Engineering the System Solution:

A Practical Guide to Developing Systems

ENGINEERING THE SYSTEM SOLUTION: A PRACTICAL GUIDE TO DEVELOPING SYSTEMS

Jack W. Hunger

Prentice Hall PTR
Englewood Cliffs, New Jersey 07632

Library of Congress Cataloging-in-Publication Data

Hunger, Jack W., 1931–
Engineering the system solution : a practical guide to developing systems / Jack W. Hunger.
p. cm.
Includes index.
ISBN 0-13-594524-0
1. Systems engineering. I. Title.
TA168.H88 1995
620'.001'—dc20 94-29420
CIP

Editorial/production supervision: BooksCraft, Inc., Indianapolis, IN
Cover art: Mike Hunger
Cover design: Ray Lundgren
Acquisition editor: Karen Gettman
Manufacturing manager: Alexis R. Heydt

Prentice-Hall, Inc.
A Simon & Schuster Company
Englewood Cliffs, NJ 07632

The publisher offers discounts on this book when ordered in bulk quantities. For more information contact: Corporate Sales Department, PTR Prentice Hall, 113 Sylvan Avenue, Englewood Cliffs, NJ 07632. Phone: 1-800-382-3419. E-mail: dan_rush@prenhall.com

Printed in the United States of America
10 9 8 7 6 5 4 3 2 1

ISBN 0-13-594524-0

Prentice-Hall International (UK) Limited, *London*
Prentice-Hall of Australia Pty. Limited, *Sydney*
Prentice-Hall Canada Inc., *Toronto*
Prentice-Hall Hispanoamericana, S.A., *Mexico*
Prentice-Hall of India Private Limited, *New Delhi*
Prentice-Hall of Japan, Inc., *Tokyo*
Simon & Schuster Asia Pte. Ltd., *Singapore*
Editora Prentice-Hall do Brasil, Ltda., *Rio de Janeiro*

SYSTEM ENGINEERING

The Eagle and the Other Side of the Mountain

There is a mountain called "Ignorance." Around this mountain are isolated valleys where small villages are established. Each village has its own language and culture and has developed many myths about the people in the other villages. The villagers have never climbed over, flew over, or tunneled through the mountain called Ignorance.

One day an eagle flew around the mountain. Seeing the different valleys he said to himself, "If these people would climb over the mountain, they could get to know each other's languages and cultures. With this knowledge they could help each other because they have developed different foods, clothes, houses, furniture, governments, educational methods, religions, laws, medicines, entertainment, songs, stories, and many skills." The eagle then said, "I will show them how they can climb the mountain called Ignorance because from my perspective I can see ways for them to travel from village to village."

The people listened to the bird and followed the eagle over the mountain to other villages, where they learned each other's languages and cultures. They worked together and combined their skills so that they kept improving. Because of this, the mountain was renamed "Knowledge."

Your villages may be called by many different names: all the various branches of engineering, government, law, medicine, and business and all the supporting organizations thereof. You speak different languages and you are part of different cultures. In addition, you are different in many other ways. Your experiences differ; your perceived missions, ages, and ethnic and educational backgrounds differ, which make it difficult for you to work together. You need to follow the eagle to learn about and from each other so you can evolve a big picture perspective. The eagle is called "System Engineering."

The first step in the system engineering process is gaining knowledge about yourselves, all the others in your village, and especially all the other villages you will have to work with to have successful missions. Gaining knowledge about yourselves is best done through an honest self-assessment. You need to know what you know and what you don't know. It is what you don't know that gets you into trouble, not what you know.

You then need to work on getting to know all the people on your team so you can work together. This includes knowing what the team doesn't know as well as what it knows. Does your team have the skills to accomplish its mission? Has your team established knowledge of and concise—no confusion—communication with all the other villages? Does your team have a single responsible eagle to lead it in performing its mission? Does your team have a mission that is clearly understood by everyone on your team and in all the villages? You must be certain that the team can answer affirmatively to these questions. If it can, then you are on your way to flying like eagles, and you have accomplished the first step to becoming successful system engineers.

Contents

* SEE refers to System Engineering Expertise

Foreword

It is most gratifying to me that system engineering is finally getting the recognition that it so richly deserves. As the complexity of our systems continues to increase, there is no more important function in any aspect of product development than system engineering. As we increasingly develop systems for global use, system engineering skills may well be the critical factor that ensures continuing improved systems.

Why is system engineering so important? It should be obvious that if we do not establish and characterize the entire system thoroughly at the start, problems will permeate through the whole process of engineering, production, and product support.

We, as engineers, talk about listening to the customer and really understanding the customer's needs. However, it falls on the shoulders of the system engineer not only to understand these needs, but to translate them into specific product requirements. We also talk about defining product costs and establishing sensible tolerances on a product to make it producible, while, in fact, this depends upon the way the system engineer partitions the system into digestible pieces that work together, with cost and tolerances assigned to each unit. We talk about the mounting complexity and costs of software development, but the work breakdown structure for the software depends on the system engineer. We talk about the importance of maintaining discipline and good documentation, but this, in turn, depends on the system engineering methodologies and procedures.

We also talk a lot about quality function deployment and its role in total quality management. If we think about it, however, in many cases our ability to do quality function deployment relies on the ability of the system engineer.

I predict that system engineering will continue to emerge as one of the most sought-after skills. You surely should recognize this as you read this book and learn more about system engineering. I believe this book will help your future.

This comprehensive treatment of systems engineering is based on decades of practical systems engineering experience and also on twenty years of teaching courses on the subject. We at Honeywell have been fortunate to benefit from Jack Hunger's teaching skills. We have a generation of system engineers to show for his efforts. Now, here in his book, Hunger offers invaluable information and advice to new and experienced systems engineers alike.

Enjoy yourself on this journey that you are about to take.

Leonard R. Weisberg
Vice President, Corporate Research and Engineering
Honeywell Inc.

Preface

This book incorporates material I originally developed with the support of Honeywell Inc. and used in the Honeywell System Engineering Manual (Minneapolis-Honeywell Inc., 1992). The material evolved as the result of my developing and teaching a Honeywell course, "System Engineering Expertise (SEE)," to over nine hundred Honeywell engineers nationwide from 1985 to 1992. The last substantial modification of the material has been its translation into a first edition textbook. I wish to acknowledge and thank Honeywell and Mr. Len Weisberg for supporting me in this work.

The information in this book can be used by a wide range of readers. In teaching this material I have, by the selection of an appropriate problem, made each specific class relate to the students' background and experience. Courses have been tailored to meet many industrial applications, covering both military and nonmilitary products, processes, and systems. The material has been applied to second-year engineering students, to nonengineers, to managers, and to multi-discipline, engineering masters level team-building courses. My course presentations have been predominantly three-day team-building workshops, but I have also taught twelve-week (once-a-week) courses with individual student projects and one-day overview offerings. The majority of the examples in this book are based on my past work experience and are related to electronic systems. Initially, I developed and presented the material as a full-quarter college course, "Electronic Circuit and System Design," with teams at Anoka-Ramsey Community College in Coon Rapids, Minnesota during 1971. The basic methodology developed for the electronic systems can apply to all systems such as government, environmental regulations, education, social programs, finance, insurance, real estate, city planning, managing any business, and other areas.

This book helps students develop generic system engineering skills that can be used to develop, improve, and manage any system. I need to emphasize that there are many excellent books on systems engineering, many of which are cited in my book. These other sources provide greater depth in specific system types and specific systems engineering analyses than those in this book. Just the same, systems engineers use a set of basic skills called "system engineering" to develop systems. In this book I am interested in these system engineering generic skills which, when applied completely in a disciplined way, have shown historically to result in successful systems. The work it takes to apply these skills is not easy—in fact, it is very demanding. When system engineering is applied to a problem, timing at the start of application is critical. The good news is that when procedures are done right at the beginning, it prevents schedule, cost, production, safety, and quality problems.

I would like to acknowledge all the students and supporters who, over a period of years, provided feedback that sustained me in my quest for understanding generic system engineering skills. Special thanks must be given to Len Weisberg, Pete Smith, Jim Lufkin, and Dr. Jay Hauge, who helped me overcome times of self-doubt. The reviews of the manuscript by Bob Sawamura, Pete Smith, Dr. Fred Zimmerman, Harold Chestnut, Dick Brown, and Dr. Jay Hauge have made them members of the development team for this book. I've used their recommendations to make it better than what I could have ever done on my own. Then there was that legion of systems engineers who kept questioning my ideas either through what they wrote, accomplished, or asked me directly: Harold Chestnut, Arthur Hall, W. Edwards Deming, Frederick Brooks, Norman R. Augustine, Dr. Henry Plummer, Kelly Johnson, John Larson, Parker Clark, Genc Timgren, Ross Schmitt, Bill Plice, Steve Sopczak, Chris Carlin, Gordon Melby, Bob Bateson, Tom Lahn, Russ Hendrick, Dick Rostal, Ron Frazzini, Don Barnhill, Paul Weasoner, Don Bailey, George Russler Jr., Ralph Pamperin, Gerald Graves, Fred Pfeiffer, Lee Carver, Keith Graham, Frank Ferrin, Sue Knipfer, Marty VanDyke, George Cole, Doug Dasgupta, Jim Grey, Marlowe Iverson, Jim Murray, Jack Lindahl, Kathie Radke, Bob Sawamura, Eric Peterson, Deb Bailey, and many others too numerous to mention. I have learned that a key, common characteristic of systems engineers is that they continually ask questions. To quote Sam Goldwyn, "For your information, let me ask you a question."

Margaret Serra must be acknowledged for her work in converting my crude writings and sketches to the text and artwork for this book. She has mastered her home Macintosh to the state of the art as evidenced by the complex figures and equations she has developed for this book. Jim Stanko provided a number of the figures by "on-the-board" drafting.

My family provided both support and the incentive to write this book. My daughter, Terri, and her husband, Eb, gave me encouragement and surprised me with a Macintosh computer. Advice from my son, Steve, helped me plan my retirement mission. My daughter, Debra, and her husband, Paul, provided me with a more complete perspective, especially in the area of quality. My son, Mike, provided the eagle on the cover and computer support. My wife, Glor, has helped me keep my sense of humor and my priorities in order. She has helped me by her systematic, common sense, loving approach to life and her daily care of me.

A prudent man leaves a legacy for his grandchildren; this is mine. My grandmother used to say, "Whoever you meet, wherever you go, make that person and place better for the fact that you were there." So taking grandma's advice, "The systems engineer helps, the systems engineer serves, consults, leads, and sometimes just stays out of the way; thus, it is better because the systems engineer was there."

It is my objective to develop system engineering skills that can be used to develop, analyze, improve, and manage any system.

Introduction—System Engineering or Mission Impossible?

A mission impossible occurs when a system fails or falls short of expectations. Systems are designed to perform missions, but a poorly designed system will demonstrate that it is *impossible* for it to consistently perform its mission.

We hear daily about mission impossibles because systems do fail or fall short of expectations: social systems bring about conflicts and abuses; political systems become politician-serving instead of people-serving. Any system that is self-serving instead of people-serving is a mission impossible. Because of poor systems engineering, planes crash, water control dams are inadequate or bring about health and delta erosion problems, so-called control systems don't control, computer systems promise much but cost too much, government systems do more harm than good, educational systems cheat the students, moral systems erode, weapon systems are too expensive, products start fires, products can't be installed, process systems function poorly, medical systems kill or injure, and new technology generates more problems than it solves. The only good news about mission impossibles is that they provide opportunities to apply system engineering to fix, improve, or replace poor systems.

Unfortunately, system engineering has been poorly applied. Because our application of system engineering has been fragmented, ad hoc, undisciplined, it has resulted in mission impossibles. If system engineering process would be properly applied when a system is first developed or improved, we could avoid most mission impossibles. The importance of applying good system engineering process at the beginning cannot be emphasized enough. It is better to pay now instead of paying a lot more to fix problems later. According to Shaw and Lacy in an *American Institute of Aeronautics and Astronautics* paper, "It has been amply demonstrated that the first 5 to 8 percent of program expenditures in front-end systems engineering effort controls over 80 percent of the product cost and unalterably defines the outcome of a product's life cycle characteristics."[1] In addition, if we continuously apply good system engineering process throughout the life of a system, we can achieve constant improvement with friendly transition to new conditions, environments, and technologies.

This book is based on a study of successful systems in which missions were accomplished, such as the highly technical landing of men on the moon and the Mayo Clinic patient process system. It has been demonstrated that systems engineering can be done, but there has been no attempt to find the critical basic generic processes embodied in the successes. This book presents and develops the basic generic system engineering skills required to develop systems capable of successful missions, and its objective is to provide students skills in the complete, orderly, disciplined application of the system engineering process.

Mission Impossibles

Mission impossibles occur continuously. We have underdesigned systems, overdesigned systems, unsafe systems which are unfriendly to people, ad hoc unanalyzed systems, poorly supported systems, deteriorating systems, outdated systems, and ineffective systems. All of these systems result in mission impossibles; they just don't do the good things they are supposed to do. Here are several mission impossible examples.

First, underdesigned mission impossibles occur when we do not study the range of mission environments. People-moving systems have traffic jams because we didn't plan for the future; computer systems with limited throughput make operators wait, and production systems have limited capacity.

Second, overdesigned mission impossibles also occur when we do not study the range of mission environments. We over design: Systems are designed for peak capacity (but practically no people show up, computer system terminals sit dormant, production facilities are idle), vibration tests apply dynamic loads equivalent to supersonic taxing on the runway, equipment is designed so it can perform at a temperature of –65°F when it is intended to be permanently stationed in the Egyptian Desert, gunfire vibration tests are applied to equipment installed in an aircraft without a gun, roll rate gyros are tested to 40°/second for a plane whose wings will come off at 10°/second.

Everyday we hear about Unsafe-System mission impossibles in which failure modes and effect analysis were not done as part of the system design. Planes crash because the engine thrust reversers came on in flight or radio interference from both inside and outside the plane cause critical flight control systems to malfunction. We seal our homes for energy efficiency and then cause a dangerous polluted environment within them. Social systems bring about crime. Gas tanks rupture in car accidents. Fires start in fire alarm equipment. Agricultural processes are injurious to our health.

Third, mission impossibles are sometimes the result of poorly controlled systems. The overstressed human has a heart attack without stress release feedback. Political systems without check or balance can bankrupt nations. When control systems are not analyzed over a wide enough frequency of operation range, they oscillate or go hard over at unexpected operational mission conditions or environmental conditions. Control systems are not designed for spurious random input or output events that cause mission impossible occurrences. We add robotics to manufacturing systems, which improve productivity greatly but cause other problems.

In the imaginary bottle factory story—which, unfortunately, has too many corollaries to real events—we see what can happen if we do not consider all the parts of a system in our mission analyses. Many times we attempt to improve a system but ignore the people involved in it, especially those people who will lose their jobs. A bottle factory was turning out 6,000 bottles a day, with 600 employees, when the market needed 60,000 bottles a day. In typical mission impossible fashion a robotic bottle factory was designed, built, and put into production. To ensure capability of 60,000 bottles daily, it produced a minimum of 100,000 bottles a day. It only required 60 people to run the factory, so 540 people were laid off. The problem of what to do with those laid off was solved by hiring the 540 people back to break and recycle 40,000 bottles back in as raw material into the factory. Of course, it appeared that the productivity had been improved—the system was providing the required 60,000 bottles—but unfortunately the new bottle factory was built with 30-year loans to be paid back by increased taxes. In addition, the 540 bottle breakers became a lower caste, the 60 who ran the factory an upper class, resulting in many social problems. We need to consider what happens to the people we lay off when we improve productivity. Furthermore, how does this improvement impact national debt, taxes, society, and the marketplace?

MISSION SUCCESSFUL

A successful mission incorporates mission analysis. There is a building at the Mayo Clinic in Rochester, Minnesota, called the "Plummer Building." It is named after Dr. Henry S. Plummer, a world-renowned doctor, who also had system engineering skills that helped him work with the Mayo brothers in developing the Mayo Clinic system.[2] In doing so he has been called an organizational genius. Dr. Plummer recognized that the mission of the clinic was to quickly, and humanely, serve patients. He engineered a patient flow process, timely data acquisition/evaluation/reporting, timely communications systems, timely record keeping/recovery, a systematic accessible historical data base that is constantly growing, and a skilled team approach to provide proper patient care.

The story is told of how Dr. Plummer in the early 1900s sketched out a diagram on how to do telephone switching for conference calls at the clinic. The telephone company had at that time told him it couldn't be done. However, it implemented his sketch and the first conference call resulted. This is just one example of how he brought about innovation to support his vision of the clinic's mission.

The continuous analysis of time, timing, proper order and sequence of events in time, the planning for contingency, the patient's experience and viewpoint in time, were key analyses to Dr. Plummer. We call this mission analysis, and it is the most important part of the systems engineering process. Successful systems are developed by the constant use of mission analyses. Unsuccessful systems (mission impossibles) are usually the result of the lack of understanding of, or the lack of, continuous application of mission time-line analyses.

What is a "system"? It is any product with a mission (an airplane, a car, a house, a computer, a gyro, a hammer). It is any process with a mission (manufacturing, educational, managerial, communication, social, medical, political, governmental). Many systems are just part of nature (the universe, solar, ecological, weather, human, cardiovascular, nervous). System engineering is the process of initial optimization and constant improvement of systems to provide better and better mission capability for greater and greater customer satisfaction.

Do you have to be an engineer to do system engineering? No. The development of successful system mission capability requires an appropriate mix of skills; it is usually a team effort. Generic system engineering skills apply to almost all professions—for example, doctors, lawyers, merchants, teachers, engineers. They can learn certain skills and develop some basic important insights. The ability to see the big picture (system perspective) reminds us that there is always a bigger system that must be interfaced to if we are to be successful. There is always a status quo, a past that must be transitioned from to evolve a better system. Mission analyses always include an analysis of all the potential ways the system can fail, the effect of the failure on the mission, and analysis of system rebaseline to eliminate all critical failure effects. Mission analyses and system perspective are two key elements of the systems engineering process. It is the objective of this book to define the system engineering process and also to help students develop skills in performing the process. The engineering of a successful system follows a process—the system engineering process.

The System Engineering Process

The system engineering process is an iterative process with multiple reentrant recursive loops operating in parallel. There is, however, a simple, definable single thread mission for each application of the system engineering process. Recognizing and managing this simple mission-driven system engineering process makes the complicated multiple parallel reentrant recursive loops doable. The system engineering process consists of defining a problem to solve, studying to gain knowledge, developing a system perspective, doing mission analysis, doing functional analysis, providing a proposal that includes a baseline and management plans, providing a specification with specific requirements, doing analyses, demanding design reviews and then doing the process all over again as required.

The system engineering process is not a committee process, but it is a team process with a single mission-driven responsible leader. It provides for consolidation of activity, not proliferation, and it follows a finite ordered sequence of events. This book shows the systems engineering process mission at its most critical time, the initial system development and documentation. It is necessary to establish the discipline and tools to perform this initial system development system engineering process for successful programs to develop successful systems. An iterative process is one that is done repeatedly over and over, where each iteration produces results that approximate the desired result more and more closely. The multiple reentrant recursive loops operating in parallel can be viewed as a group of dependent actions of actors and things, each of which strongly defends the importance of its particular sub-mission. The design of a control system, for example, requires actions of control system users, algorithm developers, electrical engineers, mechanical engineers, software engineers, specialist engineers,

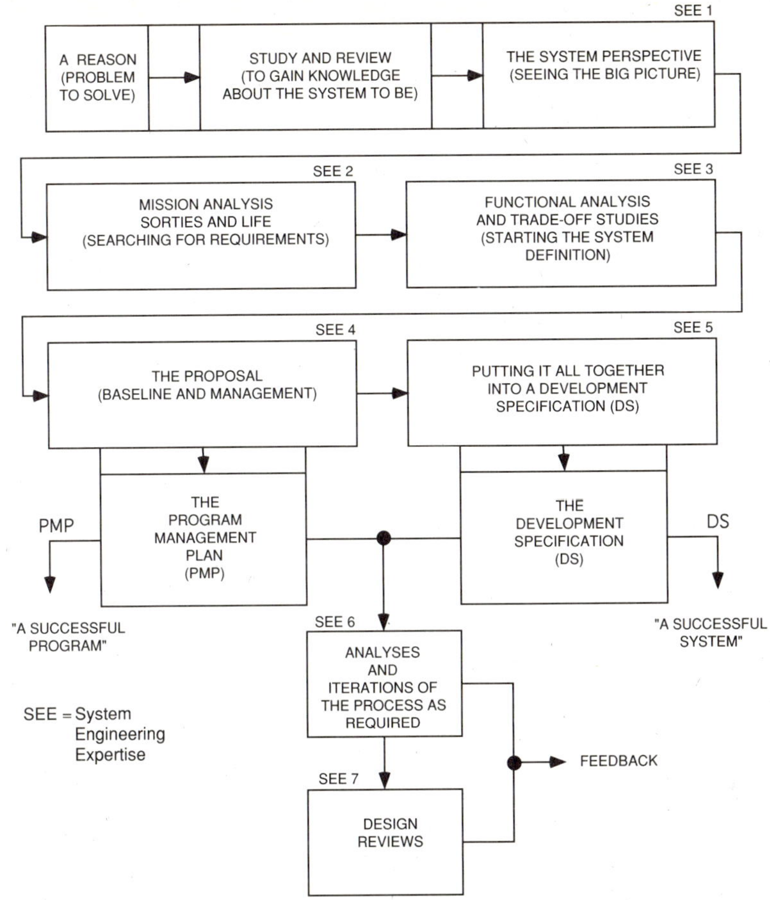

Fig. **I-1.** The system engineering process

test engineers, managers, as well as actions of system objects such as sensors, actuators, computers, mechanical package, and wires. The common dimension for all the sub-missions (recursive loops) is time. It is important to recognize this and also to recognize that no one ever gets it totally right the first time, so we develop and improve by iteration.

The system engineering process and management thereof are described in figure I-1. Given a problem to solve, the system engineering process follows this simplified flow diagram that results in the two main outputs—the development specification (DS) and the program management plan (PMP). Figure I-1 and chapters SEE-1–7 concentrate on the first pass through the process. The subsequent iterations of the system engineering process by the system engineering team will depend on the design reviews. It is the first pass that is the most important because only when it is done thoroughly and quickly at the beginning of a system development will it ensure a successful development program for a successful system.

Time Required to Do the First Pass of the System Engineering Process

A small (five to seven people) highly skilled team with prior successful experience with the specific system problem can do the first pass in three eight-hour days. They will provide as outputs two baselined structurally complete documents: a development specification that includes a complete system baseline, and a program management plan. The system engineering process and the management thereof not only can be done this quickly but must be done this quickly. A complete baseline and plans are needed to do analyses and trade-offs as the system and the development program evolve. The objective is to provide these two documents as soon as possible to establish sufficient baseline and specific requirements so communication will take place between all the people involved with the system. Everyone must understand that these documents are initially preliminary, but they better be read, questioned, and improved quickly because the cost to change them increases exponentially as the system is developed.

The Development Specification

The first pass of the system engineering process baselines the system. The baseline as recorded in the development specification must be complete and shall include mission time line definition and functional block diagrams of the system top-level, hardware, software, and detail. The baseline shall provide a specification tree that shows the flow-down of the requirements from the mission to all the required functional parts of the system. A top-level interface diagram, a mechanical sketch, an installation drawing, a top-level bill of material and an attributes table must all be provided to ensure a complete baseline. This preliminary baseline provides a basis for analyses and trade-off studies by designers, users, and all the other people involved or who will be involved with the system. The attributes table, for example, is very important because it must establish a baseline size, weight, power, cost, reliability, mission effectiveness criteria, repair time; all of these must have real numbers. We call this precision guess work, but there must not be any "to be determineds (TBDs)." Numbers are needed that reflect expected required mission capability in order to do analyses and trade-offs. When TBDs are added up, or average TBDs, or when a worst case analysis of TBDs are done, TBD is the result, which is worthless. If the numbers are preliminary estimates, then it must be indicated that they are preliminary estimates and subject to change. The system baseline is only part of the development specification needed. In addition, it is necessary to have scope, applicable documents, and especially some specific requirements so that those who are to design functional elements of our system can get started.

The "scope" statement of the development specification shall describe what the system is all about so a common, clear understanding is established of the problem to be solved. In the scope statement, the basic mission of the system shall be stated, prior associated developments shall be detailed, the players shall be determined, and the bigger system above this system shall be defined. The scope shall provide a common clear technical system perspective. "Applicable documents" shall be listed to establish a common knowledge base of prior reports, articles, and auxiliary standard or special requirements documents. The "specific requirements" shall provide a flow-down of mission requirements from the baseline so all the system designers have enough information to start designing. The specific requirements for an electronic system, for example, would include those needed by hardware circuit designers, software designers, production process designers, mechanical designers, control analysts, quality engineers, logistic engineers, and test engineers. "Quality assurance" requirements shall provide specifics to ensure a quality system. The "delivery" requirements shall provide specifics for support of the system after delivery to the customer.

It should be pointed out that the development specification is the systems engineer's document to be used in leading the development process. Thus the system engineer can modify and tailor it to the special needs of each particular system. It is intended to provide concise, clear communication, and therefore, requires a forced process of review and modification. The development specification process and its expediency are the most important part of the system engineering process.

The Program Management Plan

Not only is it important to know the technical requirements as baselined in the development specification, but it is as well to know program development schedules and funds allocated to do the work. A work breakdown structure, schedules, funds allocations, staffing with an appropriately skilled team, and assignment of responsibilities are needed. In addition, a design review plan, a software development plan, a hardware development plan, and a test plan are needed. Finally, team building communication plans and assignment of responsibilities for specific analyses and trade-off-studies are necessary to have.

A key objective in designing the development specification and the program management plan is to make them as small as possible. They will not be reviewed or used if they are long or complex so it is best to condense them. The baseline specification tree can be structured to provide a flow-down of requirements such that the individual specifications can be kept small.

In his book *System Engineering Methods,* Harold Chestnut refers to the development specification as a living document.[3] Admittedly these first pass documents at birth may be very immature, but if they are to be used as a key communication tool, as they should be, they will be alive and grow to maturity. They will be reviewed and updated in a quest for constant improvement almost daily at first, then weekly and monthly until ultimate maturity is reached.

Applying System Engineering

System engineering skills must be applied to major systems, to subsystems, and to components. They all have a mission, so mission analysis, functional analysis, customer liaison, system perspective analysis, trade-off studies, baselining, development specification process, management plans, specialty analyses, and design reviews of our biggest systems down to our smallest component can be done. Yesterday's big system is today's component that may have an embedded computer, a state-of-the-art sensor, a complicated analog conversion circuit, a multiple of software modes and algorithms, a digital serial output, built-in automatic test hardware and software, a power converter, and state-of-the-art mechanical packaging. Such a component must be system engineered.

When it's considered that components are combined into subsystems, it's not difficult to realize that the interface to the bigger system above is critical. The attributes of the components must be established so they will perform to meet the requirements of the subsystem to meet the subsystem's mission. How accurate is the component and what is its mission timing? How much does it cost? How reliable is it? What does it weigh? Is it producible? Is it a quality-constant improving design? In short, what are all its attributes? If all the subsystems are considered in the same way and their relationship to the major system and its mission, the importance of developing system engineering skills are obvious.

What the Systems Engineer Does

The disciplined systems engineer reads a lot, does a lot of analysis, asks a lot of questions, writes development specifications, allocates requirements, designs, reviews, communicates, and does it over and over (iteratively) to develop a system capable of adequately doing the desired mission. The systems engineer understands the customer and the customer's needs, works with the customer, and iteratively develops the system, subsystem and/or component development specifications. The system engineering process and system engineering management are appropriately shaped to fit the specific mission. Communication methods must be established to ensure that the systems engineer can function as the technical integrator.

The systems engineer also has to manage the team of engineers and specialists. This requires knowing a team's skills and deficiencies and balancing the team skills to achieve the appropriate skill profile to develop the specific system. This includes the application of the system engineering process, system engineering management, continuous mission time line analyses, use of the development specification process to establish "design to" goals for the system designers, and the big picture system perspective. The big picture includes "the bigger system above," which could be an organization, company, nation, the world, and even the universe. Chapter 8 of this book discusses the application of system engineering skills to government, environmental, educational, social, finance, law, medicine and other national and global systems.

The Attributes and Skills of Systems Engineers

System engineering is a discipline that involves developing skills to do the system engineering process and system engineering management. The generic skills to develop are the system perspective, mission analyses, functional analyses, baselining, development specification process, planning, specialty analyses, and design reviewing.

Chapter SEE-1 provides an exercise to show you how to do a self-evaluation of your skills to help you in your career development. This exercise is highly recommended to build teams and also to give each of you a new perspective for on-the-job and self-training.

Over the past twenty years, I have taught and worked with many excellent systems engineers. Based on my experience with them, I have come up with the following list of attributes that may give you an idea of what these eagles are like.

Attributes of Systems Engineers

- Love to solve problems
- Have an obsession for increasing their knowledge
- Ask questions when they don't understand something
- Can make almost anything work
- Can live with disappointments
- Learn from failures
- Love to study new things
- Look for correlations
- See things chronologically
- Are good organizers and enjoy organizing
- Purposeful because they have a vision and know the mission
- Know teamwork is fun
- Are persevering
- Know who the experts are
- Are sensitive to people
- Are efficient
- Have a desire to know the basics
- Have good oral communication skills
- Communicate well in writing—concise and to the point
- Don't mind being criticized
- Have excellent analytical capabilities; love to analyze
- Exhibit timing that is almost always right
- Anticipate potential problems
- Are experienced in more than one specialty

- Are quick thinkers
- Worry a lot
- Are accurate in presenting knowledge
- Understand team skills and deficiencies
- Promote growth-oriented task assignments and mentoring for team members
- Are creative
- Understand the big-picture cost and schedule impact of performing various tasks
- Generally get along with everyone
- Authoritative based on knowledge and experience
- Disciplined, they practice system engineering process
- Act as catalysts
- Don't know everything and freely admit it
- Love to help others
- See the big picture (the forest)
- Understand the small picture and its relationship to the big picture (the tree in the forest)
- Study and understand people
- Have historical perspective
- Are futurists
- Believe that everything is definable
- Visualize attributes: function, effectiveness, adequacy, size, weight, shape, power, timing, monitorability, risks, costs, maintenance, installation, aging effects, advantages and disadvantages when compared to competing systems.
- Know there is always more to learn
- Know there are always better things and better ways available
- Know that sometimes "new" is not better
- Avoid hysteria
- Have a sense of humor

How Do You Get to Be a Systems Engineer?

This book's first seven chapters lead you through the system engineering process. Chapter 8 is intended to start you flying on your own in the application of the system engineering process.

This book is intended to help you improve and expand your system engineering skills so you can become a systems engineer or become a better systems engineer. The application of system engineering skills not only can help you in your career, but, more importantly, can have an impact on society. The health of people, businesses, countries, and the world can be improved by your use of system engineering skills. Ross Perot, the successful Texas businessman, has a motto: "Eagles don't flock, you have to find them one at a time."[4] The question is how did they get to be eagles in the first place? What skills did they develop to get the attributes of an eagle? Who taught them and mentored them? Will the eagles pass on their knowledge and skills by mentoring other potential eagles?

References

1. Shaw, Thomas E. and Lacy, James A. *An Overview of Systems Engineering Management Methods and Processes,* Washington, D.C.: American Institute of Aeronautics and Astronautics, Inc, (AIAA), 1992.
2. Clapesattle, Helen. *The Doctors Mayo,* Minneapolis, MN: University of Minnesota Press, 1941.
3. Chestnut, Harold. *System Engineering Methods.* New York: John Wiley & Sons, 1967.
4. Gross, Ken. *Ross Perot—The Man Behind the Myth,* New York: Random House, 1992, page 84.

The System Perspective

(They shall SEE as an eagle)

In this chapter the goal is to develop your system perspective skills; that is, your ability to perceive the big picture. In the Introduction "a reason" was established—there are too many mission impossibles; and the goal is to improve system engineering skills in order to develop successful systems. It is now time to study and review some definitions pertaining to system engineering in order to better understand what a system is, what system engineering is, what the roles and responsibilities of a systems engineer are, what a development specification is, what a system engineering plan is, and what system engineering skills are. In addition, at the end of this chapter there is a list of recommended reading for those who wish to further expand their knowledge of system engineering.

A very important part of your development of a system perspective requires that you establish your relationship to the system and to other people involved with the system. You will take a system engineering self-assessment skill survey that will help you become more aware of what skills you have and what skills you lack. It is highly recommended that a team building exercise be done after the personal self-assessment. This exercise will help each team member learn very quickly about the other team members at the beginning of a project. It's important to know what the team's composite skills and deficiencies are in order to avoid conflicts among people later in the system engineering process. Do the team's composite skills meet the needs of the system to be engineered? You need to be especially sensitive to deficiencies because it is what you don't know that gets you in trouble, not what you know. It is the aim of the skill survey and team building exercise to develop an incentive to improve the breadth and depth of your skills. Many will realize, after completing these exercises, an obligation to mentor those less experienced in order to develop a winning team.

Based on your knowledge of system engineering definitions and your team's skills and deficiencies, you will be ready to start the system engineering process. Given a specific problem to solve, which gives a reason to apply the system engineering process, you can now study and review the problem to gain knowledge about the system to be. This chapter shows you, the student, a number of examples that use the system perspective cube to develop a big picture view of the actions required in order to solve the problem. This system perspective action plan is the initial step in the system engineering process (fig. 1-1).

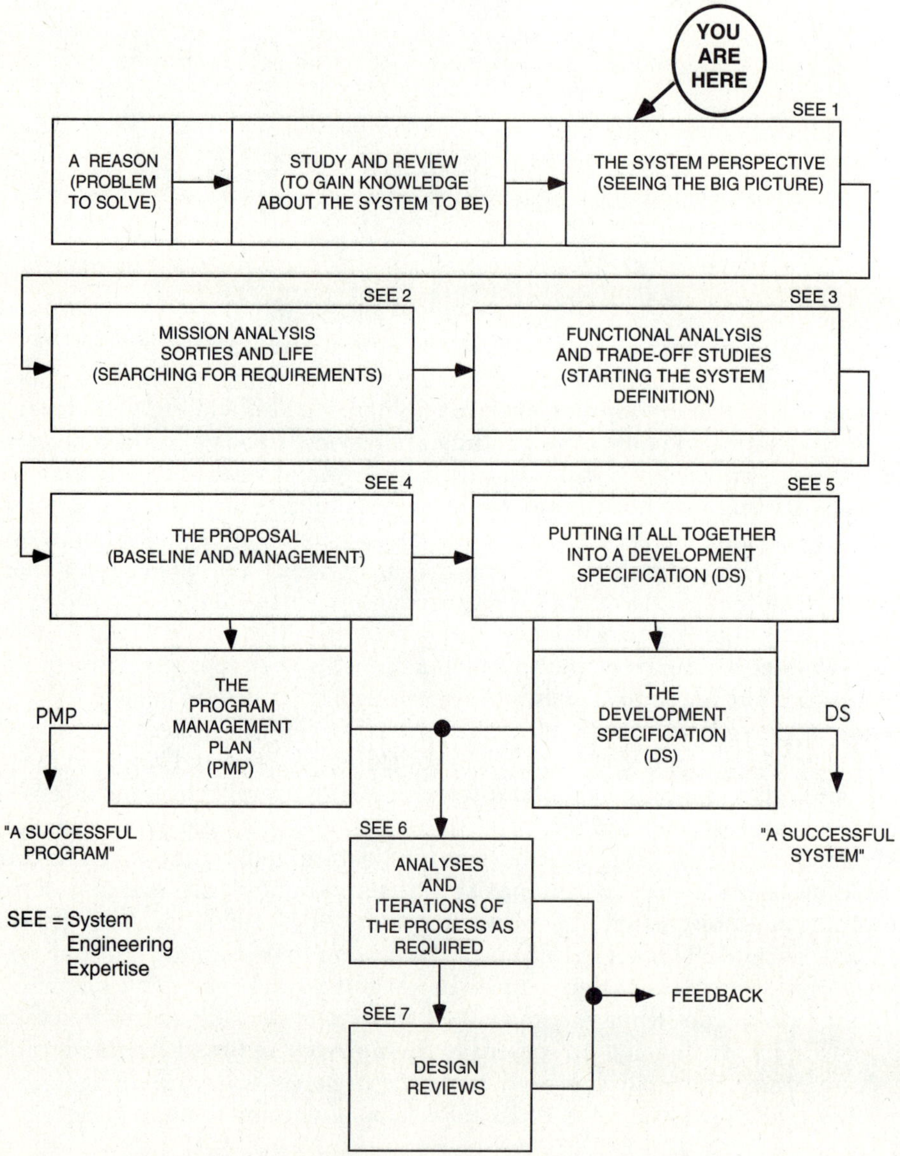

Fig. 1-1. The system engineering process

1.1 A SYSTEM

Some basic definitions can help you understand "system," "systematize," and "systematic." A *system,* for example, is defined by five separate classifications (*American Heritage Dictionary,* 2nd ed., 1986), which can all be related to system engineering expertise (SEE).

1. **A group of interrelated elements forming a collective entity**—This is descriptive of each of the products we design, develop, build, sell, and support, such as airplanes, automobiles, homes, flight controls, automotive controls, home controls, navigation, and space systems, and all products. It also includes a lot of other systems we don't design but study, monitor, learn to work with, and sometimes attempt to control, such as the cardiovascular system, the nervous system, the ecology system, the weather system, and governmental systems. We have learned that we can develop *process systems* such as product producing process, production process, chemical process, and total quality management process.
2. **A network such as communication, travel or distribution**—The interconnects we have with each other and our customers and suppliers. The design and management of communication networks are needed for all human missions and are especially critical to our product development programs.
3. **A set of interrelated ideas, principles, rules, procedures or laws**—Engineering operating procedures, for example, embody a set of interrelated principles, rules, and procedures that establish an engineering process system.
4. **A social, economic or political organizational form**—We live in a democratic form of government. This could also include the form of our specific engineering organization, how are we organized, do we have a matrix organizational system or a product line organizational system?
5. **The state or condition of harmonious, orderly interactions**—This should define how our systems work and also how we work together. It should be our goal to develop system engineering skills so we can establish a condition of harmonious, orderly interactions in our business, our programs, our projects, our products, and our lives.

The dictionary also defines *systematize* as to formulate into or reduce to a system.

Systems engineers systematize. If they are going to design, build, or buy a house or a new control system, they study and do analysis and trade-off studies in a very systematic way. They intuitively systematize at home and on the job.

The word *systematic* is also defined three ways in the dictionary as:

1. **Carried on in a step-by-step procedure**—For example, a systems engineer will use the System Engineering Process for the development of a system.

2. **Characterized by purposeful regularity**—I know the purpose of each step in the System Engineering Process and I do them in order on a regular basis for each system I develop.

3. **Methodical**—I follow each step in the System Engineering Process methodically, never leaving a step out.[1]

1.2 System Engineering

A key difference between a successful program and an unsuccessful one is the adequacy of the system engineering. The system engineering process is an iterative process used to formulate a system by using a disciplined procedure. A systems engineer is a person who has both the authority and the responsibility to apply the system engineering process. The systems engineer is disciplined, systematic, communicative, suitably technically competent, and has the system perspective. The systems engineer develops such skills as system perspective analysis, mission analysis, functional analysis, trade-off studies, baseline system design, development specification process, technical communications, engineering management, system analysis, and control system design. Most important, there must be a reason to do system engineering, a clearly stated problem to solve. This is the first question the systems engineer must ask: What is the problem?

1.3 System Engineering Process and System Engineering Management

Systems go through a natural evolutionary process, or life cycle, in which actions taken (or not taken) in the very early stages can mean the difference between mission accomplished (success) and mission impossible (failure).

The purpose of system engineering process is to prevent failure by completely defining all requirements on the system and establishing a system configuration that is proven early on to be capable of meeting those requirements. System engineering is often referred to as a "front-end" process. That is, the majority of system engineering tasks are completed in the initial life cycle phases of the systems life.

System engineering management is the management of the system engineering process. Program success is measured in terms of cost, schedule, and technical performance. This includes the organization of the people (skill elements), the organization of the effort (work elements), and the organization of the system (functional elements). This is referred to as "we three-tree," the staffing team tree for skill elements, the work breakdown structure tree for work elements, and the specification tree that establishes the functional elements and requirements thereof in order to perform the required mission successfully. These three trees evolve in the following sequence: definition of the system hierarchy by use of system engineering process permits allocation of mission requirements to defined functional elements; this in turn permits allocation of functional requirements to defined program work elements; this in turn permits assignment of responsibility for development work to specific skilled people elements.

1.4 SYSTEMS ENGINEER'S ROLES/RESPONSIBILITIES—THE DEVELOPMENT SPECIFICATION

System engineering process begins with the system perspective study process, followed by mission analysis, functional analysis, trade-off-studies, baseline definition, and system engineering management. These all lead up to the key systems engineer's responsibility—the development specification—which is always done as part of the system engineering process. It is the most important output. If you are not generating a development specification, you are not doing the system engineering process.

The systems engineer is the lead technical interface to the customer; therefore, he or she is the translator of customer technical needs to development specifications for the various system designers. This transition must be *timely* and *structurally complete.* The systems engineer uses the system engineering process and system engineering management to lead the development process. Starting with the system engineering perspective, it is the systems engineer's goal to establish a development specification to lead the technical development, and to construct a program management plan to manage the development.

Timely means that a preliminary development specification is provided by the systems engineer during the pre-proposal time frame. It need not be complete or neat, but it should provide sufficient information to adequately (not perfectly) define the requirements for the system. The development specification is *structurally complete,* however, in that it provides the outline (checklist) for the system, even if some interfaces and requirements are to-be-determined. To-be-determined requirements should be stated in this way: "The accuracy shall be to-be-determined (1.0%—tentatively)." To-be-determined cannot be analyzed, but the "1.0%—tentatively" offers a number that can be used to get started. A systems engineers' goal will be to remove all to-be-determined's from the development specification as soon as possible.

The initial purpose of a development specification is to guide the development of the system. It is called the "part I" and is provided early in the program; a system definition and requirements for the system are the main objectives for the development specification at this time. This specification, at the beginning of the program, will be adequate (not perfect) to define the requirements for the detailed design. At the start of the contract the system engineer must immediately work very hard and quickly in liaison with the customer to establish all the detailed data and requirements necessary to ensure that when detailed design is started, sufficient definition is provided to the various designers. For an electronic system for instance, this could be analog hardware, digital hardware, software, mechanical and analysis design engineers.

The development specification must provide a system functional description, algorithms, physical definition, installation requirements, partitioning, block diagram, a specification tree (see fig. 1-2), accuracy requirements, environmental requirements, power requirements, input/output interface specifications, and any other special requirements. Chapters SEE 1–5 will lead you through the first pass of the system engineering process that results in the development specifications process. The specification tree is a must requirement for defining a system because it shows the top-level partitioning down of requirements of the system to lower-level requirement development specifications.

Subsequent to the development specification part I and prior to system testing, a development specification part II will be generated. This part II will provide inspection, test, and analysis procedures to be performed to prove system compliance to the requirements of part I.

Subsequent to production of the system, the development specification may be updated and enhanced to fulfill new objectives such as aid to production, deployment of the system, marketing aid, and a guideline for subsequent similar system developments. While the initial specification part I was provided for the

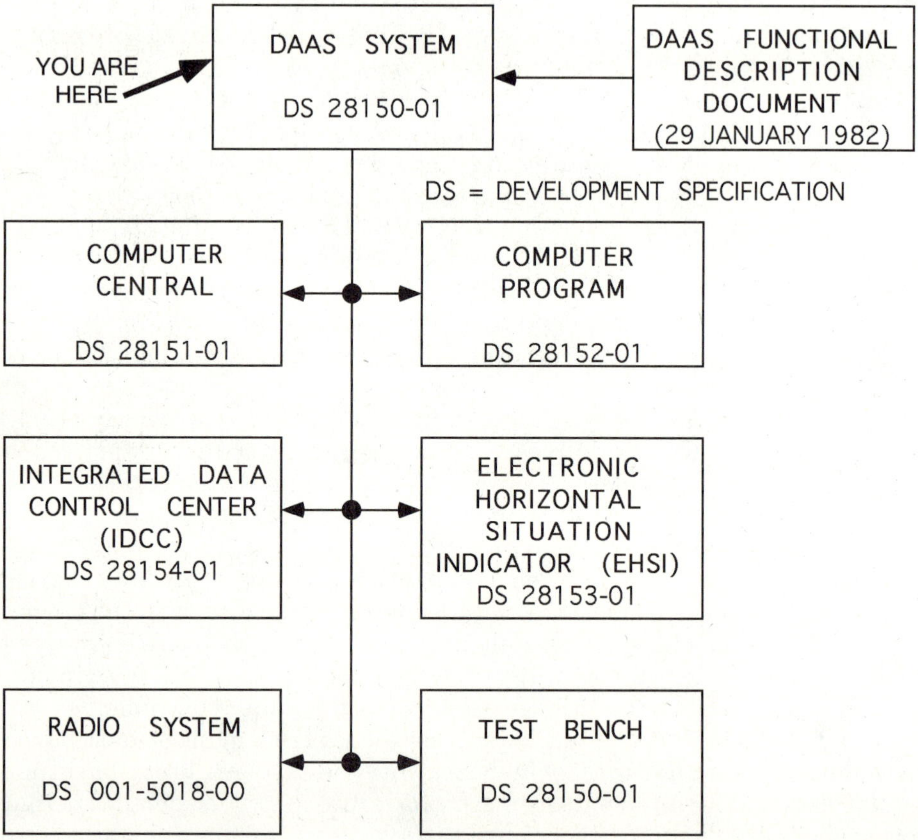

Fig. 1-2. An example of a specification tree (one of the required figures for system description; others are interface diagram, functional block diagrams, physical sketch, installation sketch, and attributes table)

design engineers and part II was for the test evaluation and data analysis engineers, this update is now to be provided with additional data as required for production engineers, field service engineers, marketing engineers and trainee engineers. Chapter SEE-5 is intended to help you develop your skill in doing the development specification process.

Analysis: The development of the system development specification may require these analyses:

- Market analysis
- Mission analysis
- Life cycle analysis
- Control system analysis
- Functional analysis
- System trade-off analysis
- Allocation (partitioning) analysis
- Accuracy and error analysis
- Failure mode and effects analysis (FMEA)
- Safety analysis
- Attributes balancing analysis
- Thermal analysis
- Legal issues analysis
- Government regulation analysis
- Educational issues analysis
- Power system analysis
- Testability analysis
- Producibility analysis
- Cost analyses
- Matrix reduction analysis
- Data analysis
- Logistics analysis
- Technical risk analysis
- Structural analysis
- Electromagnetic interference (EMI) analysis
- Maintainability analysis
- Mean time between failures (MTBF) analysis

The systems engineer must be sufficiently familiar with these so that he understands the need and timing for specific analysis. The systems engineer does not necessarily do them all but calls upon the appropriate analyst when timely. Familiarity with these various analytical requirements must be of great enough depth so the systems engineer can do some of them and do a cursory job on all the others. It should be the goal of systems engineers to continuously improve the scope and depth of their analytical skills. Chapter SEE-6 will expose you to many of the critical analyses that, if overlooked or are poorly done, will result in system problems.

1.4.1 System level verification

The development specification part II is prepared by the systems engineer to provide test and analysis procedures to be performed to prove system compliance to the requirements of part I. The systems engineer is then responsible for actively supporting the verification of the design. That the system meets the requirements of the part I requirements is verified by using the test and analysis procedures established in the part II of the development specification. This may also include support of a flight or field test program. During the verification of the system, the systems engineer's responsibility for analysis is key to proving to the customer that the system does fulfill the customer's needs.

1.5 The System Engineering Management Plan and the Program Management Plan

Management is a key part of system engineering. You must combine the management of the system engineering process by system engineering management with all of the program management. Program managers must have system engineering skills in order to do their mission. They know that every program needs a plan to be successful. This plan is referred to as a "program management plan." In some large system programs there may be a large number of systems engineers and associated system engineering tasks to be done. In such situations a system engineering management plan may be required as a subset of the program management plan. If so, refer to chapter SEE-4 for assistance in developing a system engineering management plan. In smaller programs, the conventional program management plan will include plans to ensure that the various system engineering tasks are planned into the program.

1.6 Skills

In order to do system engineering process and system engineering management, you need to continuously work on skill improvement. The standard I.Q. test provides a very limited measure of skill potential in only the areas of mathematics and language. There are really seven skill areas that must be developed to continuously improve your basic system engineering skills: 1) mathematical-logical (an obvious engineering skill); 2) language (we need to develop our ability to communicate); 3) musical (to provide a rhythm and timing to your control systems); 4) spatial (we must be able to envision and depict things in three dimensions including distance relationships); 5) bodily kinesthetic (the bodily capabilities and limits of those who will use our systems); 6) understanding others (do we have the skill to put ourselves in the other persons shoes not only in body but also in wants, needs, and desires?); and 7) self-knowledge (do we honestly know what skills we have and what skills we do not have, are we constantly trying to improve our skills? How do we deal with our desires, fears, and frustrations?)

Systems engineers need to recognize that they need to strive to improve all of these seven basic skill areas so they can effectively serve the public. In developing your skill of knowing yourself you may have to openly admit that you have very poor skills in some of the basic skill areas.

In 1962 Arthur D. Hall identified five traits of the ideal systems engineer.[2] These traits are: 1) an affinity for the systems point of view (in this book we work on the system perspective, mission analyses, and balancing attributes to develop this trait); 2) faculty of judgment (we work on trade-off studies, design charts, and analyses); 3) creativity (we do mission analyses, brainstorming, and iteration for constant improvement); 4) facility in human relations (we do team building, people systems, skills analysis, and working with others); and 5) a gift for expression (we develop both oral and written communication skills, development specification, program management plan, and design review processes).

Tracy G. Peck, a systems engineer with IBM, stated in a 1985 paper, "To be a successful systems engineer requires that one achieve a blend of technical and business understanding, personal relations, and effective communications."[3] This restates Hall's traits in a slightly different way, but both are correct.

This chapter provides a self-assessment skill survey example and a team building methodology to learn more about yourself and your team. This will give you the incentive to develop your skills and to mentor others.

1.7 SYSTEM ENGINEERING SKILLS SURVEY—AN EXAMPLE

When you are about to engineer or re-engineer a system you must make an honest attempt to assess the skills required. What skills are required for this type of system? What skills do you have? What skills do you lack or are weak in? What skills does your team have or lack? What does the team have to do to overcome its lack of specific knowledge, experience, and skills?

The following system engineering skills survey example is intended to help you in a self-assessment of your skills for engineering electronic control systems. A self-assessment skills survey similar to this must be established for your specific system type in order to establish that you and your team have the skills needed. Chapter SEE-8 shows you how you may develop a skill survey for your special system type, it may be any number of fields—judicial, civil engineering, transportation, government, education, medical, or social. Note that the following example is a good starting point for your specific system type because many of the skills are generic to any system. Most mission impossibles are directly traceable to arrogant or naive people who are unaware of their skill deficiencies. You need to establish your present understanding of your skill level and profile against a wide range of sample skills.

Your ability to communicate in electrical, mechanical, software, analytical, management, and basic systems engineering is important for you to become skilled in systems engineering an electronic control system. You need to also establish yourself as an expert in at least one skill area of electrical, mechanical, software, or analytical, engineering or in management or as a mission specialist. Note that to be an expert does not necessitate a degree in engineering. The mission specialist, for example, is usually the user of the product or system.

The analysis of the SEE-1 skills survey will help you develop your present skill level and profile. With this information you can plan your study of SEE and other subjects to raise your overall skill level, establish an expert area, and flatten your skill profile.

You should now take the following skills survey. This is not a test! It is merely an honest personal assessment of certain current engineering skills you have. *Note:* You may be submitting your skills survey to your instructor for analysis. If so, it will be returned to you along with an analysis report. Chapter SEE-8 provides the instructions and templates if you wish to analyze your skill survey manually yourself.

Also, do not be discouraged by this skills survey. Many accomplished engineers with years of experience have an overall skill level of less than 50 percent on this survey. The objective is to find your areas of skill deficiencies so you can work to strengthen those areas and also to assist in building teams. If you have questions regarding the skills, refer to the skill definitions in chapter SEE-8.

This survey can be analyzed numerically to provide your overall skill level and a normalized skill profile of six skill areas: common system engineering skills, mechanical skills, hardware skills, software skills, analysis skills, and program manager skills. The ideal system engineering profile can be characterized as "a jack of all trades and a master of at least one." A subjective analysis may also be done by your instructor who will then provide a report of the numerical and the subjective analyses with career development recommendations.

SYSTEM ENGINEERING SKILLS SURVEY

Name: ______________________ Address ______________________

____

____

SURVEY OBJECTIVES:

- To determine your perceived electronic control type system—system engineering capability.
- To assist in developing system engineering training needs.
- To provide a communications medium (a survey analysis report may be provided by your instructor to those who request it).
- To thyself be true and also to those who work with you.

Do you consider yourself: (Check all that apply)

A Systems Engineer ____________

A Development Engineer ____________

A Program Manager ____________

A Lead Engineer ____________

Other ______________________ Years Experience ________

Grade the skills you feel you have developed (0 to 9)
9 = Expert 5 = Average 1 = Interested 0 = I don't know.

Analog Circuit Design ____________

Digital Circuit Design ____________

Software Coding ____________

Software Design ____________

— Development Specification Writing ____________

Engineering Specification Writing ____________

Breadboard Building ____________

Use Test Equipment ____________

— Data Analysis ____________

SYSTEM ENGINEERING SKILLS SURVEY

— Functional Analysis __________
— Design Review __________
— Customer Liaison __________
— Reliability Analysis __________
— Unit Product Cost __________
— Finishing a Job __________
— Human Factors __________
Acoustics __________
Worst Case Design __________
Value Engineering __________
— Failure Mode & Effects Analysis (FMEA) __________
— Maintainability Design __________
— Safety Analysis __________
— Life Cycle Cost __________
— System Spec Writing __________
— System Partitioning __________
— Standards (Industry, Military, etc.) __________
Error Analysis __________
Error Apportionment __________
Testability Design __________
Accuracy Enhancement __________
Power Supply Design __________
Power System Design __________
— People Systems __________
Mechanical Packaging __________
Stress Analysis __________

SYSTEM ENGINEERING
SKILLS SURVEY

	Electromagnetic Interference (EMI) Analysis	________
	Thermal Analysis	________
	Integrated Circuit Technology	________
—	Time Analysis	________
	Microprocessors	________
	Computer Technologist	________
	Computer Hardware	________
	Computer Software	________
	Probability	________
	Statistics	________
	Hydro-Dynamics	________
	Variance Analysis	________
	Propulsion	________
—	Mission Analysis	________
	Computer-Aided Design (CAD)	________
	Computer-Aided Manufacturing (CAM)	________
	Drafting	________
—	Proposal Writing	________
	Control System Analysis	________
	Data Busses	________
	Architectures	________
	Sketching	________
	Company Procedures	________
	Tools Knowledge	________
—	Working with Others	________
—	Training	________

SYSTEM ENGINEERING SKILLS SURVEY

	Risk Analysis	________
	Travel	________
	Software Systems	________
	Production Design	________
—	System Design	________
	Wire Lists	________
—	Program Management	________
	Lead Engineer	________
—	Solve Difficult Problems	________
—	Planner	________
—	Cost Estimator	________
—	Innovator	________
—	Report Writer	________
—	Communicator	________

Other Skills ____________________________________

____

Education: ____________________________________

Your Job Title: ____________________________________

I'm interested in becoming a Systems Engineer ________

— I'm interested in improving my System Engineering Skills ________

Name: ____________________________________

Present this survey to your instructor for analysis or utilize the manual analysis method outlined in chapter SEE-8.

This survey will be returned to you by your instructor after analysis for specific SEE class skill needs and for team building.

1.8 TEAM BUILDING

The self-assessment skill surveys may be analyzed by a computerized skill score sorting process coupled with an objective analysis by your instructor. Or if you wish, you may manually score your own by referring to chapter SEE-8. You may also use your mission analysis skills from chapter SEE-2 and functional analysis skills from chapter SEE-3 to develop a skill survey and do team building for your special system type problem.

Instructors can build skilled systems engineering teams by using the datum from skill surveys. The overall applicable skill level, number of years of applicable experience, skill profile, degrees, plus various subjective criteria, have been used to build successful workshop teams. This team building process can be done prior to meeting the team members and can be based on the skill survey only. The goal is to establish a work load balanced, high skill level, no critical skill deficiency team with sufficient prior successful applicable experience. In SEE workshops, given twenty-eight students' survey datum, it has been demonstrated repeatedly that they can be formed into four teams of seven each before the class convenes. These teams do work together and perform the system engineering process and system engineering management successfully. When these pre-selected teams first meet, a fifteen-minute team attributes workshop session is set up. This very vocal, fun, fifteen minutes helps the team to quickly get to know each other prior to working together on a project. We tell them after this session that with the total team weight and height, they should be able to walk through walls and leap tall buildings with a single bound.

1.9 TEAM ATTRIBUTES WORKSHOP INSTRUCTIONS

The team will be seated around a table. They will ask: What is the problem? They then will *orally* collect and *record* the following data for each team member on a form similar to the electronic control systems example shown in figure 1-3.

- Name and address
- School and degree
- Years experience (include applicable prior to degree)
- Experience (single most prominent)
- Specialties (pick top 3):

  H = Hardware

  SW = Software

  A = Analyst

  P = Program management

S = Systems engineer

M = Mechanical

O = Other

- WT (record your weight)
- HT (record your height)
- New and good (what is new and good for you within the last year? Just married? New baby? Discovered you are part Norwegian, etc.?)
- Total the teams: (# years experience, most predominate experience type, top three team specialties*, WT, HT, most predominate new and good)
- Comment: Is your team "right brain" (intuitive), "left brain" (rational), or "composite?"

  *Does your team have any skill deficiencies? If the team has a skill deficiency, what are you going to do about it?

  You should discover that your team is made up of (or has the skill potential for):

  — A mechanical systems engineer

  — A hardware (circuit design) systems engineer

  — A software systems engineer

  — An analytical systems engineer (preferably a control system analyst)

  — A program manager systems engineer

  — A high skill level, flat skill profile, very experienced system engineer—mentor

  — Other specialty engineer

  — Some inexperienced learners

Now as you work on solving a problem by using the system engineering process, you will find it is fun to work on a highly skilled system engineering team. Team building by skill profile matching and establishing mentoring relationships can make your job fun and rewarding. The inexperienced learners must ask questions to ensure mentoring is taking place. These questions will force the mentors to think as well as the mentored to learn. You must learn to respect each others skills, communicate, and work together toward a common goal. Chapter SEE-8 provides sample workshop problems and workshop worksheets for the team to perform system engineering process and system engineering management.

1.10 A QUICK SELF-ASSSESSMENT SKILL SURVEY

Occasionally, it is necessary to expedite the team building process because time does not allow for analysis of the seventy-five skills survey. The short form similar to that shown in fig. 1-4 can then be used to develop a quick self-assessment. The left side pictorial plot on this form provides a quick profile; the right side shows system engineering skill, and the bottom shows the overall skill level.

SEE TEAM # __________ ATTRIBUTES DATE: __________

NAME MAIL STATION	SCHOOL DEGREE	YRS EXPR	EXPERIENCE	SPECIALTIES (PICK TOP 3) H/SW/A/P/S/M/O	WT / HT	NEW AND GOOD
TOTAL						
AVERAGE						

H = HARDWARE
SW = SOFTWARE
A = ANALYST
P = PROGRAM MANAGER
S = SYSTEMS ENGINEER
M = MECHANICAL
O = OTHER
WT = WEIGHT
HT = HEIGHT

Fig. **1-3.** Team attributes workshop recording form

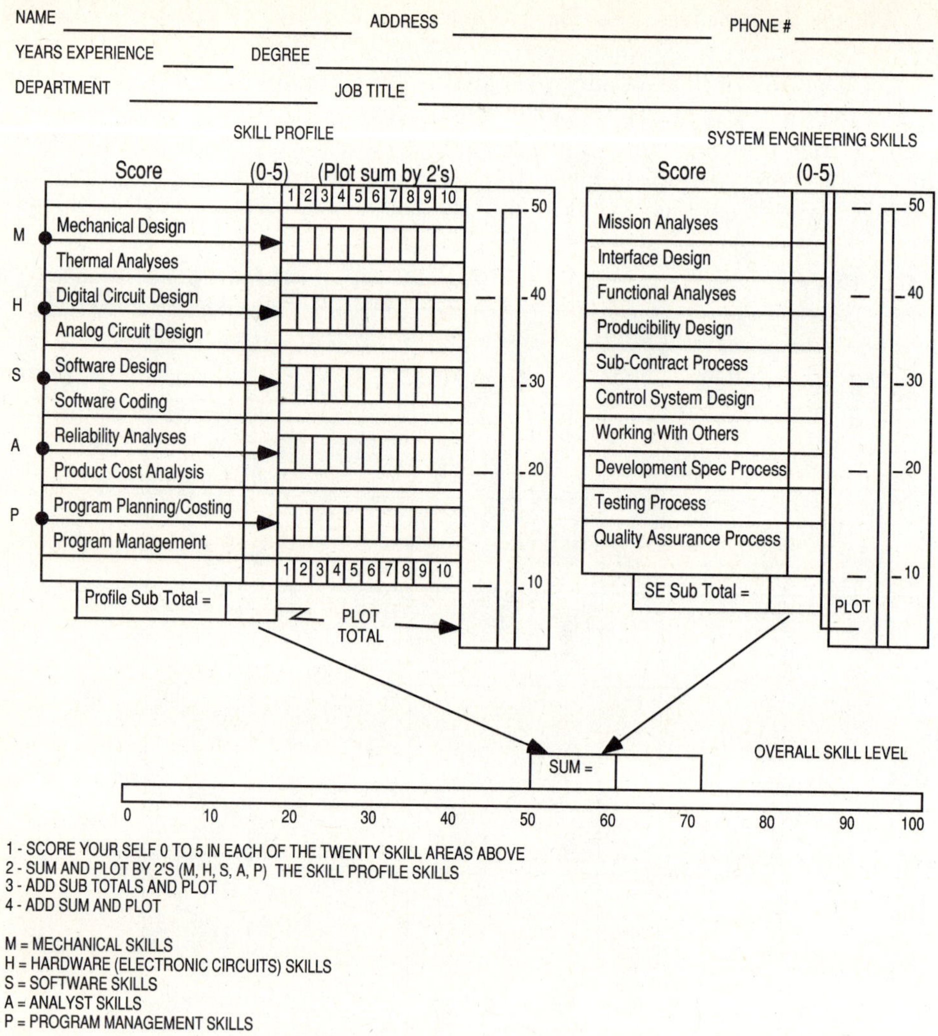

Fig. **1-4.** Short form—self-assessment skill survey

This data only takes a few minutes to obtain and, when coupled with the data on the top of the form, can be used to start to build a balanced systems engineering team. The goal is to establish a highly skilled team with prior successful applicable experience. This form is obviously not as effective as the long form survey because it does not collect as much data.

1.11 The System Perspective

The system perspective is the ability to see the big picture, that is, the development of an understanding of the system's life and purpose. In "IEEE Transactions for Systems Science and Cybernetics," A. D. Hall III used a three dimensional structure of systems engineering.[4] If you wish to further study the genetic structure of systems engineering as postulated by A. D. Hall III, his book *Meta Systems Methodology* is recommended.[5] Through a series of modifications to the original concept, I have evolved a six-sided cube to use as an aid in developing a big picture perspective for specific systems. The cube in figure 1-5 helps show the relationship of six system viewpoints embodied within the system perspective. These six ways to look at a system are: 1) its relationship to other bigger systems above it (hierarchial); 2) its relationship to past/present/future (time); 3) its physical characteristics—what is the system made of? 4) its constraints (for example, limited technical capability required and limited funding to develop); 5) actions it requires (plan for what the team must do), and 6) its specific time and cost goals (how much does the system cost to buy and to own, and how much time and money is available for the team?). The summation of these six viewpoints onto a cube on a single piece of paper provides the big picture system perspective.

The systems engineering team has established a clear problem statement, done a skills survey, and some team building. Now it needs to develop a system perspective.

When you apply the systems perspective cube there is a recommended order to follow:

1. Hierarchial structure—Your system is always only a subsystem. This is a very important perspective because you need to recognize and interface to the bigger system above.
2. Time line—What has been done in the past? What is the market? What must be done in the future? Where are we on the time line? (Note that the deploy/recycle phase of the system's life has been further broken down to show that recycle should also include concern over disposal. What do we do with the excrement from our system? The disposal of garbage, nuclear waste and sewage are part of the initial system perspective concerns.)
3. Physical attributes—All systems involve people, so who are they? Also what is the system hardware, software, documentation, tools, and facilities?
4. Constraints—What are the technical constraints on the system (its capability, weight, power usage, etc.)? What are the program constraints—how much time and money have been allocated to do this specific development program?
5. Actions to perform—What must be done now to fulfill the development program mission? What is the plan?
6. Specific time and cost goals—What are the dates development is to be started and completed? How much money has been allocated to do the development? When is the system (subsystem), or product to be ready for initial use by the customer? How many are to be produced? How much is the subsystem or product to cost?

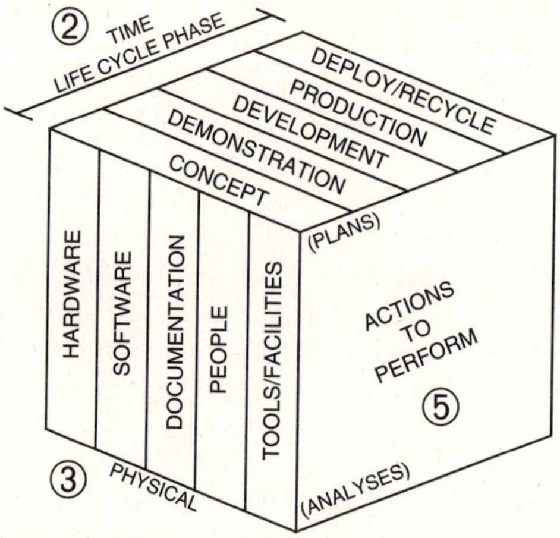

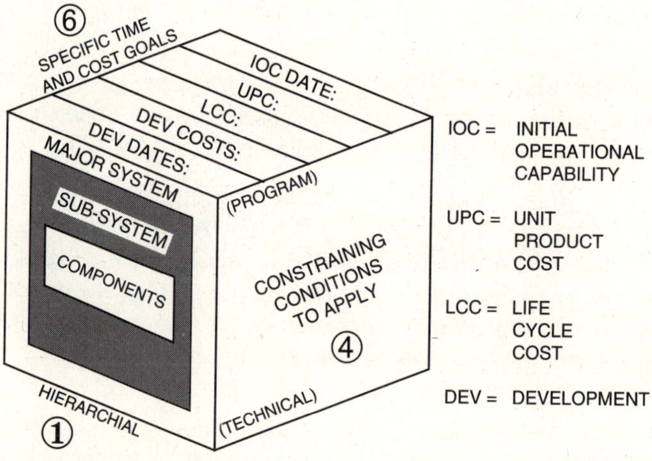

Fig. **1-5.** The system perspective cube

1.11.1 Examples of the System Perspective

The three following figures show the cube used as a worksheet to develop a system perspective on different types of missions. Study these figures by looking at the sides of the cube in the following order: 1) hierarchial, 2) time, 3) physical, 4) constraints, 5) actions, and 6) specific time and goals. This logical sequence has been shown to expedite the use of the cube regardless if you are developing your own or reviewing someone else's system.

1. The Nutcracker Sweet
   We have been asked to develop an automatic nut processing subsystem for a chocolate candy factory. The nuts are purchased in the shell with a mixture of various types of nuts at a low price in January every year. The problem is that they must be handsorted and cracked by manual labor. We need to study this problem to better understand how nuts are sorted and cracked and how fast this is done. We need to know if the people now doing it have any ideas to make the job easier or better and if we will have to retrain and find other jobs for some people. Finally, we want to know what equipment already exists to help perform automatic nut processing, and what the skills of the present workers and the development team are. With the information we have gathered through study and questioning, we need to develop a system perspective. Figure 1-6 provides a worksheet for the Nutcracker Sweet system perspective.

2. The Flight Control Computer—Unit Product Cost Reduction
   It has been discovered that the cost of a flight control computer now being designed is higher than the unit product cost goal. A system engineering team has been assigned by the program manager to develop ways to reduce the unit product cost (UPC). This "Green Team's" system perspective is shown in figure 1-7.

3. Engineering in the 1990s
   A systems engineer has been asked to make a presentation at a technical symposium. He uses the system perspective cube to develop a big picture perspective. Figure 1-8 shows how the cube helps him systematically approach a ten year mission.

These three examples demonstrate that the cube can be used to help you quickly develop a big picture perspective for a wide range of problems to solve and products and processes to develop. Anytime a systems engineer is to do some system engineering, the cube is an excellent tool to use to get started.

When you use the cube, follow an orderly methodology. For example, always consider your project with respect to the system perspective cube. First, on the hierarchial side, you are always working on a subsystem; no matter what your project, there always is a major system above your project and components of your project. It all depends upon your perspective. For instance, a resistor is a subsystem to a resistor manufacturer. His resistors are used in electronic assemblies that constitute major systems. The components to his subsystem are leads, resistor body material, marking, and lead bonding. In like manner, your project is a subsystem. Your flight control (sub)system is really only a (sub)system of the avionics (sub)system, which is really only a (sub)system of the aircraft, and so on. It is a lack of this subsystem understanding that leads to system integration interface problems. Remember there always is a bigger system above.

Next, establish which phase of the time life cycle your subsystem is presently in. Then, list the five basic physical attributes of your subsystem and also note

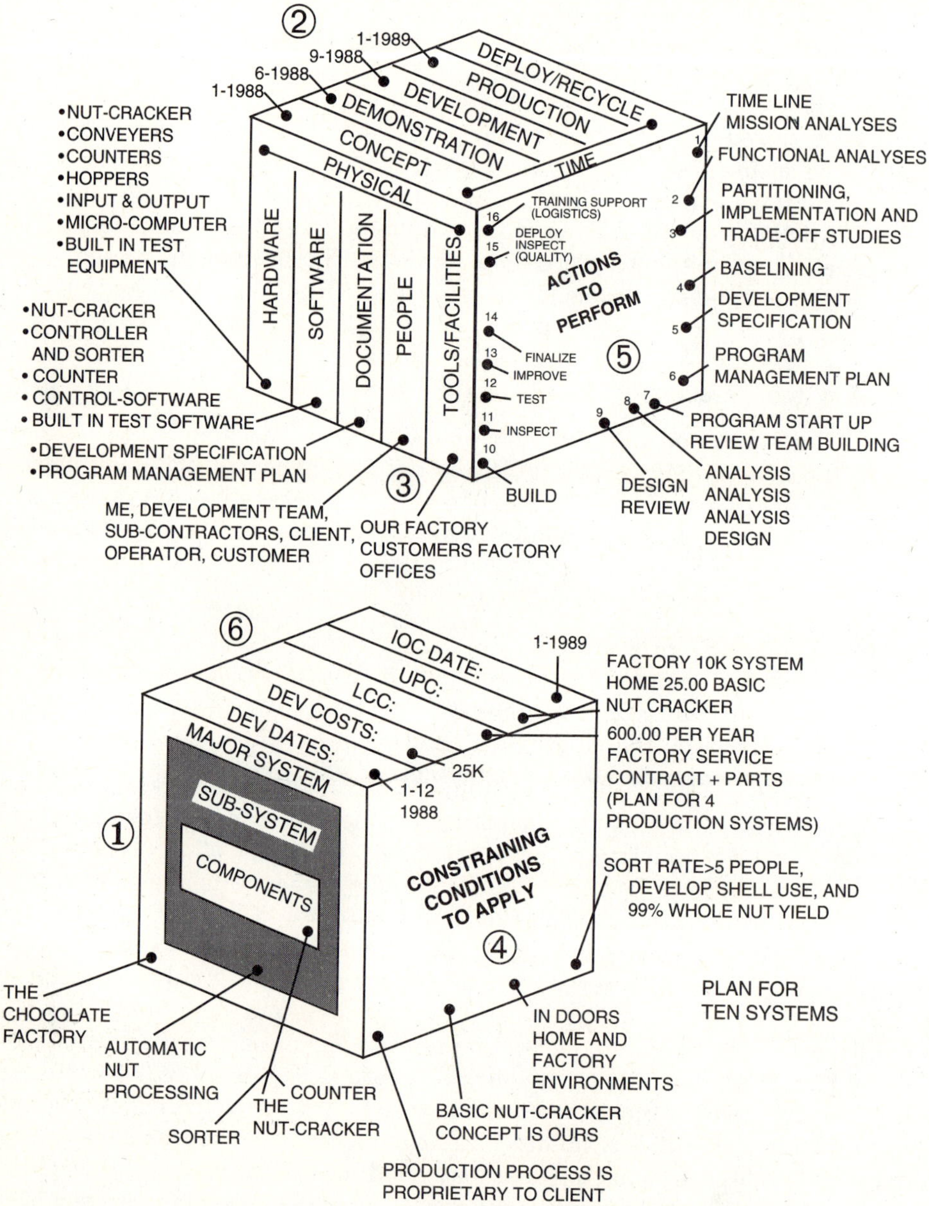

Fig. **1-6.** An example worksheet for the system perspective (the nutcracker sweet)

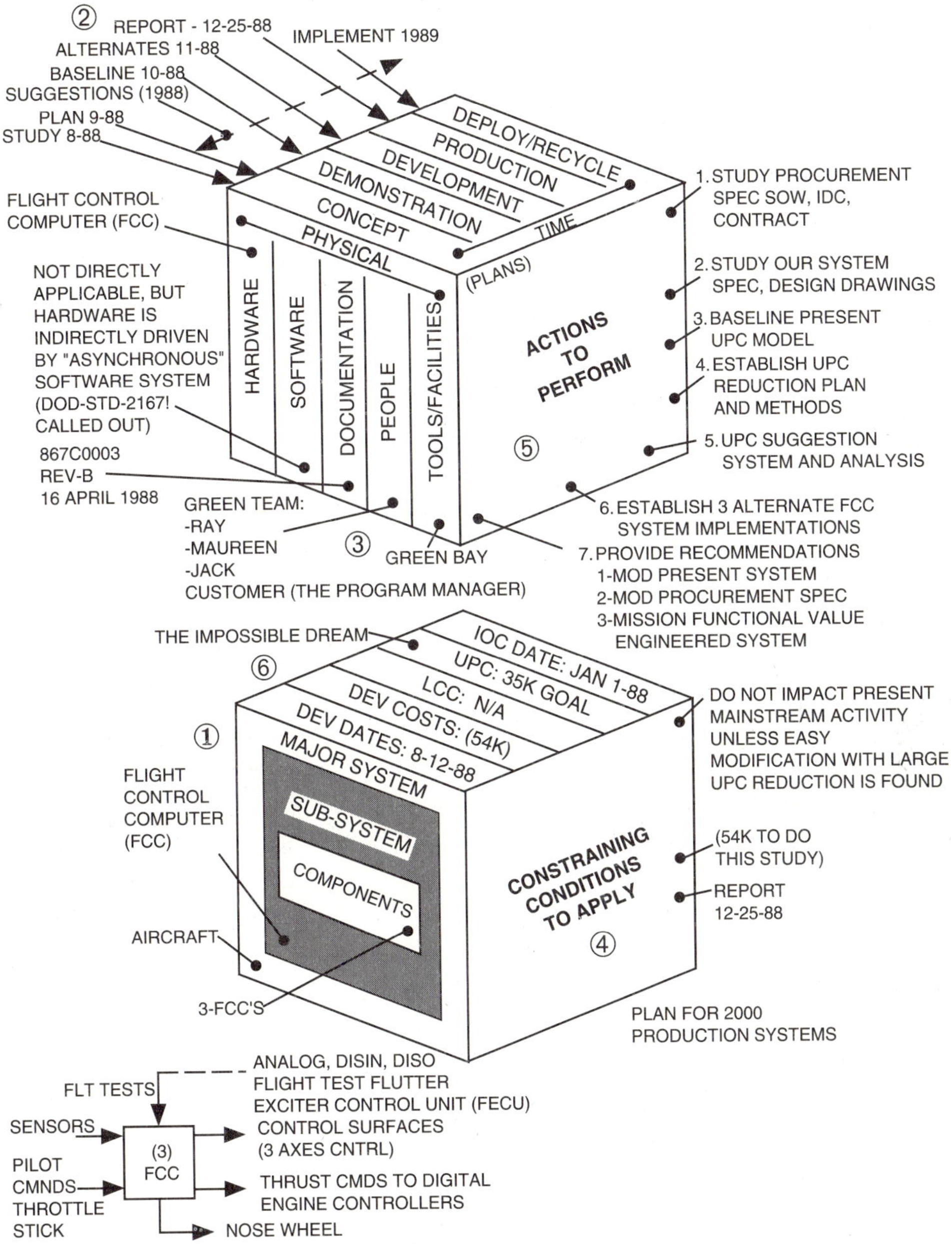

Fig. 1-7. An example worksheet for the system perspective (FCC—UPC reduction)

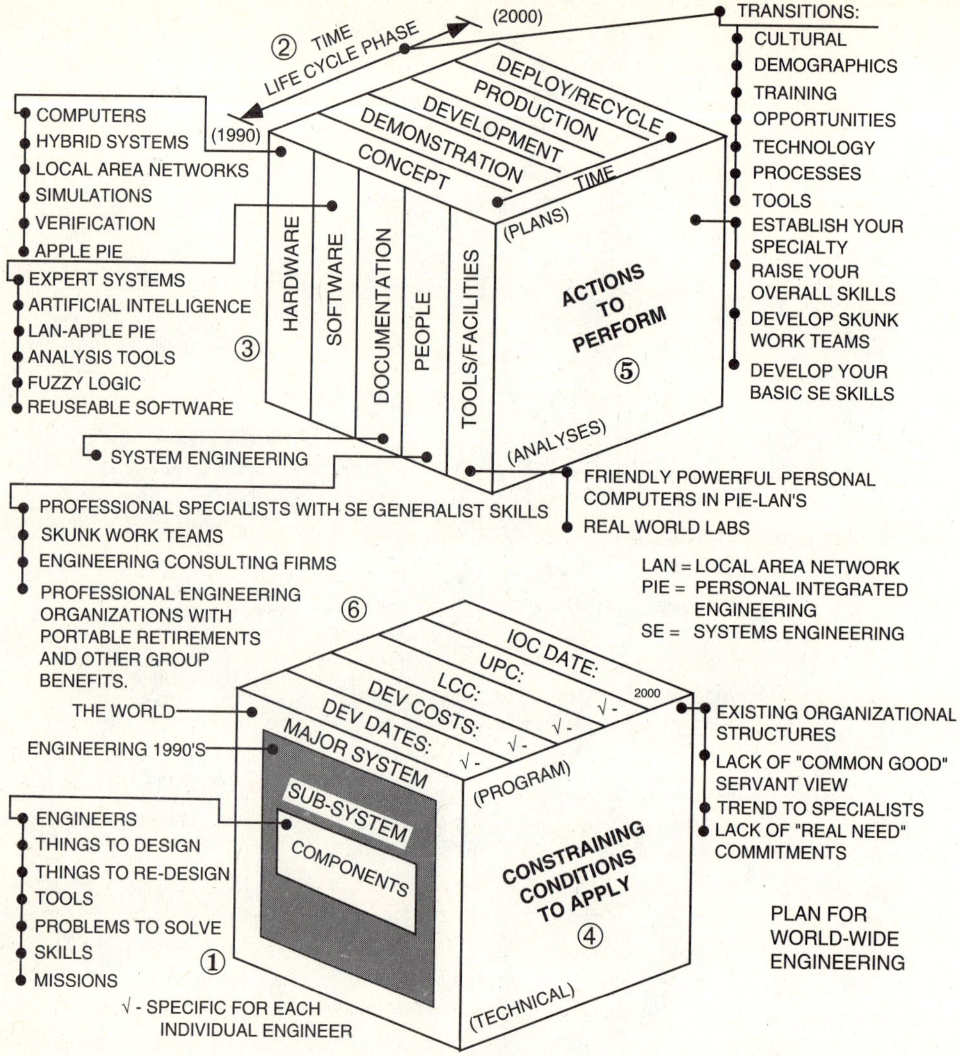

Fig. **1-8.** An example worksheet for the system perspective (engineering in the 1990s)

the actions to be performed by the personnel, any constraining conditions, and specific costs/dates applicable to your subsystem.

Look back (and forward) to prior and future phases of the time life cycle. Does your subsystem take full advantage of what was learned from prior phases? In your present phase, is consideration being given to future phase requirements? Is your subsystem going to be producible with a profit when it reaches the production phase?

Applying the cube in an orderly methodology to study your subsystem helps you develop your system perspective; you are beginning to see the big picture. You are improving your chances that your subsystem will interface to its "major system" and will successfully provide its intended function for the desired mission. On the "actions to perform" side you are starting your management planning.

The application of the systems engineering perspective is required at all points in the hierarchy for a system. The aircraft designer, the aircraft avionics designer, the aircraft flight control designer, the navigation system designer, the digital-to-synchro card designer, the digital-to-analog converter designer, *all* must develop the system engineering perspective for their particular subsystem.

1.11.2 Applying the Cube to a Matrix Format

Let us apply the system perspective to the design of a digital-to-synchro converter printed circuit card assembly. A digital-to-synchro converter converts digital data from a computer to alternating current (AC) voltage that drives a synchro motor. Our subsystem is a printed circuit card assembly (which will be baseline defined in a specification by a detailed block diagram that shows all the electrical interface, and a mechanical assembly sketch showing the physical hardware, size, weight, power, and transfer function equations). The major system, my subsystem, is part of a navigation system. Our subsystem receives digital signals from a computer (which is another subassembly of the navigation system). My subsystem provides outputs to other major systems, the cockpit displays and the flight control system, thereby requiring output interfaces from my major system (the navigation system).

We are designing this subassembly for a demonstration flight test program. Our subsystem consists of: 1) Hardware (digital computer bus buffers, digital-to-analog converters, Darlington power output drivers assembled on a 6″ × 6.25″ × 0.5″ printed (6 layers) circuit card assembly); 2) Software (computer program sequences that properly load the digital-to-analog converters); 3) Procedural data (a development specification that defines the subsystem, its requirements, and test procedures. Schematic and assembly drawings are also provided); 4) Personnel (the designer, a test technician, a build technician, a systems engineer, the customer, a software programmer, our boss and an analyst—all of whom we must interface with); 5) Facilities (desk, data files, a lab to build in, a lab to test in, actual displays to simulate aircraft installation, environmental test chambers, test equipment (scopes, meters, etc.), build tools, programming facilities, test jigs).

The actions to be performed by the personnel are design, analysis, testing, building, interface analysis, communication with all the interfaces, specification development process, software programming, test jig design and test, data taking and analysis. The constraining conditions are –40° to +75°C operating temperature. Unfortunately, the customer considers the interface to the displays a low priority issue, and the customer procurement specification is very poor. It is questionable whether the required circuitry will fit on the 6″ × 6.25″ × 0.5″ printed circuit card allotted. Our interface must work with a dozen ill-defined aircraft type interfaces, and our schedule is only three months to design, develop, build and test prior to flight test. We do not have all the key personnel or facilities we need

TIME	PHYSICAL	CONDITIONS	ACTION	DATES & COSTS	HIERARCHY
CONCEPT **GROUNDED Z-LEG,** **NEW DIGITAL TO ANALOG (D/A) CONVERTER**		NO SCOTT TRANSFORMERS, NO TRANSFORMER ISOLATION NO SECOND SOURCE	STUDY AND TRADE-OFF ANALYSIS	COSTLY	
DEMONSTRATION **DIGITAL TO SYNCHRO CONVERTER SUBASSEMBLY**	**HARDWARE:** 6X6.25X0.5 IN. PC CARD **SOFTWARE:** PROGRAMMED SEQUENCER **DATA:** SPECIFICATION SCHEMATIC ASSEMBLY DRAWING **PERSONNEL:** DESIGNER TECHNICIANS **FACILITIES:** DESIGN TOOLS LABORATORY	POOR CUSTOMER SPEC, 3 MONTHS, TIGHT FIT, INTERFACE DEFN. POOR, -40°C TO +75°C OPERATE, KEY PERSONNEL NEEDED	DESIGN, ANALYSIS, TEST, DATA ANALYSIS, SOFTWARE, SPECIFICATION PROCESS, COMMUNICATE, BUILDING, RELIABILITY ANALYSIS	UPC = 3K, LCC = NO FIELD COSTS DEV $50K, 3 MONTHS	MAJOR: NAVIGATION SYSTEM (AND INTERFACE TO COCKPIT DISPLAYS)
DEVELOPMENT	AS ABOVE	2ND SOURCE FOR D/A	WORST CASE DESIGN, PRODUCIBILITY, FAILURES, DESIGN REVIEWS	UPC = 3K COST REDUCTION	
PRODUCTION	AS ABOVE PRODUCTION TEST EQUIPMENT		PRODUCTION READINESS REVIEW BUILD 1000		
DEPLOY RECYCLE	AS ABOVE	REMOVE THIS ASSEMBLY IN THE FUTURE BECAUSE OF DIGITAL DISPLAYS?			ALL DIGITAL DISPLAYS

Fig. **1-9.** System perspective example—worksheet

to do the job. The unit product cost we are designing for is $3000 at a quantity of 50; life cycle cost is not defined; no field maintenance will be allowed (send back printed circuit card assembly to factory); development costs are to be less than $50,000 with completion in three months.

Looking back, prior work has indicated that a "grounded Z-leg design approach" may provide adequate room to get all the circuitry on a

6″ × 6.25″ × 0.5″ printed circuit assembly. Also, a new combined buffer, digital-to-analog converter, and integral output power driver component has been studied.

Looking ahead, a second source must be found for the key special parts required in order to keep unit product cost down. Producibility, testability, fail safe, and reliability features must be designed in *now*. Interface testing with all possible external display interfaces should be analyzed now and planned for testing soon to ensure useability on a long list of aircraft. Figure 1-9 places the above data on a worksheet for the system perspective cube.

The matrix worksheet method of doing the system perspective allows you not only to develop your big picture perspective at a specific point in time, but to better study past and future system life cycle phases. This method can be used to construct very large and highly detailed total life time system perspectives. One of the objectives of the systems engineer should be to put the total system definition on a single piece of paper so system developers and users can get the big picture system perspective. These single pieces of paper may get to be very large, but they can be powerful tools.

1.11.3 The System Perspective Worksheet Conclusions

The cube worksheet is recommended for getting a quick big picture perspective. It helps generate a three dimensional perspective of the system and is especially good at pointing out that the system is really only a subsystem of a bigger system above. The matrix worksheet is slower but provides greater depth in developing a system perspective. It is very helpful in accentuating the system changes through the various time phases of concept, demonstration, development, production, and deploy/cycle. The matrix is helpful for total system life program planning by pointing out changing physical, actions, conditions, and dates and costs.

1.12 STUDY PROBLEMS

1. Provide a worksheet that shows application of the system perspective to your present job.

   Note: If your present job does not seem appropriate, apply the system perspective cube against buying a car, a house, installing a swimming pool in your backyard, or other similar personal project.

2. Twelve comments that relate to the system perspective cube follow below. Establish how each is related to the cube. Does the comment relate to: activity prior to the cube, the entire cube, the hierarchial side, the time side, physical side, conditions side, action side, dates and costs side, physical-people, physical-tools/facilities, physical-hardware, physical-software, physical-documentation, action-analyses, action-plans, etc.?

   a) A lack of system engineering perspective is like using a needleless mirrored compass. When you look at it, it will show you who is lost, but you will not know what direction you are going.

b) You need to know the history of an existing system in order to improve it. You also need to know how it is presently functioning.

c) Let me ask you a question so you may know what to do.

d) The system perspective can be shown by a six-sided cube.

e) I prefer teams over committees.

f) The team has a great attitude, but I only wish they could clearly communicate.

g) Quality starts with a complete baseline, specific requirements, and a good design that can perform the system mission.

h) The system specification is a living document.

i) Continuously find and solve problems.

j) System engineering requires planning.

k) Your system is always only a subsystem.

l) The systems engineer is smarter than the collection of inanimate objects that make up a system; unfortunately the systems engineer does not always take the time to prove it.

3. A team building exercise is highly recommended utilizing self-assessment skill surveys and the team attributes workshop. Remember that it is the skills the team is deficient in that will get you in trouble. As a team you should worry about what your team doesn't know instead of trying to impress each other with what you do know. Worksheets and workshop problems are provided in chapter SEE-8. Your team will do the team attributes exercise and also will do the system perspective cube for the problems provided.

4. Provide a generic total life time system perspective matrix. Fill in each box of the matrix with elements to be considered for any system.

5. Make a list of all the generic system engineering skills in the 1.7 "Systems Engineering Skills Survey" example.

6. Develop a skill survey for your specific system-type—start this exercise by selecting all the generic systems engineering skills in the 1.7 "Systems Engineering Skills Survey" example. Then brainstorm to include a sufficient number of skills (plus knowledge and experiences) required for your specific system type. Apply the survey to yourself and your team.

1.13 References and Recommended Readings

1. *The American Heritage Dictionary of the English Language*, 2nd ed. Boston: Houghton Mifflin, 1986.

2. Hall, Arthur III. *A Methodology for Systems Engineering.* New York: Van Nostrand Reinhold, 1962.

3. Peck, T. G. "Worldwide Systems Engineering." *IBM Systems Journal,* 24, 1985, p. 186.

4. Hall, Arthur III. *IEEE Transactions for Systems Science and Cybernetics.* New York: IEEE Press, April 1969, p. 156–160.

5. Hall, Arthur III. *Meta Systems Methodology: A New Synthesis and Unification.* Tarrytown, New York: Pergamon Press, 1989.

1.13.1 Recommended Readings

Augustine, Norman R. *Augustine's Laws.* New York: Viking Penguin, 1986.

Bateson, Robert N. *Introduction to Control System Technology.* Columbus, OH: Merrill Publishing, 1989.

Blanchard, Benjamin S., and Walter J. Fabrycky. *Systems Engineering and Analysis.* Englewood Cliffs, NJ: Prentice Hall, 1981.

Brooks, Frederick P. Jr. *The Mythical Man Month.* Reading, MA: Addison-Wesley, 1982.

Chambers, George J. *The System Engineering Process: A Technical Bibliography.* IEEE Transactions On Systems, Man, and Cybernetics. New York: IEEE Press, 1986.

Chestnut, Harold. *Systems Engineering Tools.* New York: John Wiley & Sons, 1965.

Chestnut, Harold. *Systems Engineering Methods.* New York: John Wiley & Sons, 1967.

Deming, W. Edward. *Quality, Productivity, and Competitive Position.* Boston: MIT Press, 1982.

Eisner, Howard. *Computer-Aided Systems Engineering.* Englewood Cliffs, NJ: Prentice Hall, 1988.

Engineering Management. Military Standard 499A, Department of Defense, May 1, 1974.

Goldratt, Eliyahu M., and Jeff Cox. *The Goal—Excellence in Manufacturing.* Milford, CT: North River Press, 1984.

Hall, Arthur III. *A Methodology for Systems Engineering.* New York: Van Nostrand Reinhold, 1962.

Specification Practices. Military Standard 490A, Department of Defense, May 18, 1972.

System Engineering. U.S. Army FM 770–78, Department of Defense, April 1979.

System Engineering Management Guide. Defense Systems Management College, Fort Belvoir, VA, October 1983.

Transition from Development to Production. Department of Defense Directive 4245.7, 1983.

Truxal, John G. *Introductory System Engineering.* New York: McGraw Hill, 1972.

Mission Analysis

(They shall follow the right path)

Writing this book is a mission. The fact that you are reading this book is an indication of mission accomplished. If no one ever reads the book, then I have had a mission impossible. So what? Who cares? It doesn't exist. The fact that you are reading the book does not tell how effectively the mission has been accomplished. If, however, you practice and apply the basic system engineering skills to develop and improve systems, then the mission of the book has been accomplished.

When I wrote this book, I followed the basic systems engineering process. I first needed to establish a *reason* for doing it. This was easy. I recognized a need to reduce the number of mission impossibles. This book is therefore dedicated to all those who have been killed or seriously hurt physically or socially by poor system engineering. The application of good system engineering can help us avoid a lot of human grief.

The second thing I needed to do was *gain knowledge*. What is a system and what has been published to help me do system engineering? I have read and studied hundreds of books, papers, and abstracts on the subject of systems, system engineering, and many other associated subjects such as quality. The search for knowledge has become an obsession, so I continue to read and study.

I then needed to establish a big picture perspective, the *system perspective*. This is based on my reason and my knowledge gained. I have learned, for example, that all systems are control systems; they have inputs, activities, outputs, and control feedback. A means is established to obtain feedback so my work can be controlled. Therefore, manuscript reviews, book reviews, educator and student feedback methods were established. This book writing is a subsystem of a bigger system. It is developed, produced, deployed, and improved to a defined schedule. Physically, the book is hardware, software for both text and graphics, and permission forms that were required to use prior published references. It required tools and facilities, and I needed to recognize all the people involved—students, educators, reviewers, publisher, and authors. I needed to establish constraints, actions to perform, and how much money and time could be expended to write and publish the book. I asked how many pages will the book have and what will the book cost?

Mission analyses were then done. The major system sortie indicated that the student purchases the book, reads it, is lead by the instructor, does the work assignments, and thereby develops knowledge and skills to do the systems engineering process successfully so mission impossibles are avoided or overcome. In addition, the life mission of the book itself requires analysis. The book was baselined, a publisher found, a contract established, much re-writing was done, the manuscript was submitted, reviewers reviewed; the manuscript was rewritten, finalized for publication, published, advertised, sold, used in major system sorties, evaluated, feedback received, revised to improve next edition published, and so on. Every mission was analyzed for failure modes and effect (FMEA = failure mode effect analysis). What if no student purchases the book? What if I become ill and cannot work on the book? I needed to modify the missions to compensate for these possible contingencies. Mission analysis included developing mission requirements to overcome or provide fallback missions to prevent failures from causing mission impossibles.

Given the missions for writing this book, I then did *functional analysis.* I brainstormed the question: What are all the actions of persons and things required to successfully perform the book's missions? This resulted in functional diagrams showing all the functions required to ensure that the missions will be accomplished. *Trade-off studies* were required to select the optimum implementation of computerization of the manuscript. I then *baselined* the book. This was easy because I had previously developed an eight-chapter workshop course manual, the *Honeywell System Engineering Manual.* Specific requirements had been established by the publisher regarding formatting, content, form, indexing, and permissions. These along with the baseline constituted a *development specification.* I had also established a development plan with a schedule, methods, a team, and costs. This development plan had a goal to provide the initial manuscript to the acquisitions editor by June, 1993, a three month manuscript review process, incorporation of reviewers comments, resubmitted manuscript prior to 1 January 1994, a production process with a publish date of late 1994 or early 1995. This plan constituted a *program management plan.*

If I can use system engineering process to write a book, I or you can also use it to accomplish any mission. In this chapter you will develop skills in mission analysis. Figure 2-1 shows where we are in the system engineering process.

2.1 MISSION ANALYSIS

What is mission analysis? Mission analysis is the development and analysis of time line scenarios. Mission analyses start with a major system sortie. It is

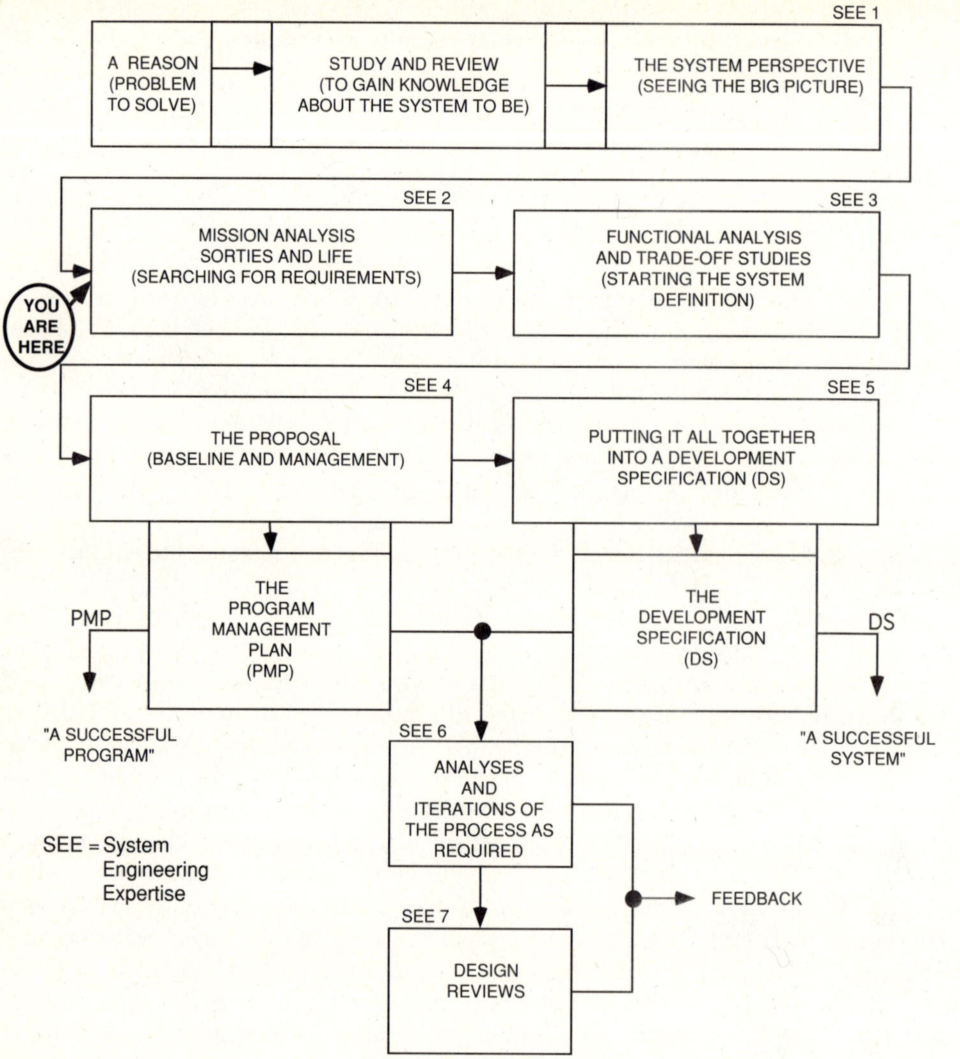

Fig. 2-1. The system engineering process

then flowed down to subsystem missions, component missions, proposal effort missions, program management missions, design process missions, production process missions, customer support missions, etc. Anytime there is a mission to accomplish, there are many time lines and, consequently, mission analyses must be done.

Mission analyses are a very important part of the system engineering process. You must work very hard to develop your mission analysis capabilities because this skill is used for the following system applications.

1. Developing a new system. Mission analysis is a tool used to establish the system requirements and to ensure you have not forgotten any-

thing. Sortie missions, mission life of a system family, life mission of a specific system family member are examples of these types of analyses.

2. Managing the development of a system. Management planning time lines are also "missions" to be analyzed. Proposal, concept, demonstration, verification, full-scale engineering development, production, and deployment programs all need program mission time line development and analyses thereof.

3. Analyzing an existing system. Mission analysis can be used to study and improve existing systems. Existing systems not only are systems we produce or use but are also process systems such as engineering, production/manufacturing, education, retailing, service—process systems. All existing systems must be continuously analyzed by time line mission analyses in order to continuously improve quality and safety of systems.

4. Subsystem development. Subsystems can be analyzed for requirements using mission analysis. Mission analyses flow down from major systems to subsystems and to components. The mission of the system above us must always be analyzed and it must be recognized that everybody and everything has a mission that is related to the bigger system.

5. Working with a customer to develop a system by doing mission analysis in support to the customer. Mission analyses skills can help the customer (and you) to develop successful constantly improving quality systems that perform the customer's missions.

In the introduction to this book, I talked about mission impossibles where systems and products failed to perform their missions. The main reason these systems failed is because there was a lack of in-depth serious mission analysis coupled with failure mode and effects analysis (FMEA), and with safety analysis. The more dangerous or difficult the mission, the more reason to do serious in-depth mission analyses. I had experience working on the mission analysis of the Apollo man on the moon mission. This experience, along with many subsequent system developments, show that if you don't do mission analysis, you will have problems.

The following examples illustrate how the neglect of mission analysis can create problems. Customers don't buy from you anymore, and you don't know why (you are not working on their mission). Students are not getting an education they can use (you are not relating to the students' life mission). Your manufacturing system has poor yield (the product you are trying to manufacture was not designed to meet a manufacturing mission time line). Your management mission fails (you did not do a software development mission analysis). Your equipment cannot be installed because it will not fit through the door (you did not do a life mission analysis of installations). Your ship shoots down a commercial airplane by mistake (you did not do a good mission analysis of what the system operator had to do manually or mentally with the data your system provided).

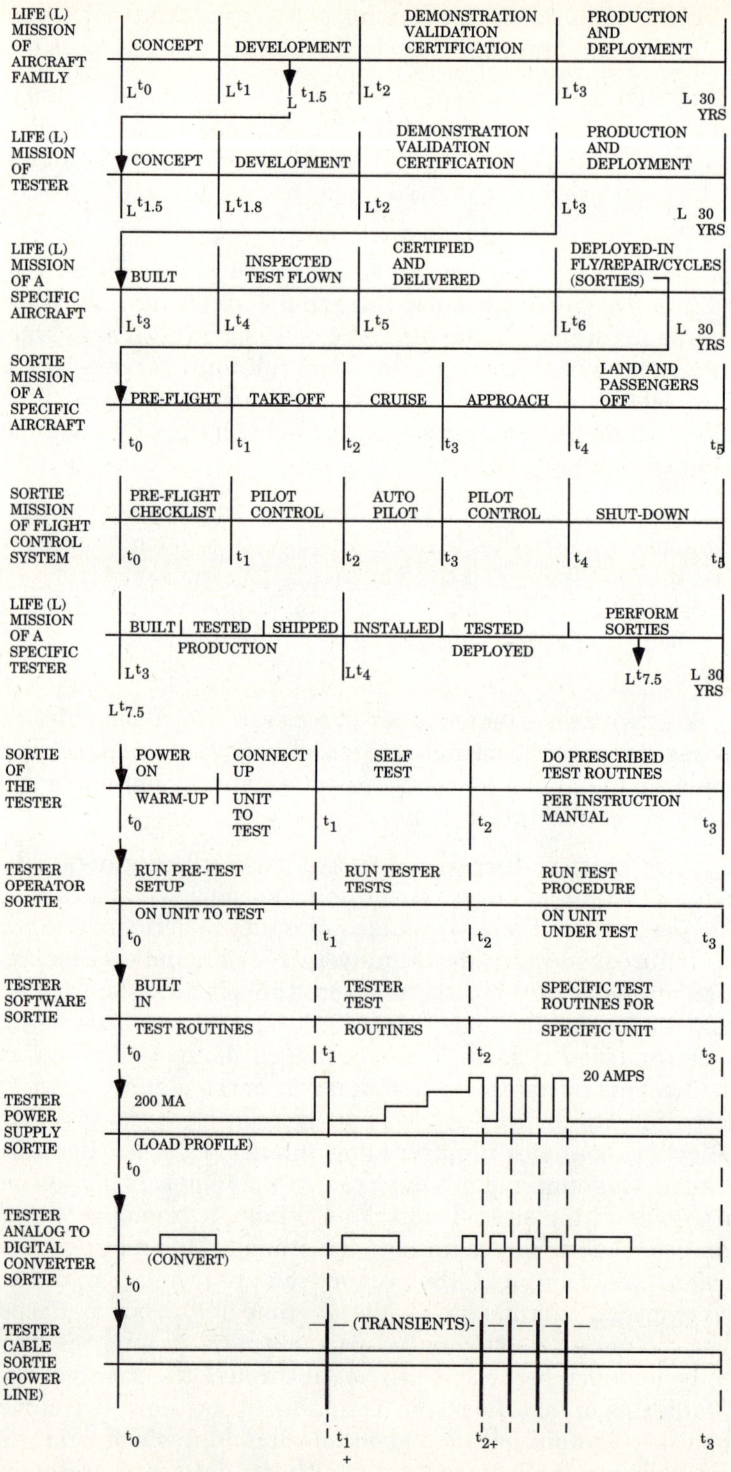

Fig. 2-2. Relationship of missions and sub-missions: mission relationships of tester equipment that is to test the electronic equipment on an aircraft.

Software programming causes spurious hardware dynamics that cause your system to shut down (your software designers and hardware designers did not do mission analyses of hardware and software timing). All three channels of your triple redundant system turn off at the same time (your software designer did not do mission analysis of common built-in software's impact on the bigger system mission). Three airplanes crash in three separate events when they fly by a powerful radio tower (mission analysis did not consider the electromagnetic interference environment this electronic fly-by-wire triple redundant system would be exposed to). Your new system causes fires (the system timing dynamics along with long installation cables generate large transient voltages that cause component failures at the interface to your system). The time spent early on in the development of a system doing mission analysis pays off by reducing the number of mission impossibles.

Mission analysis applies to everything and every person involved in a system. The following example shows the relationship of missions and sub-missions in a system.

The life and sortie mission time lines of figure 2-2 show that all system, subsystem, component, and people associated with a system are related in time. It is very important that the system engineer is capable of developing and analyzing all these time lines. The system engineer must analyze all events occurring along all time lines. Breaking down the life mission to lower-level sortie missions is an effective method to facilitate analysis. The remainder of this chapter is devoted to introducing mission analysis tools and subsequently showing you how mission analysis is performed. It should be obvious that everything and everybody involved in a system has a mission, all of which are related to the sortie of the major system.

2.1.1 Sortie Missions

Sortie missions are the basic reason the system exists. Systems are developed so they are capable of performing their various sorties. Likewise subsystems and components have sorties that are sub-missions they must perform to support the purpose of the major system or subsystem above them. Usually, a sortie mission is described as a scenario. For example, the helicopter will take off in any weather at any time of day from an aircraft carrier, fly 100 ±20 feet over the ocean for 60 ±40 N miles, pick up two passengers and gear not to exceed 600 lbs

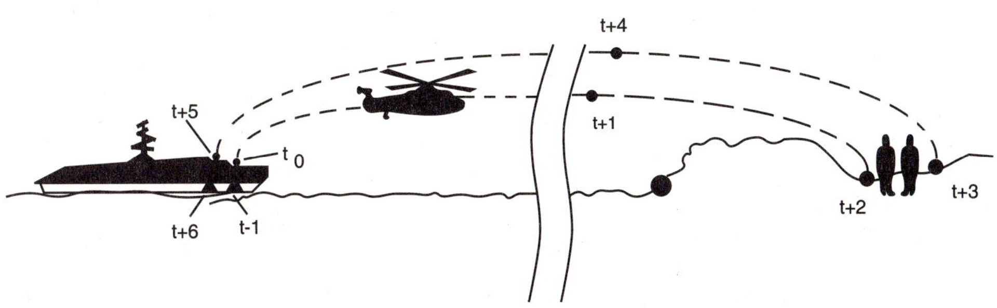

Fig. 2-3. A sortie mission

from a ball field 10 ±10 N miles inland, fly evasive nap-of-the-earth maneuvering back to the ocean, then return to the aircraft carrier by flying at 520 ±480 feet altitude. This sortie mission can be drawn as a time line pictorial (see fig. 2-3).

The analysis of this sortie mission will include the following:

1. What are all the possible environmental weather conditions for this mission?
2. What are all the bigger system interfaces?
3. What are all the bigger system interface variables?
4. What is the impact of mission scenario variables such as passengers not there yet, 80-foot waves, ball field in the middle of four 20-story buildings, a ball game in process, someone shooting at you, passengers are the President/Vice President and it is being reported on national television?
5. What is the impact of equipment failures?
6. What personnel training is required?
7. What alternate implementations are needed to perform the basic mission?

Examples of sortie mission scenarios are given in figures 2-4, 2-5, 2-6, and 2-7.

Mission analysis is done to make sure the mission can be successfully performed; that is, it must be ensured that any contingency that could result in a failed mission hasn't been overlooked. It is a mental exercise in which the mission is thought through step by step many times. It is like rehearsing the mission over and over to get it right. It helps define a system and its requirements that will be capable of performing the mission. When mission analysis is seriously performed, along with the subsequent functional analysis and total system engineering process, we successfully perform helicopter sortie missions, win proposals, solve system problems and develop successful subsystems. Sortie mission analysis can also be used to develop a new jury selection process, improve medical test data timing or develop a work-at-home system to reduce rush hour traffic. All the branches of engineering, government, law, medicine, and business have sortie missions.

The output of mission sortie analyses is the establishment of major requirements for a system that can perform the desired mission. These requirements will reflect all of the concerns found on the system perspective cube. These requirements will appear as the output of system engineering process and system engineering management in precise/concise easily understood development specifications and program management plans.

2.1.2 Life Missions

Life missions relate to the system family in total and also to specific family members. System family life mission analysis tells the life story of a system from

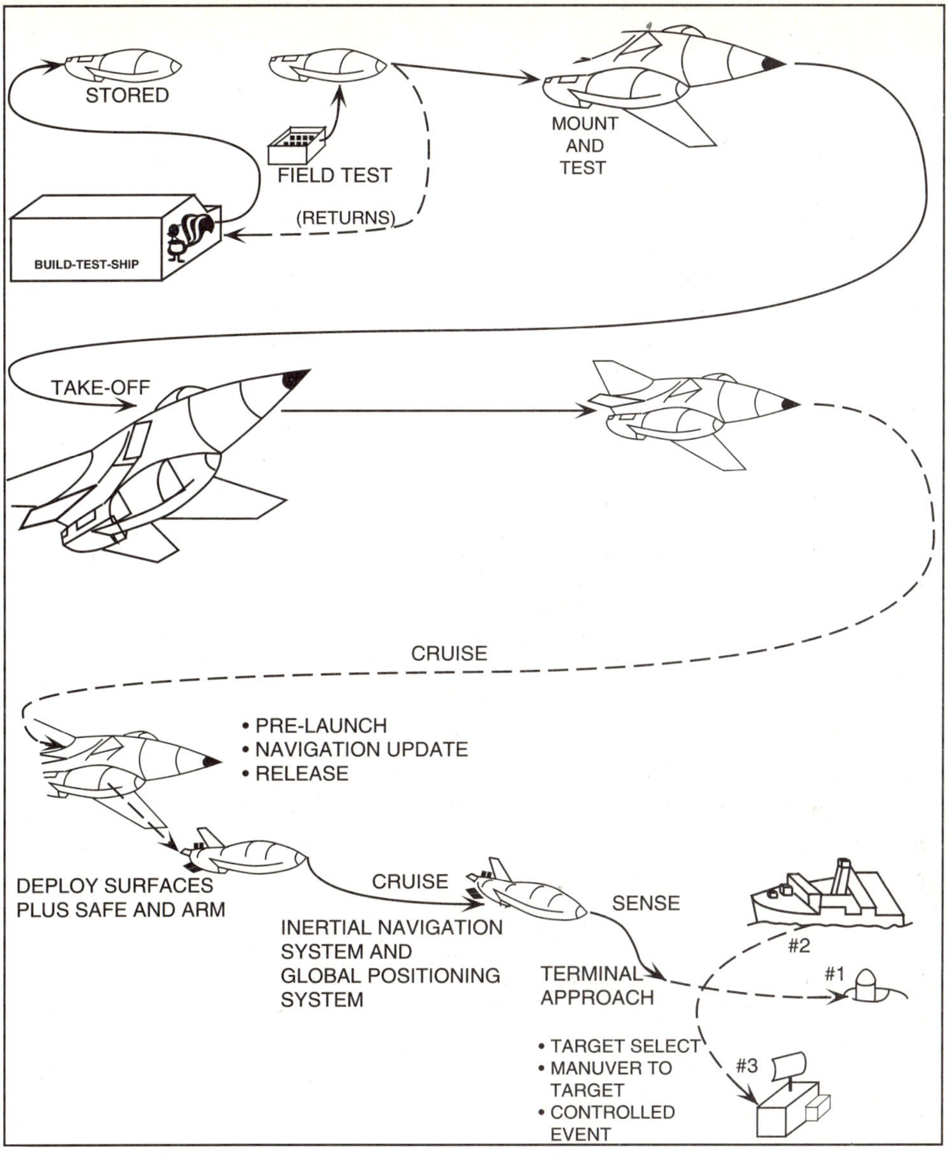

Fig. **2-4.** A sortie mission for a missile system

concept to disposal, and it includes all the family members. A family member life mission analysis, in contrast, might deal with a specific serial number of the system: Serial #12 was built in production in 1986; it was shipped; it was removed from service after 52,784 hours and disposed of by being scrapped out. Family member life can also relate to the generic system member: The family members of this system family are designed to be produced in two weeks, have ten years shelf life, a 5,000-hour mean time between failures (MTBF), and an ultimate life of 50,000 hours.

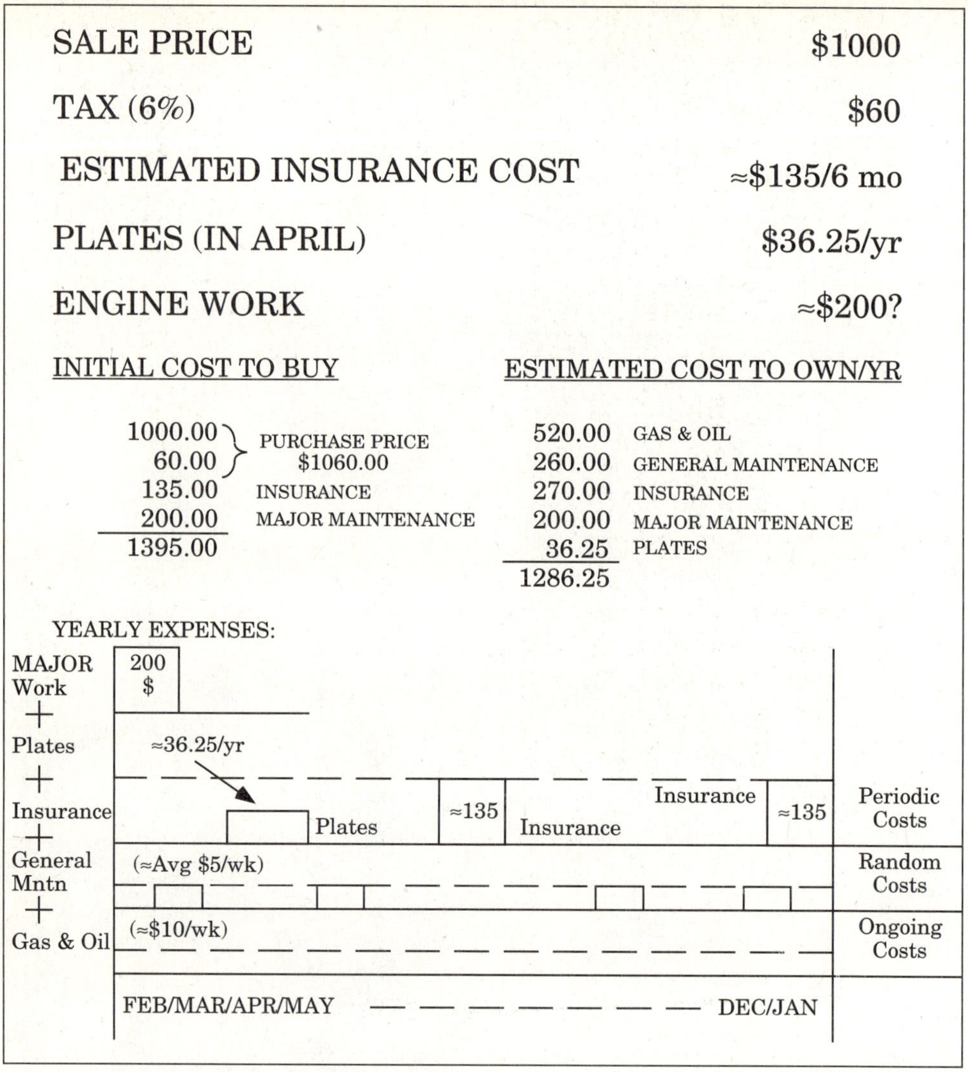

Fig. **2-5.** Used car: mission analysis (example)

System family life mission analysis

When you look at the system perspective cube you will see time life cycle phases on one side of the cube. If you would draw a line diagram showing the five phases, you could use this as a worksheet to do system family life mission analyses.

Figure 2-8 can be used as a worksheet to perform specific analysis of the family life of the system. Four sides of the system perspective cube should be analyzed: physical, action, specific time/cost goals, and conditions. Physical attributes for example, will require the five worksheets: hardware, software, documentation, people, and tools/facilities.

EXAMPLE: PARKING RAMP SYSTEM

- MISSION: PARK YOUR CAR, TO DO YOUR THING

  1 — YOU ARE GIVEN A TICKET

  2 — YOU MUST TAKE IT

  3 — THE GATE OPENS

  4 — YOU DRIVE IN

  5 — YOU PARK, DO YOUR THING (WORK, SHOP, DOCTOR, ETC.)

  6 — YOU DRIVE OUT

  7 — THE ATTENDANT TAKES YOUR TICKET

  8 — YOUR MONEY—YOU MUST PAY

  9 — THEN OPENS THE GATE

  10 — YOU LEAVE

- MISSION ACCOMPLISHED!
- NOW DO A FAILURE MODE AND EFFECTS ANALYSIS (FMEA)

  What can go wrong, the gate doesn't open - - - so we put in an attendant call button. How can we design the system so it can still accomplish the mission even if something goes wrong or fails?

- NOW USING THIS SAME BASIC METHOD, ESTABLISH THE FOLLOWING PROCESS MISSIONS WITH SUBSEQUENT FMEAs.

  1. GET A CONTRACT TO DESIGN A PROJECT
  2. DESIGN A PRODUCT
  3. PRODUCE A DESIGNED PRODUCT
  4. TEST A PRODUCT
  5. MARKET A PRODUCT

Fig. **2-6.** A systematic process—applicable to parking your car or to developing a new product

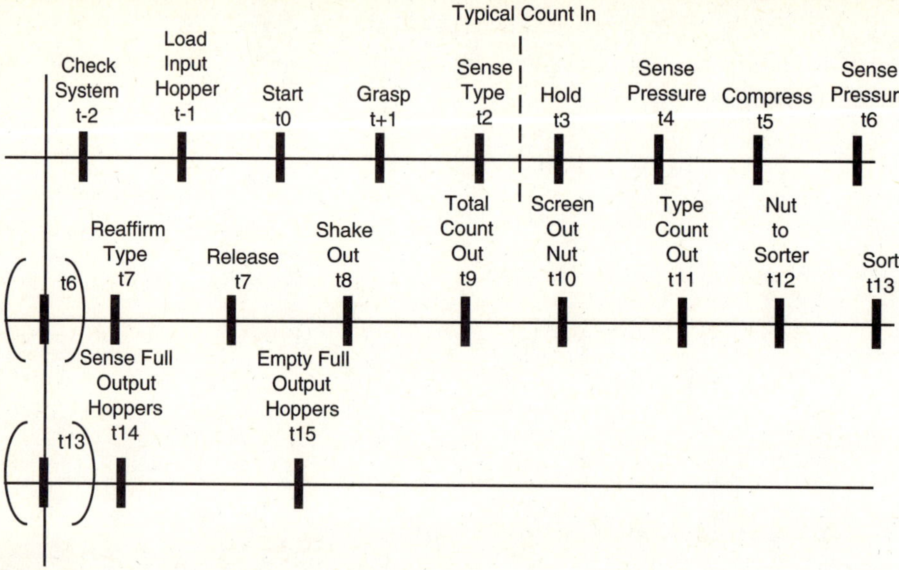

Fig. 2-7. Sortie mission analysis nutcracker sweet

Equipment and documentation family life missions for our helicopter system of the sortie shown in figure 2-3 could look like figures 2-9 and 2-10. The Nutcracker Sweet sortie of figure 2-7 could have a life mission as shown in figure 2-11.

Obviously, these family mission analyses will be brought down to lower levels. That is, each subsystem will be analyzed, and then each component of each subsystem will be in like manner. You see, the sixth side of the cube, hierarchial structure, was not overlooked.

The family life mission analyses include development "actions" as well as specific hardware vs. time line (fig. 2-9). There will be five subcubes whereby the hardware development actions (program management requirements) for each phase will be analyzed. Figure 2-12 illustrates the twenty-five "action" subcubes.

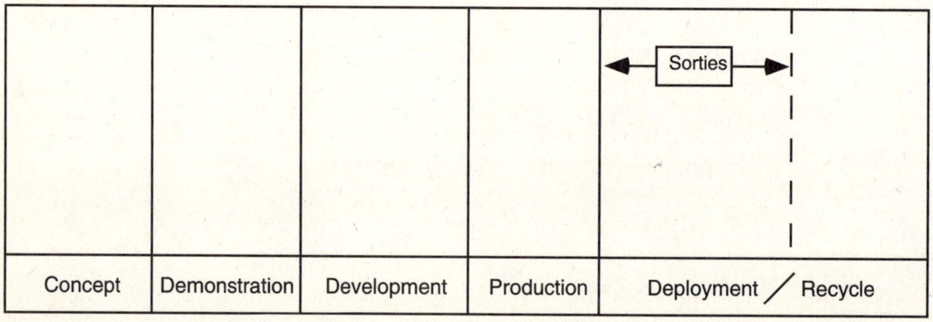

Fig. 2-8. A family life mission

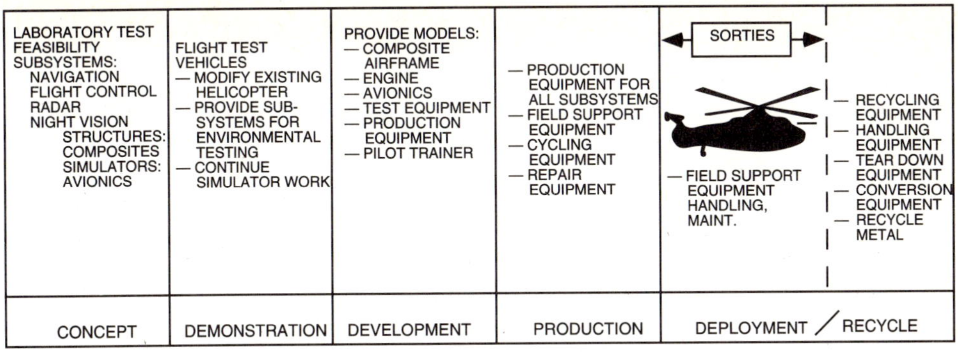

Fig. **2-9.** A family life mission analysis for a helicopter—hardware

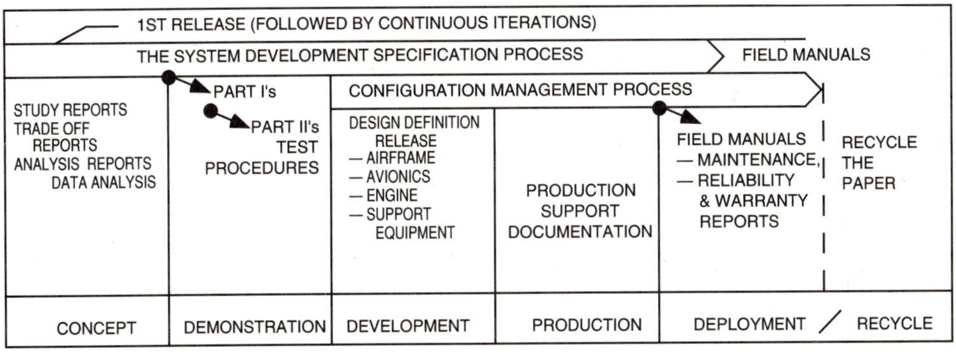

Fig. **2-10.** A family life mission analysis for a helicopter—data

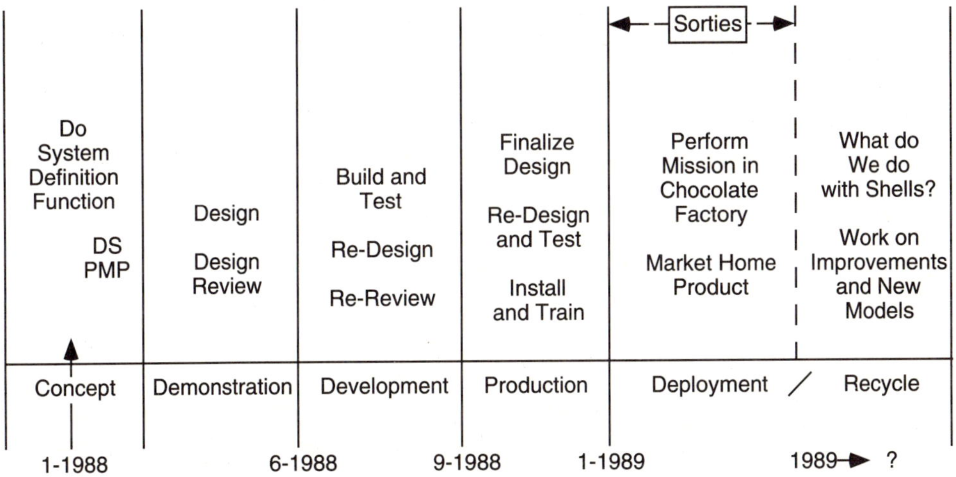

Fig. **2-11.** Life mission analysis example —nutcracker sweet

Mission analyses start with a major system sortie that is flowed down to subsystem missions, component missions, proposal effort missions, program management missions, design process missions, production process missions, customer support missions, etc. Anytime there is a mission to accomplish, there is a time line, and consequently, mission analyses must be done by everyone.

LIFE MISSION ANALYSES

THE CUBE ACTION SIDE

WE MUST NOT ONLY DO

HARDWARE VS. TIME LINE BUT MUST ALSO DO

HARDWARE DEVELOPMENT ACTIONS VS. TIME LINE

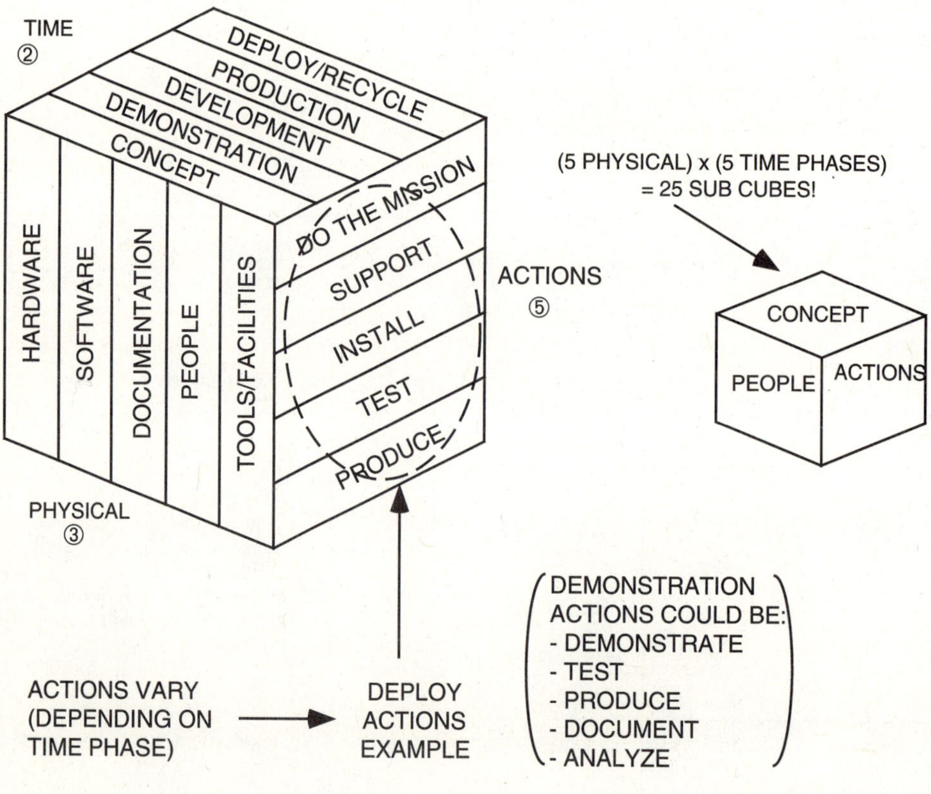

Fig. **2-12.** "Action" subcubes

2.1.3 The Proper Sequence for Mission Analysis

Sortie missions are analyzed because they are the reason for the system. Systems are developed to perform sorties during the deployment of the system; therefore, it is necessary to always start with the sorties.

The following sequence of missions is the proper order to follow when doing mission analysis.

1. Do your major system sortie analysis.
2. Do your major system life mission analysis.
3. Do your subsystem sortie analysis.
4. Do your subsystem life mission analysis.
5. Do your components sorties.
6. Do your components life.

Always do the major system above you first. You need to do this to ensure you establish your interfaces to the major system and also to guarantee that you properly flow down major system mission requirements.

In conclusion, life mission and sortie missions must be compatible. If you do the sortie mission analysis and then the life mission analysis with an output that defines the desired system roughly (a baseline) and provide the requirements for that system, you have a good start in developing a system that will perform the sortie missions desired.

You must do both sortie and life mission analyses. In addition, their relationship to each other must be recognized. The helicopter hardware system in figure 2-9 was developed so it could perform sorties. This hardware life mission then must be combined with the associated software, documentation, people, and tools/facilities life missions to ensure the major system sortie can be performed. The purpose of the life missions is to provide capability to do the sortie missions. The car purchased only for its life mission shown in figure 2-13 resulted in a poor system sortie capability. Don't forget the sortie!

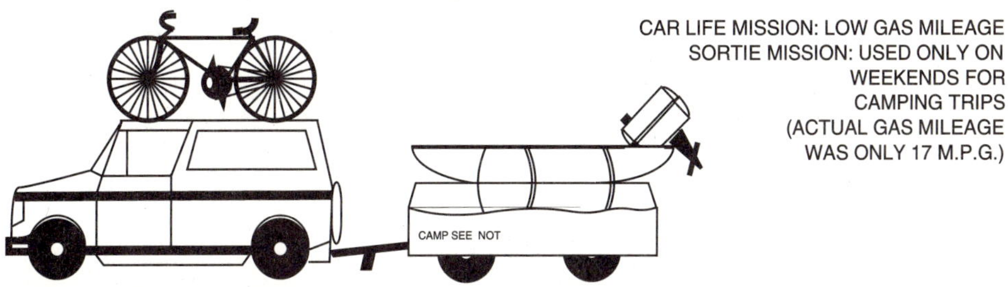

Fig. **2-13.** Don't forget the sortie!

2.1.4 The System Engineering Process as a Sortie

If you develop systems so they can perform sorties you can also use sorties to develop a system. Your time lines to perform development programs or projects are sorties. They require mission analysis and system engineering process. Every subactivity used to design, improve, produce, or support a system must be viewed as a mission. Development specification process sorties are done as well as design sorties, design review sorties, laboratory test sorties, trial run sorties, flight test sorties, rehearsal sorties, and other process sorties. These sorties help us develop the system. System engineering process can be considered a disciplined sortie methodology. You iteratively use this process sortie methodology over and over to develop and redevelop systems. This relationship is shown in figures 2-14 and 2-15. Figure 2-15 shows mission time lines for replacement of a product, the engine pressure ratio (EPR) device, that had been deployed over forty years and was reaching the end of its life mission. A preplanned product improvement (P^3I) replacement product, the digital engine pressure ratio transducer (DEPRT) had to be system engineered. System engineering process is a sortie mission used over and over during the total family life mission.

2.2 Mission Analysis Methods

The methods of performing and recording mission analysis can vary all the way from a fleeting time line pictorial in your head to teams generating voluminous reports with a multitude of formats for a major system, its subsystems, and all its components. A family picnic is an example of the fleeting pictorial, and the Apollo moon landing mission is an example of the latter complex analysis. You do mission analysis to make sure you don't forget anything or make a silly mistake. Both of these reasons apply to a family picnic or Apollo, but the impact of a poor mission analysis is much greater on an Apollo mission. If you run out of fuel on a picnic, it is not too serious, but on the moon? It is also comforting to know when you ride a commercial airline that sufficient mission analysis has been done to give you confidence that your mission (flight) will be successful despite practically any contingency. Therefore, the greater the desire to have a successful mission, the more important the mission analysis. Consequently, a wide range of methods is used to perform and record mission analysis. All missions should be preceded by a mission analysis, but the depth of the analysis can vary depending on mission criticality. You now will be exposed to seven possible methods that you may use to do your mission analyses.

2.2.1 The Scenario Sketch with Time Line Notation

Figures 2-3 and 2-4, "A Sortie Mission," are examples of the scenario sketch method used during the system conceptual phase as a mission analysis tool. These sketches are the working diagrams used for a multitude of walk-throughs. A walk-through is mentally thinking through the scenario with some specific concern in mind. The impact of potential winter environments on the mission and

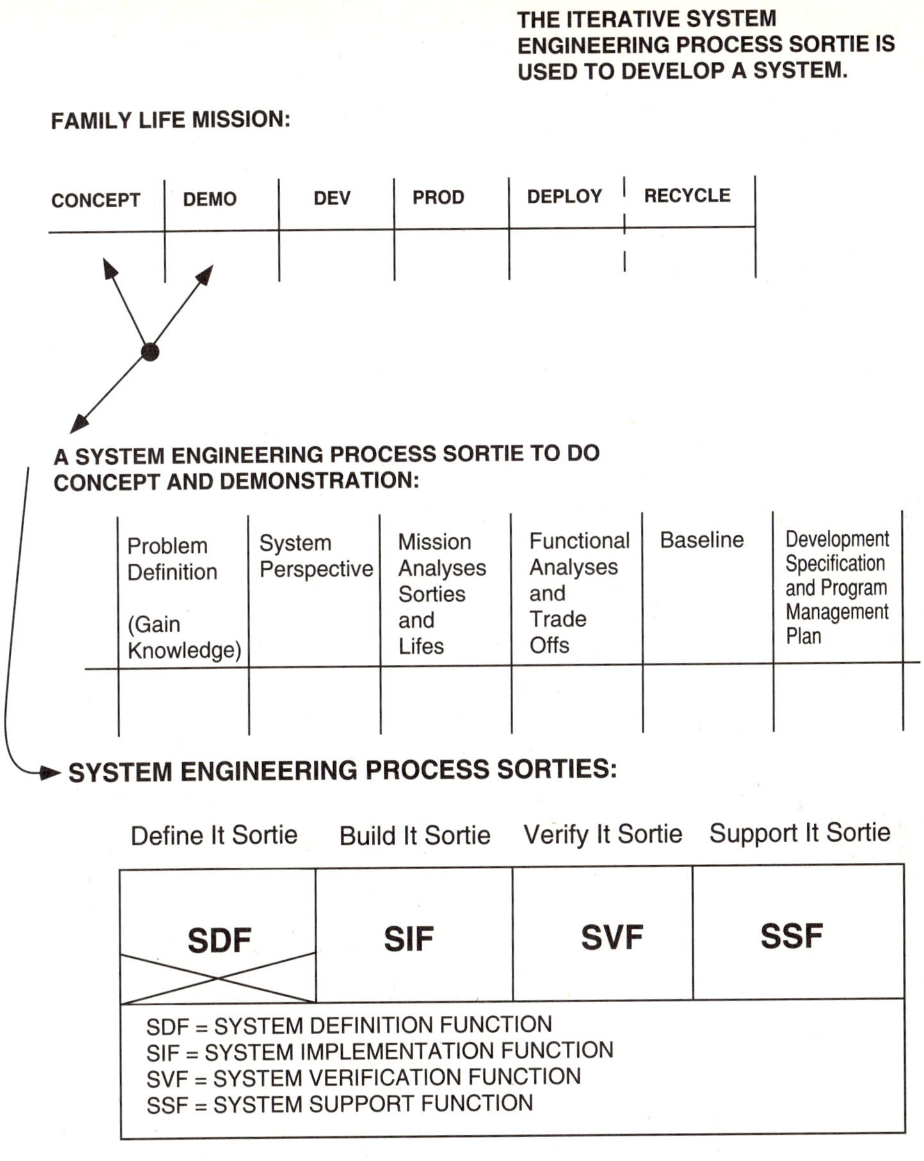

This figure shows the relationship between the system engineering process sorties and the family life mission analyses.

Fig. **2-14.** System engineering process is a sortie

the system requirements for the helicopter sortie are part of one possible walk-through: It is a winter night with a snow storm in process; at t – 1, deicing equipment is required; a radar altimeter is required in the system; a 50-watt frequency-modulated-radio control between ship/passengers/helicopter is required; land lighting is required at t2, t5; etc.

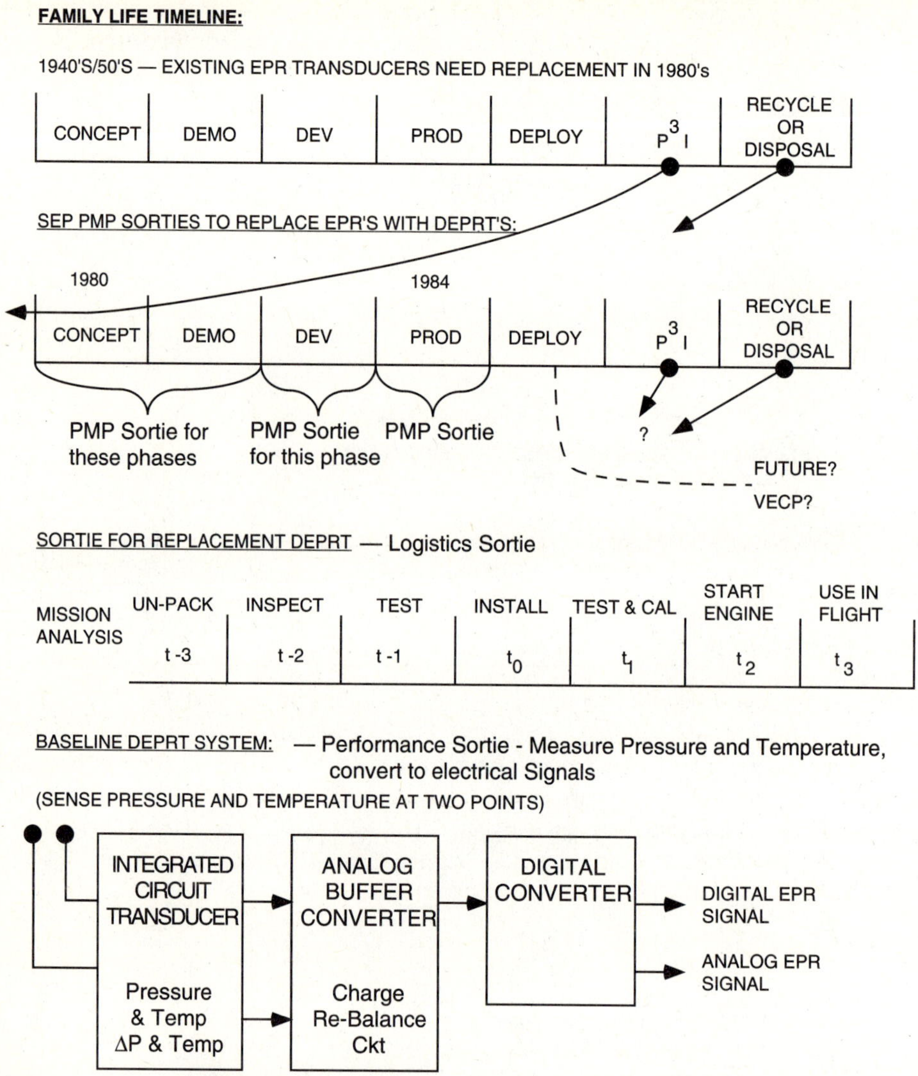

EPR	=	Engine Pressure Ratio
P^3I	=	Pre-Planned Product Improvement
DEPRT	=	Digital Engine Pressure Ratio Transducer
PMP	=	Program Management Plan
VECP	=	Value Engineering Cost Proposal (Cost Reduction)

Fig. **2-15.** Sorties applied to the development of a product improvement

One of the walk-throughs for the Apollo moon landing sortie was an assessment of impact of equipment failures at any time throughout the total sortie. This was done with the objective of establishing a system architecture and repair methodology that would ensure a successful mission despite equipment failures. You may have seen some of the animated scenarios that illustrate Apollo and

Space Shuttle missions. These, too, are examples of time notation pictorials in which the sequencing of the scenes relate to time segments.

2.2.2 The Time Line Diagram

Figures 2-2, 2-5, 2-7 through 2-11, 2-14, and 2-15 are examples of time line diagrams used for mission analysis. This method is probably the most used, because time lines are easy to draw. The time base can be in years, months, days, hours, minutes, seconds, milliseconds, picoseconds, or various combinations of time bases. The time base may also be derived from a crystal-controlled clock in a computer, a sequence of random nonsynchronous events, a heart beat, or combinations of multiple clock sources and random events.

The impact of the occurrence of a random event on a mission time line is one of the most important analyses done as part of a mission analysis. The impact of random power down on a system is a good example of this type of analysis. Consider a computer data acquisition system, for example, that periodically stores blocks of data in nonvolatile memory whenever a data block is full. The data are accumulated in a volatile memory (a memory that will lose data if power is turned off) prior to storing the data on nonvolatile memory. The transfer of data from volatile to nonvolatile occurs at a variable rate depending upon the rate at which data are generated. The time to transfer the block of data from volatile to nonvolatile is one millisecond (0.001 second). A time line diagram of this data acquisition system can be done, where one of the mission analysis walk-throughs is to access the impact of random power loss.

Figure 2-16 is a time line diagram (sometimes called a timing diagram) that depicts part of the mission for the computer data acquisition system. To do an analysis of random power loss, you mentally walk the power down cursor across the time line, stopping at selected different conditions 1, 2, 3, 4 to determine impact on the system at these specific points in time.

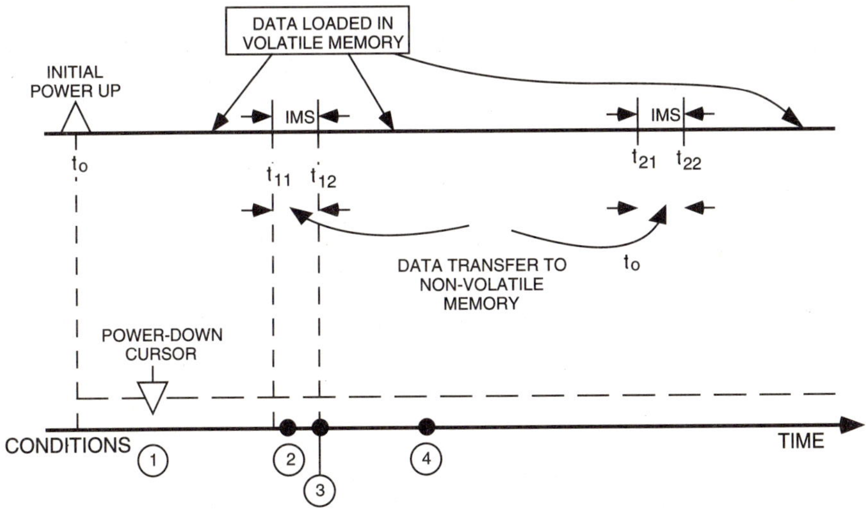

Fig. **2-16.** Computer data acquisition system — time line diagram showing the impact of a random event

The result of this specific analysis will be a definition statement or a requirement for the system that will ultimately appear in the system development specification provided by the system engineer.

The System Development Specification may say:

Power Interruptions

1. In case of power interruption during the loading of data into volatile memory, data loss is acceptable.
2. In case of power interruption during the transfer of data into nonvolatile memory, the system shall be designed to set a flag upon successful completion of transfer of a block of data. If at repower up this flag is not set, the data in this nonvolatile memory block shall be deemed invalid and the entire block shall be set to all zeros.

The preceding power interruption time line analysis incorporates a timing diagram usually generated by digital logic designers to determine the "impact of a random event." Random event analysis must be done by everyone! Random-event analysis also can be applied to life mission time lines. If a "no funding" cursor is moved across the time line, what is the impact of random loss of funding for random periods of time? What would be the impact of new technology? Will it make your present product under development obsolete? How about the impact of nonavailability of key material? The nonavailability of key skilled people? Random event impact also may be applied to other sortie sub-missions such as the impact of a chattering set of interrupt request lines into your software; random high energy high frequency interference fields to your hardware; a sudden random thermal shock from a blow torch to your hardware? Mission analysis applies to any mission so all associated with the system must be mission conscious. Mission analysis must be done and the impact of random events must be included so systems can be developed that will not be seriously impacted by random events.

I have been discussing mission analysis methods. So far I have discussed two methods, the scenario sketch with time line notation and the time line diagram. I have also discussed multiple walk-throughs, which is applicable to all mission analysis methods and the random event cursor that is also applicable to all mission analysis methods.

2.2.3 Time Segmentation

Time segmentation is another system analysis methodology. The mission is broken up into a sequence of time segments; then each segment is analyzed separately. Given a 16-second mission sortie, each second could be assigned to a page in a sequential 16-page analysis notebook. Given a 30-year life mission, we could write 6 mission analysis reports, one for each life phase, a report for 1 year of concept development, a report for 3 years of demo/validation, 3 years of full-scale engineering development, 6 years of production, 20–30 years of deployment, and 1 year of recycle (note deployment overlaps production and recycle).

2.2.4 The System Perspective Cube Checklist

Life mission analysis can be accomplished by systematically checking each subcube of the system perspective cube. In figure 2-12 the cube had 25 subcubes with time, physical, and action dimensions. A life mission analysis checklist can be developed using these subcubes. There are over 50 checklist items, which follow this question pattern:

I. Cube time side-2 vs. physical side-3 (25 questions)

1.1 During concept time phase:

1.1 What hardware is needed for the concept phase?

Under what physical and environmental conditions?

1.2 What software is needed for the concept phase?

With what computer program development conditions?

1.3 What documentation is needed for the concept phase?

With what procedural development conditions?

1.4 What people are needed for the concept phase?

With what availability and training conditions?

1.5 What tools/facilities are needed for the concept phase?

With what physical and environmental conditions?

2.0 During demonstration phase:

2.1 What hardware is needed for the demonstration phase?

Under what physical and environmental conditions?

(Repeat the prior questioning sequence)

3.0 During development phase:

(Repeat the prior questioning sequence)

4.0 During production phase:

(Repeat the prior questioning sequence)

5.0 During deployment/recycle phase:

(Repeat the prior questioning sequence)

II. Cube action side-5 vs. side-2 time (25 questions)

1.0 During the concept phase what actions must now be performed to ensure that the system will:

1.1 Accomplish its mission?

1.2 Be supported when deployed?

1.3 Be deployable?

1.4 Be testable?

1.5 Be producible?

2.0–5.0 (Repeat this question sequence for demonstration, development, production, and deployment/recycle phases.

III. Cube specific dates side-6 vs. side-2 time questions. Other questions on the time line relate to specific dates for such events as design reviews, product cost report dates, and field service process (logistics) report dates. These questions will establish the life mission time line in detail for system engineering management of the system.

IV. Cube hierarchial side-1, specific costs side-6, and program/technical constraints side-4 questions. Questions that help define and establish requirements for the system and the system development program.

NOTE: The exercise of mission analysis by applying the system perspective to a project is worthwhile for all people working on the project, not just the systems engineers.

2.2.5 The Action Function Element and PERT Diagrams

The use of combining action function elements (fig. 2-17) and Program Evaluation Review Technique (PERT) diagrams (fig. 2-18) form another mission analysis methodology. The system is looked at as an "action function," the system's purpose or function can be defined by what inputs are needed to perform the function, what action activity constitutes the function, what the outputs are from the function, and what the feedback is. In many cases the feedback can be human or some other outer loop. For example, it could be the detection of no outputs from the element. Figure 2-17 illustrates the basic action-function element.

This action function element basically shows that all systems are control systems with a time dimension. In a conventional control system there are inputs

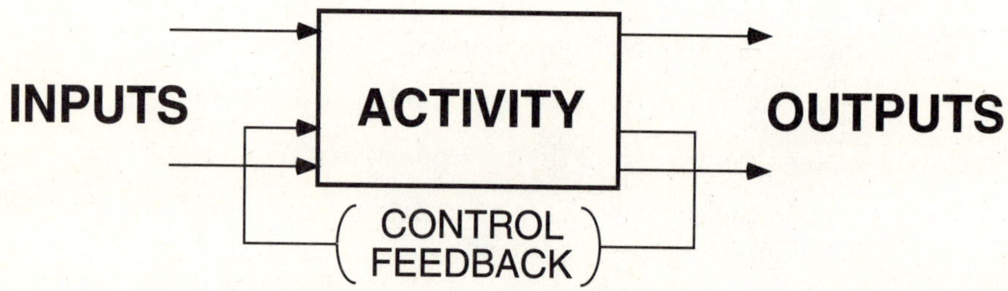

Fig. **2-17.** The "action function" element

from sensors that have a certain "latency" or age in time, there are control algorithms as activity that takes time and outputs are provided to control or to become an input to a different action function or to provide information. Feedback from positional sensors, humans, our own action function element, or a multitude of possible sensors exist to "close the loop" and establish a controlled transfer function. These action functions can be and are analyzed for timing, stability, gain, frequency response, dynamic response to a step input, and range of control. These control system action function elements can represent any control system or any process system. This includes manufacturing process, quality control process, educational process systems (students in, students out with more knowledge/skills/capability), medical process systems, management systems, retail systems; that is, all systems.

The action function elements also can be applied to define all the lower-level functions the system must perform within the "activity" part of our system. All the action function elements are strung together in a mission chronological as parts of the analyses of the systems missions.

The PERT method was originally devised as a way to evaluate engineering development program plans and progress, but the fundamental technique is applicable to the analysis of all types of missions. Once again the time dimension is the key feature that relates it to mission analysis. The beauty of the PERT method is that it gives us a method of generating a time sequence linking of action functional elements. Outputs of functions become inputs of other functions; the latter functions do not become correctly functionally active until all the required inputs have occurred; thus, there is a time line analysis tool. It must be ensured that feedback is identified, especially where it is human (this could be a survey at class reunions to evaluate how effective an educational system has been). The monitor of the outputs for credibility as a function of the outputs can also be a feedback method. A test case (Does this action function give the desired output when known inputs are provided?) is an example of forced feedback.

Conventional PERT for program review uses circles and annotated time

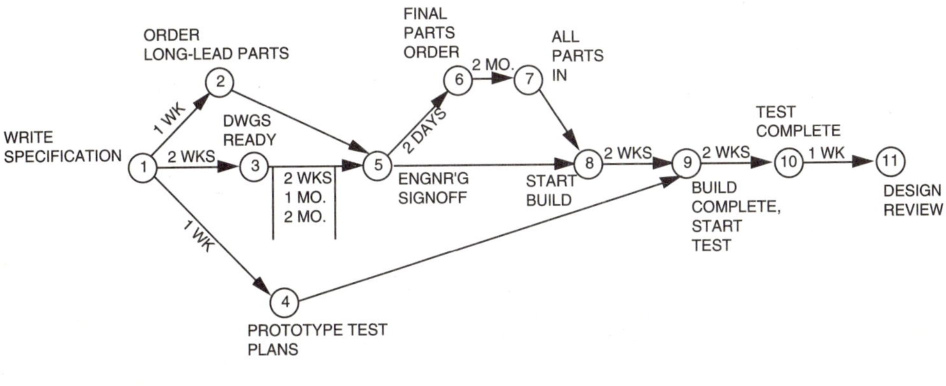

Fig. 2-18. Program evaluation review technique (PERT)

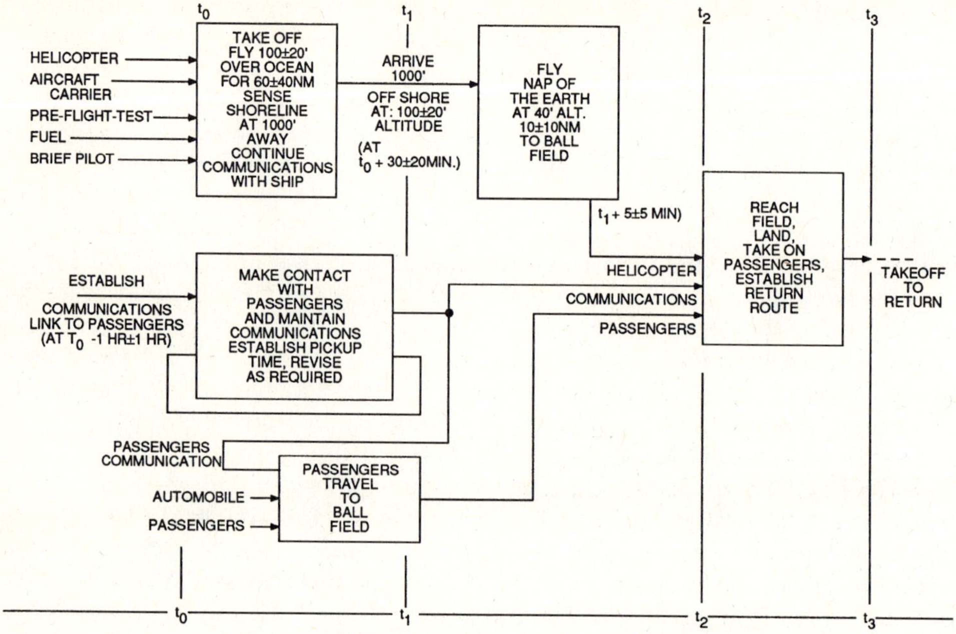

Fig. 2-19. Doing mission analysis using PERT and action function elements

lines. Figure 2-18 is a simplified example. What happens if it takes us two months instead of two weeks to get the drawings ready for sign off? In expanded use of the PERT diagram, each time line will have three numbers (best case, nominal, and worst case times); analysis will determine the critical timing path and the critical action functions.

We can combine the action function elements and the PERT method to further enhance mission analysis methods.

Go back to the figure 2-3 sortie mission for a helicopter. Using the action function and PERT method, we will start a mission analysis.

At t_0, what inputs are required to perform the activity to get to t_1? What is the activity and what is the output? In like manner, t_1 to t_2 activity will be analyzed, and also t_2 to t_3. Figure 2-19 shows how we can do mission analysis using PERT and action function elements.

2.2.6 Flow Diagrams and State Machine Diagrams

Flow diagrams provide a timing sequence; therefore, they can be useful for mission analysis. The input/activity/output feature of the application of PERT diagrams to mission analysis can be used, plus two other additional features of value: annotation and decision. Look at the prior helicopter scenario example.

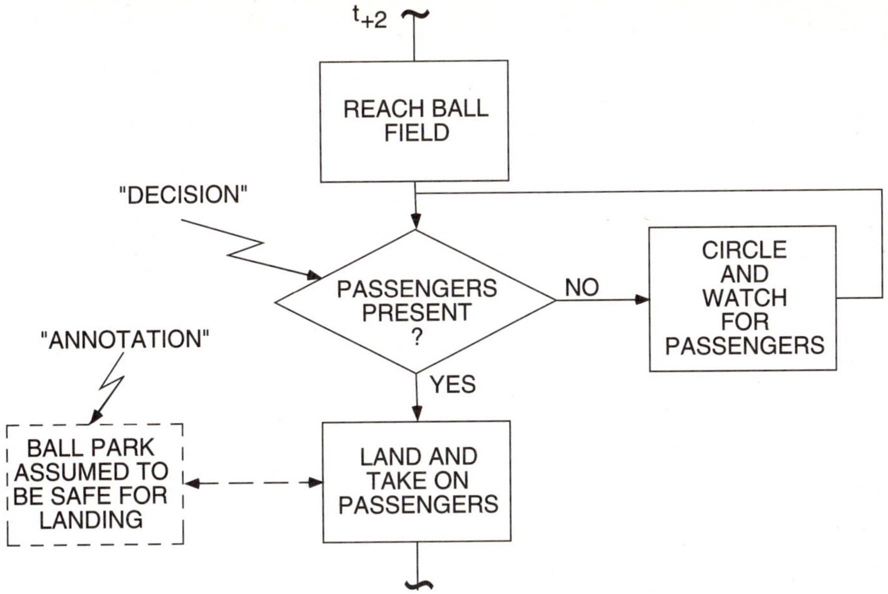

Fig. **2-20.** Using flow diagram techniques for mission analysis

These software engineer's tools, "flow diagrams" and "state machine diagrams," can also be used as an aid for mission analysis.

2.2.7 Quality Function Deployment (QFD)[1]

The quality function deployment process also can be used as a tool to support mission analyses. QFD provides a means by which to translate the cus-

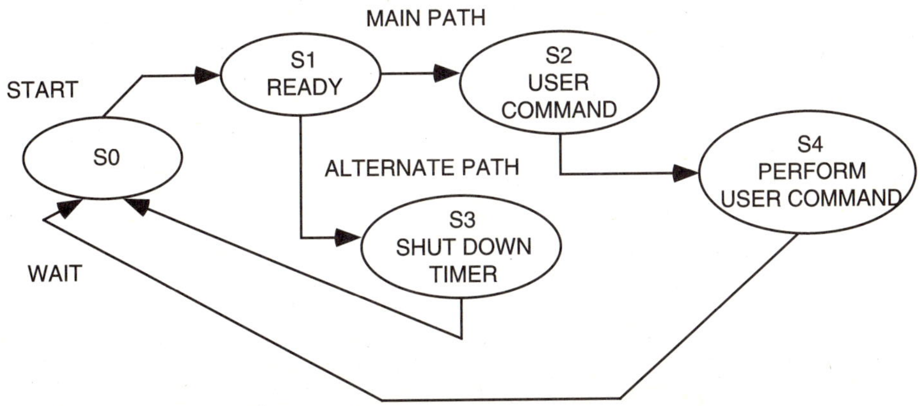

Fig. **2-21.** Using state machine diagram techniques for mission analysis

tomer's requirements into the appropriate technical requirements needed to provide the customer's wants (in QFD these are called "WHATS"). The first goal of QFD is to establish who the customers are and then determine what the customers want (WHATS). The next part of the QFD process is to determine how to satisfy the customer's WHATS. In QFD these are referred to as "HOWS." The WHATS and the HOWS are then placed on a matrix as shown in Figure 2-22. This matrix helps put the requirements at the top level and attributes of the implementation on a single piece of paper. Correlation analyses can now be done (do the hows really correlate with the whats?). Attribute balancing and prioritization can be done (how much does it cost to provide a how to meet a what? Is it worth it? What is the most important what? What is the least what?). These analyses are really a subset of mission analyses. It still is important to provide time line diagrams of the missions to ensure the very important time dimension of mission analyses are included. Then it is also very important to ensure the translation of the requirements established by the use of QFD to the development specification. If the output does not become part of the development specification, then you are not doing the system engineering process.[1]

2.2.8 Method to Use and How Much to Do

The following lists some of the potential methods you can use to do mission analysis.

1. *The Scenario Sketch with Time Line Notation.* This is neat for a quick, concise, single picture of sortie missions. This method is good for quick, multiple, visionary walk-throughs.
2. *Time Line Diagram.* This straight line method is simple and allows for random event impact analysis. It can be used for sortie, life, or specific critical time period mission analysis. It provides a methodology to show relationships between all the missions—major system, subsystem, component, people, life and sortie.
3. *Time Segmentation.* Partition the mission into specific periods of time and perform detailed analysis of each period. Document each time period analysis in separate pages or reports. Do this for highly detailed critical time phase sequences, similar to that of my experience working on the mission analysis of the Apollo man on the moon mission.
4. *The System Perspective Cube Checklist.* Use this method for family life mission analysis. Although a question checklist like this may take a few days, it is a highly recommended exercise. It should be remembered that it does provide hierarchial structure, so it is applicable at the major system above you. It also flows down to the subsystem and the component level.
5. *The Action Function Element and PERT Diagrams.* Apply PERT diagram techniques to mission analysis to provide insights of the input/action/function/feedback/output interrelationship in time

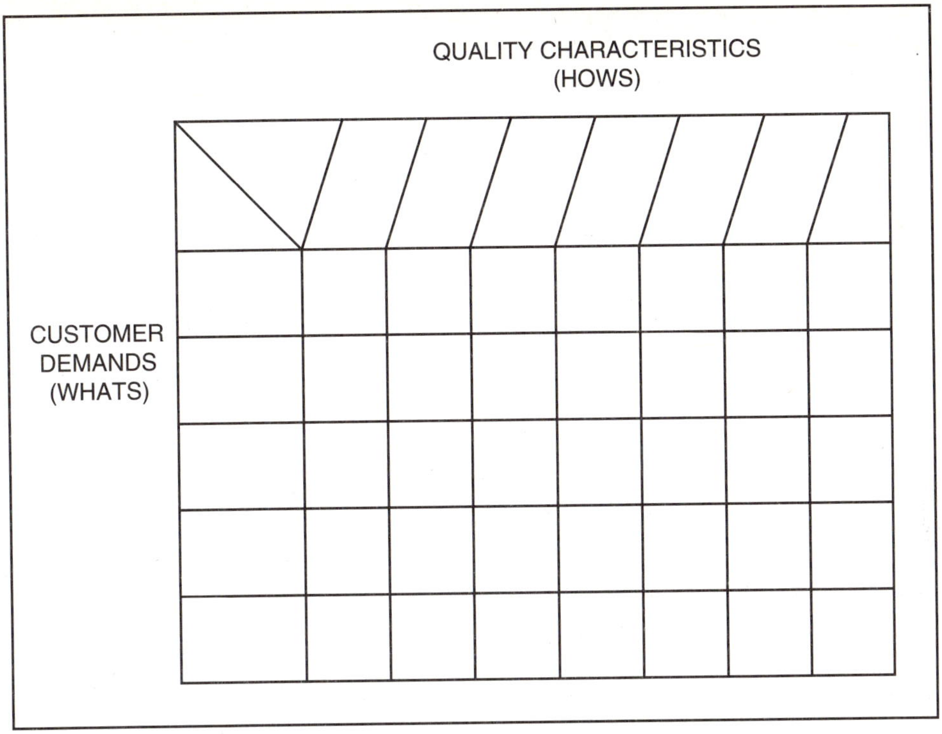

Fig. **2-22.** Basic QFD matrix

sequence. *Use this method when the system has multiple paths with dependent relationships*. This method was initially used for program management missions, but it is also very powerful for all other missions, especially process and top-level control systems.

6. *Flow Diagrams*. Flow diagrams provide a means for further enhancing the PERT diagram method by adding "annotation" and "decision." Flow diagramming methods may also be used for chronological sequential type system analysis. They are, for example, applicable to the scenario sketch, time line diagram, or time segmentation methods because flow diagrams can provide greater detail to the somewhat cursory nature of these other mission analysis methods.
7. *Quality Function Deployment*. QFD is a good method to ensure you are providing the customer's wants ("whats"), but this method must be supported with time line methods to ensure the system you evolve has the proper timing and also incorporates all the life, sortie, major, subsystems, and component sorties, with the required interrelationships.

One recommended method to use is to start with a mission scenario sketch and some time line diagrams, and next do a life mission time line diagram. Then stop, and consider your risk areas in the system; use whatever method that seems appropriate in the high-risk areas. You will probably come up with mission analysis method combinations and other new methods of your own.

Experience has shown that uncontrolled iterations of mission analyses increase development costs and complexity of the system being developed. So when have you done enough mission analysis? Use your common sense. At a minimum, a mission scenario sketch and a system perspective checklist against a life mission time line should be adequate. However, in high-risk systems, additional mission analysis should be done. You must always analyze your missions for the impact of random events and failures so you can be confident your system will not put people at risk of injury.

2.3 Summary of SEE-1 and SEE-2

SEE-1 introduced you to the system perspective cube. SEE-2 said let us take a good look at the time dimension of the cube and learn how to do time line scenario analysis—mission analysis. It was said you do mission analysis to ensure you don't overlook something that could result in a failed mission. The output of your analysis contributes to a baseline definition of the system and its requirements for that system. In mission analysis you have applied methods that result in scenario sketches, time line diagrams, and flow diagrams. Mission analysis is time line analysis. *All* time lines reflect a mission that must be analyzed. The degree of analysis depends on the desire for a successful mission and the number of risks involved.

In summary, mission analysis is the development and analysis of time line scenarios. The outputs of mission analyses are preliminary system requirements, plans, schedules, and they include innovation. The development of your system engineering skill, mission analysis, will lead you to think in the time domain. You will ask yourself questions: Is there a quicker way? A safer path? An alternate method, process, route, or way? Why do we do this? Wouldn't it be neat if at this point in the mission we . . . etc. This leads to innovation. Better and more innovative cars, computers, processes, for instance, have been developed because designers made believe they were users with missions to accomplish, in time.

You will now take your mission analysis scenario and ask: What are all the actions of persons and things required to perform this mission? SEE-3 will show how to use your mission analysis datum to perform "functional analysis," which will lead you closer to your ultimate goal: a system design that will perform the system missions.

2.4 Problems, Questions, and Things to Think About

1. On your present project, do at least two specific mission analyses—one as a sortie and one as a life mission. Now indicate the impact of a random event on the sortie and also a random event on the life.

2. Match the following:

a) Scenario sketch	A. By software engineers
b) PERT diagrams	B. By developing a 16-page notebook
c) Flow diagrams	C. Never
d) Time line diagrams	D. By system engineer
e) Time segmentation	E. By project administrator
f) System perspective cube checklist	F. By program evaluators
g) PERT diagrams and time line diagrams	G. To determine random event impact
h) None	H. By digital logic designers
i) Cursor on time line diagram	I. During conceptual phase
j) All methods	J. All people on the project

3. Review the following sixteen cases and record where mission analysis would have avoided the problem. Your response should indicate which mission analysis would have first prevented each problem, S (sortie) or L (life).

   a) The tester for the USN that would not fit through the door on the ship for shipboard installation. (L)

   b) The energy efficient house that had unhealthy air in the winter. (S)

   c) Temperature testing of a major system to -65°F for equipment intended to be permanently stationed in the Egyptian Desert.

   d) The testers that randomly burst into flame at the customer's facility, due to large load profile changes when used to test.

   e) The gunfire vibration test for navigation equipment installed in an aircraft without a gun.

   f) A vibration test that applied dynamic loads to a sensor component that were equivalent to supersonic taxiing on the runway.

   g) The deployed equipment that has no repair manual for a repairable component.

   h) Power up control of a major system that doesn't always power up because the power up built-in test equipment (BITE) did not consider that power would be applied by a slow-acting relay during flight line test.

   i) The helicopter that was designed to Mil-Spec-Rqmt of Mil-E-5400 Class 2 (required to go to 70,000 feet altitude).

   j) The Calibrator system that was tested to 25 g and failed 25 times before it was discovered that the aircraft mounting was only good to 2.5 g, so it was tested the 26th time at 5.0 g and passed.

   k) The C-5A airplane roll rate gyros that failed roll rate tests of 40 deg/sec. (The wings will come off the C-5A at 10 deg/sec.)

   l) The analog-to-digital converter that was mounted coresident with a microprocessor on the same printed circuit board. The sharing of common power supplies and ground returns gave untenable noise conditions when mounted in the system.

m) The wirewound resistor that dissipated so much heat it overheated a memory chip on an adjacent card.

n) The multiaircraft use radar altimeter that oscillates at certain altitudes when coupled with certain display interfaces on certain aircraft.

o) The aircraft flight control system software engineer who decided to add a test at the beginning of an iteration cycle to determine whether or not the software completed the previous cycle. This test was intended to catch latent software problems during system test and debug, if a failure was detected the system was shut down. The test, however, was placed in the software such that it could remain a part of the software during actual aircraft flight test, where a system shutdown would have been a catastrophe.

p) The ASWAN Dam on the Nile—An example of lack of mission analysis where a man-made system, the Aswan Dam, resulted in a set of problems on the Nile River.

Construction of this massive dam ensured that the Nile will never flood again, solving an age-old problem. However, several new problems arose: 1) The food chain was broken in the Eastern Mediterranean, thereby reducing the fishing industry. 2) Rapid erosion of the Nile Delta took place, introducing soil salinity into Upper Egypt. 3) No longer limited by periodic dryness, the population of Bilharzia (a water-snail-borne parasite) has produced an epidemic of intestinal disease along the Nile.

Mission analysis for the Aswan Dam included flood control, power generation, artifact preservation, etc.

It did not include impact on fishing industry (river and sea), river bank stability, aquatic life, insect life, water purification, social structure, etc.

4. True or False?

a) Mission analysis is done only at the beginning of the project.

b) Mission analysis can be applied to buying a house.

c) Mission analysis must be applied to all projects.

d) Mission analysis must be recorded on all projects by the system engineer down to the component level.

5. Provide action function elements for the following: The skill survey in chapter SEE-1, your present project, test equipment, a helicopter, the overall mission in the helicopter scenario picking up passengers—as a medical emergency, this class, an electronic analog audio amplifier with a gain of 10, a software functional skill survey scoring element. What is the feedback mechanism in each case?

6. Three things to think about:

a) You are flying on a commercial jet about to land at an airport where a summer storm is in progress. You are comfortable because you worked on the mission analysis for this particular airline and airplane. A set of the walk-throughs in the analysis included the bad weather conditions at landing, one of which was a bad summer storm with wind shear conditions. You think through the landing portion of the scenario again. What are you thinking? What will the pilot do? What will the ground controller do? Is the system fail-safe?

b) (See figure 2-23.) A cost estimating system is under development. The system is based on a procedural structure. The hardware and software used to implement this system is purposefully specified to allow usage by people in different parts of the country with a wide range of personnel computers and software skills. Study the mission analysis for the cost estimating system (computer-aided). What other walk-throughs should be performed in order to develop this system?

c) Complete the PERT/Action Function method mission analysis diagram (figure 2-19) for the mission sortie shown in figure 2-3. Add flow diagram annotation and decision features to your diagram. How would you do random event analysis? (Consider lightning, gunfire, power loss, engine failure, communication interruption, flight control system hardware failure.) How do you expedite this analysis? How do you document the analysis? Does your analysis serve a purpose? Does your method of documentation serve a purpose? Why?

7. Everything has a mission. Draw a major system sortie time line for your car to drive it to school. Now, on the same piece of paper show time lines for the subsystem sorties of engine, driver, brakes, transmission, computer. Also, show component sorties for battery current level, driver's foot, brake disk function, transmission switching, and computer software.

8. Draw on a single piece of paper the following: a major system sequential sortie for a 24-hour home heating system. Then, show a cyclic temperature measurement sortie as an integral part of the home heating control. Now, show a random event time line such as the home owner's resetting the control temperature. Remove power from the house. Plot expected room temperature over the 24-hour period.

2.5 REFERENCE

1. Adams, Tamara J., Richard W. Amos, Alan R. Wyskida, and Gary A. Maddux. *Utilization of Quality Function Deployment as a Tool for the Systems Engineer.* Sunnyvale, CA: National Council on Systems Engineering, July 1992.

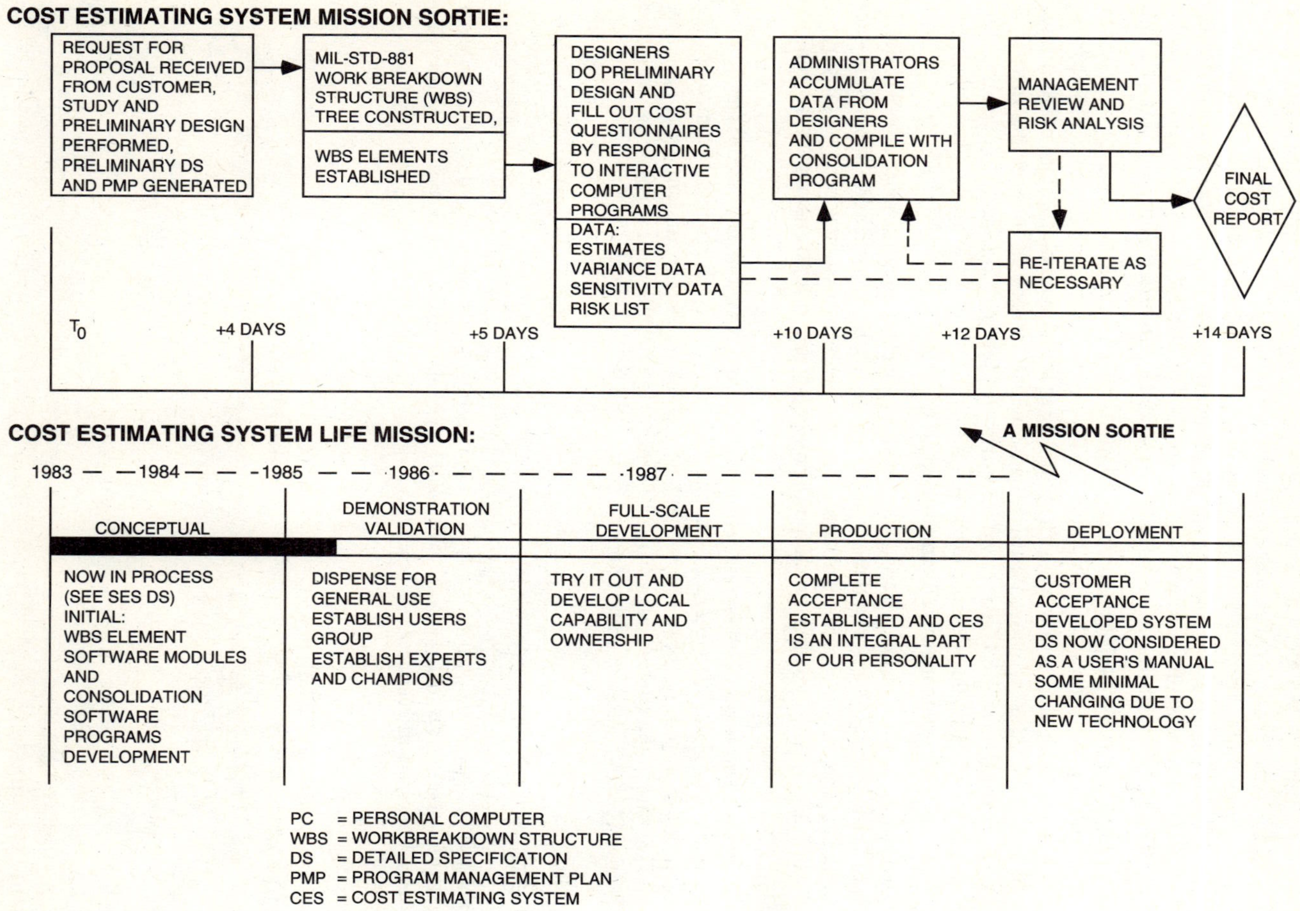

Fig. **2-23.** Cost-estimating system (computer-aided)

Functional Analysis

(What are all the Actions of Persons and Things)

3.1 Functional Analysis

In SEE-1, you learned some system engineering terminology, established your skill profile, did some team building and related the system engineering perspective cube to your present job, project, or mission.

In SEE-2, you expanded upon the perspective cube "time" axis by exposure to mission analysis for system sorties and mission analysis for various system lives. You studied the time axis of the system perspective cube and developed skills in doing both sortie and life mission analysis. You also developed familiarity with numerous analysis methods and applied these to your present job. You learned that time line analysis skills can be and must be applied to *all* systems. You also sensed that embodied within your sortie and life mission analysis scenarios there are *functions* that must occur in order to successfully perform the desired missions. This leads to a natural transition to the next system engineering skill of interest, "functional analysis."

In SEE-3, you will learn how to use the mission analysis datum to perform "functional analysis" that will lead you to the ultimate goal: a system functional design that will perform the system mission. You will also learn how to do tradeoff studies, which are used to optimize system implementation for specific functions of the system. Figure 3-1 shows where you are in your system engineering process skills development.

Functional analysis is a method for organizing the determination of actions and things required by a system to perform its sortie and life missions. Functional analysis is performed as part of the process used to develop the definition and the requirements for a system. Functional analysis is done after mission analysis. Functional analysis also is done to lower levels; that is, you perform functional analysis of systems, then of subsystems, and then of components. This analysis methodology is used throughout all phases of the system development and at all system levels. The functional analysis activity is shown in figure 3-2 as an action function element that requires your prior mission analysis time lines as inputs. This activity follows a disciplined methodology that when properly done, will result in the typical outputs shown. These outputs become the inputs to the system baseline definition and the plans for its development. When done with a team, functional analysis also provides for team building by developing a common understanding of the system. When you study the functional analysis methodology in this chapter, you will realize that functional analysis

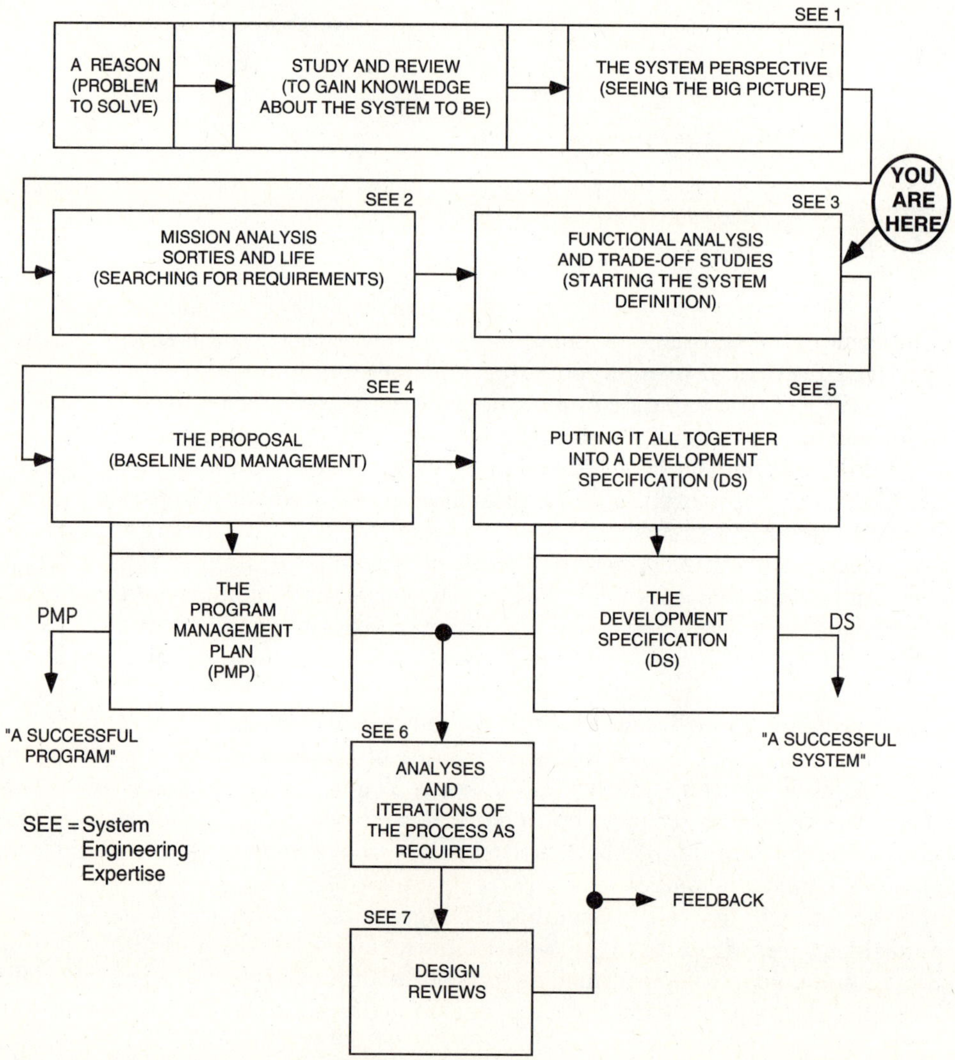

Fig. 3-1. The system engineering process

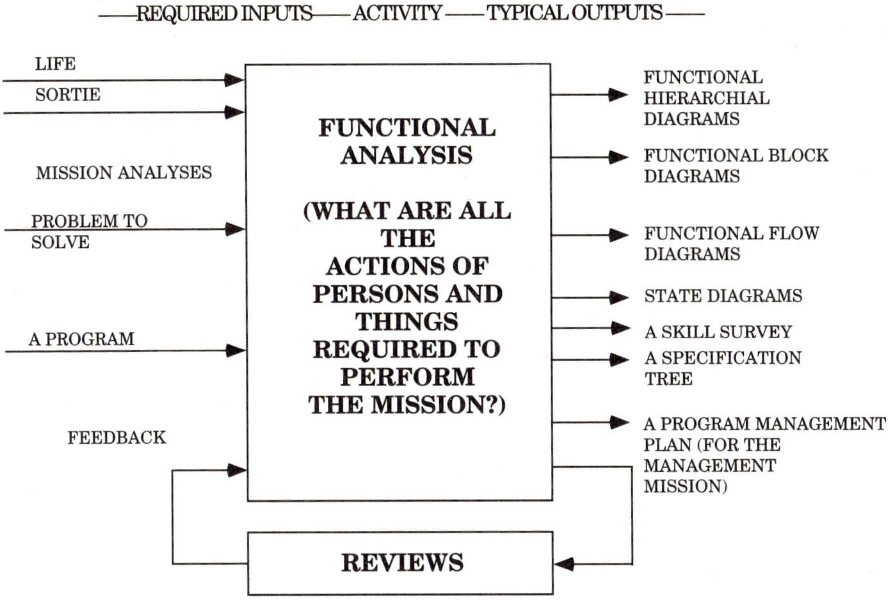

Fig. **3-2.** Functional analysis input/output activity diagram

will promote innovation when brainstorming is done properly and also will help avoid conflicts within the team. When a difference of opinion occurs during the functional analysis, avoid the potential conflict by quickly setting the differences of opinion aside. Deal with them at a later date and assign a specific appropriately skilled individual the responsibility for a special study or a trade-off study. The functional analysis activity is also the first step in baselining the attributes for your system. In chapter SEE-6, "Analysis Topics," I also will show you how the functional analysis methodology can be used to write "function of" equations that can be used to solve difficult problems, partition out requirements, analyze error contributors, and determine the functions for existing systems.

3.1.1 Functional Analysis Methodology

1. First, select the major system above your subsystem and do a functional analysis of its sortie mission. This makes you aware of the interfaces and relationships to other subsystems of the major system.
2. Brainstorm—Listing all the actions of persons and things required to perform the major system sortie mission. Brainstorming is an open, friendly, and comfortable exercise, where you are expected to say anything that comes to your mind as you study the mission time line. There are no right or wrong comments; all are acceptable and every comment is recorded. Wild and crazy comments are expected and

humor is required. Brainstorming is a catalyzing activity where one thought triggers others leading to some unique and innovative ideas. Make sure that you find actions of persons and things from the beginning to the end of your mission.

In the helicopter sortie mission of SEE-2, figures 2-3 and 2-19, some of the brainstormed action functions could be:

- Flying
- Navigating
- Controlling altitude
- Communicating
- Transporting 600 pounds of passengers and gear
- Evasive maneuvering
- Training
- Testing
- Flight controlling
- Propulsion controlling
- Equipment powering
- Having the pilot go the bathroom before takeoff
- Radar altimeter
- Automobile
- Passengers
- Fuel
- Pilot
- Aircraft carrier
- Preflight test
- Brief the pilot
- Radio
- Coffee and cake after the mission
- Night vision helmet sight subsystem
- Life vests
- Lights
- Helicopter

- Landing gear
- Helicopter speed
- Helicopter agility
- Lunch
- Drinking water
- Gimbaled forward looking infrared radar
- Pilot helmet controlling the gimbal, line of sight
- 45% field of view
- Trade-off system architecture alternates
- Do cost of mission analysis
- Study helicopters available for the mission
- Attributes of the helicopter: reliability, weight, payload capability, power, speed, agility, availability, logistic needs, etc.
- Pilot cockpit controls and displays
- A good book to read

(Do this until you feel you have covered most of the required functions from beginning to end of the mission sortie or until your team gets tired of brainstorming.)

3. Collect the brainstormed functions into a hierarchial structure. Check off the brainstormed items as they are placed in the hierarchial. Make sure you find a place for all the items of your brainstormed list. Do not omit any of them! If you have ever been in a brainstorming session where one of your ideas was voted out or discarded, then you know how quickly you will lose interest in the system, so make sure all the brainstormed items find a place in the hierarchial diagram. Your hierarchial should have boxes marked "other," "studies, analyses, and trade-off studies," and "attributes."

4. Sort out and list attributes such as size, weight, speed, agility, reliability, installation, power, etc. Place these items with your "attributes" box on your hierarchial diagram. Also sort out and list analyses, trade-off studies, and special studies. Place these items with your "analyses" box. If you have some items for which you can't find a place, put them with your "other" box. Remember, do not omit any items on the brainstormed list. If other things come to mind as you are working on your hierarchial diagram, add them to your list. You may, for example, think of all the various flight control modes that may be required to accomplish the mission, so you will add them under flight control automatic hover and other control modes.

5. Repeat the process for each of the hierarchial boxes on your diagram. What is the sortie mission for each? Then brainstorm, collect on a lower-level functional diagram, and work to lower and lower levels as appropriate to the complexity of the system. This is known as functional decomposition. It is important to perform this decomposition so you can identify the "control" function or functions of the major system. All systems are control systems. There is feedback, and you must find and identify it! A system that is defined only as a hierarchial is doomed if it never discovers its control and feedback. You need to realize that even the most seemingly mundane functions on your hierarchial have some sort of feedback or control. Even "other" functions must be reviewed and feedback reported on. "Trade-off studies" require feedback reports. "Attributes" require monitor and analysis with feedback reporting. "Fuel" requires a supply and feedback monitor control system, and the people of the system require training with testing to provide feedback of the effectiveness of the training subsystem.

   The most important control and feedback of the system that you must identify is the major system outerloop to the customer's mission. When the customers stop buying from you, it may be too late to realize that "no more customers" is the feedback that indicates your product or service is poor, or your competitor has taken the business away from you, or the market no longer exists. You better establish a control function with constant feedback and a response activity to that feedback to ensure your system is healthy, performing the customer's mission, and is constantly improving. Your control function diagrams of system and subsystem shall be patterned after the basic action function element from our mission analysis methods of chapter SEE-2. You need to identify inputs, control activity, outputs, and feedback for these control functions of your system.

6. Perform the prior five steps on the major system sortie mission, the major system life, your subsystem sortie, subsystem life, component sortie, component life, ad nauseam. The amount and depth of your mission and functional analyses depend on the criticality of the mission. Simple low-risk missions do not require a lot of lower-level functional analyses. Usually, in these cases a single major system hierarchial, your subsystem hierarchial, major system top-level control, and your subsystem control functional diagrams are sufficient for the sortie mission of the major system and the sortie mission of your subsystem. The life mission also will be functionally decomposed by functional analysis at the top level, then down to lower levels as your common sense deems necessary.

3.1.2 Functional Analysis Examples

The process of functional analysis is best shown by examples of actual analyses. The following examples should be carefully studied in order to see the power and the beauty of this part of the system engineering process.

Example 1. "A Functional Analysis for the Total Avionics System for a Commercial Aircraft" (see figs. 3-3, 3-4, 3-5, and 3-6). This example shows the aircraft major system sortie functional analysis with subsequent flowdown to the flight control system sortie functional analysis.

Example 2. "A Functional Analysis of the Engineering Department" (see figs. 3-7, 3-8, 3-9, 3-10, and 3-11). Given a sortie mission, this example shows the major system, the engineering department life functional analysis with subsequent flowdown to the "people subsystem" life functional analysis.

Example 3. "How the total system functional diagram is derived from functional analysis" (see fig. 3-12). Given a sortie mission, this example shows how the sortie functional hierarchial and lower-level functional diagrams are combined into a single drawing to develop a complete system functional diagram.

Example 4. "How Functional Analyses flow out of multiple mission types: sortie mission, life mission, a specific program" (see fig. 3-13). This figure shows that all missions require functional analysis: the user's sortie mission during deployment of the system; the program manager's demonstration program three-year mission to develop and demonstrate the system to the user-customer; the life mission of the system to be developed to ensure it can be sold at an acceptable price, produced, installed, and serviced so it can perform the user-customer's mission. All three of these mission analyses (sortie, life, and program) must be done at the very beginning of the system development process, and then they must be functionally analyzed.

Example 1 (transport airplane)

Our mission is to provide the avionics for an aircraft that is to transport passengers in a soon to be developed commercial aircraft from Minneapolis to Boston.

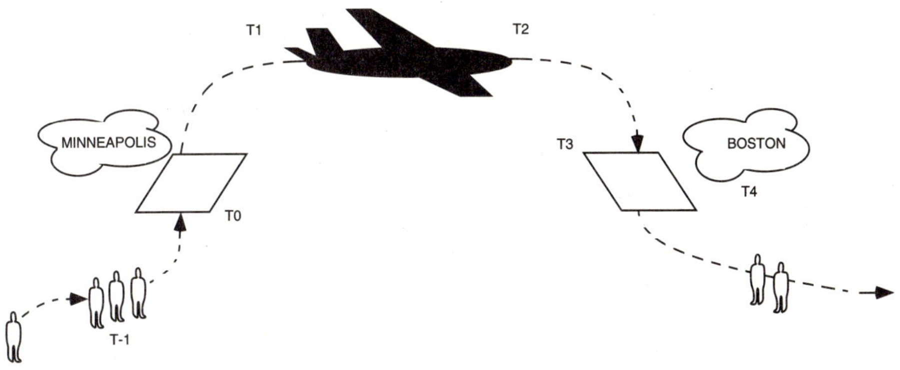

Fig. **3-3.** Major system sortie mission

Step one. Brainstorming: we brainstorm to determine all the actions of persons and things required to perform the major system sortie.

List functions (actions of persons or things)

- Attitude control
- Integrated test equipment
- Communications
- Maintenance equipment
- Propulsion
- Flight planning
- Lighting—internal and external
- Landing gear
- Safety interlock
- Toilets
- Navigate
- Ground support equipment
- Directional control
- Instruments
- Personnel training
- Passengers get a lei on debarking (in Hawaii)
- Passenger control
- Get passengers
- Thrust control
- UPC (Unit Product Cost)
- Internal environmental control
- Safety
- External environmental control
- Reliability
- Spatial orientation
- Maintainability
- Fuel control
- Logistics support
- Preflight test
- Ice and rain protection
- Collision avoidance
- Toilet training
- Weather radar
- Paint plane in attractive color

- Taxiing
- Remove all hi-jackers before boarding
- Power system—electrical
- Explode all bombs before loading
- Pneumatic hydraulics
- LCC (Life Cycle Cost)
- Man-machine interface
- Architecture

Step two. Separate into top or lower level: collect your brainstormed items into a hierarchial structure.

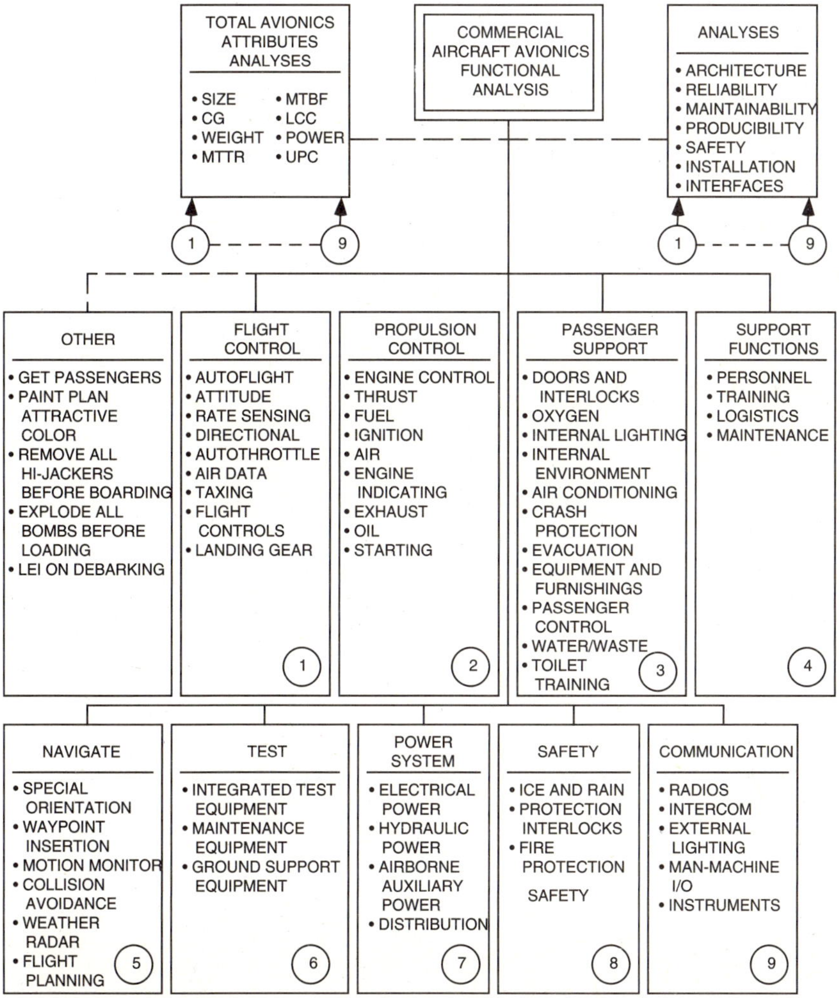

Fig. 3-4. A functional analysis for the total avionics system for a commercial aircraft

Step three. Brainstorming: work to lower levels by brainstorming to determine all the actions of persons and things required to perform the hierarchial subsystem function (flight control for example).

List functions (actions of persons or things) for flight control subsystem.

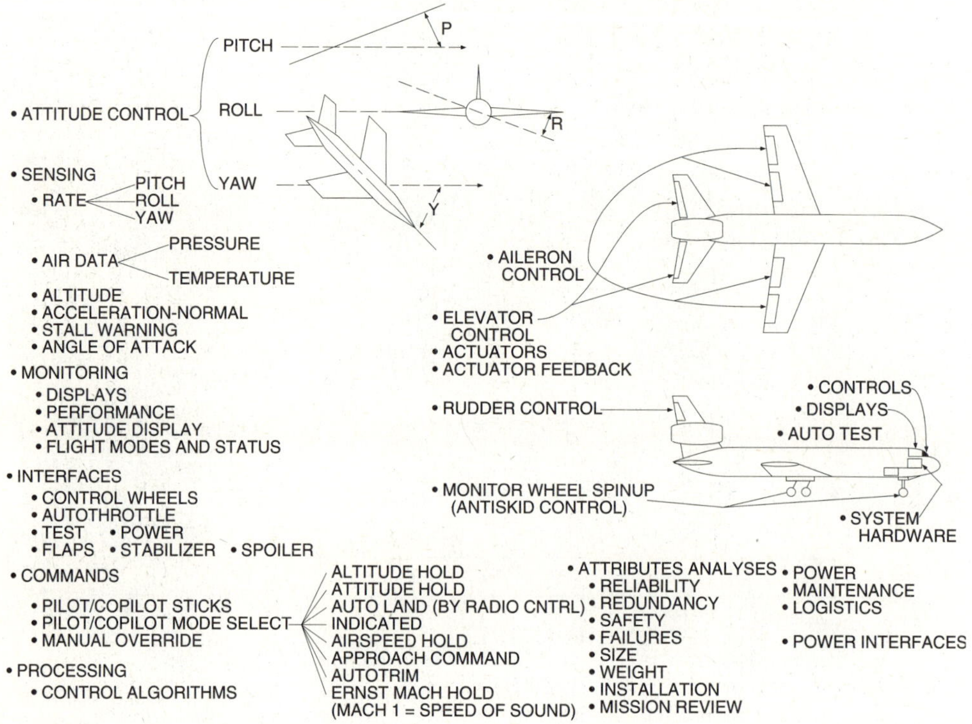

Fig. 3-5. Functional analysis—brainstorming the flight control subsystem

Step four. Collect the brainstormed items into a "control" functional diagram utilizing action function elements because this particular function is to be a control function.

Separate into top level or lower level.

Step five and on. Now repeat step three for each of the hierarchial functions developed in step two.

Example 2 (engineering department):

A functional analysis of the engineering department results in a defined and controlled engineering processing system. Given a life mission to process the inputs into the outputs as shown in figure 3-7, action function element, you end up with a hierarchial development specification tree (fig. 3-9) and a simple program management control process diagram fig. 3-11.

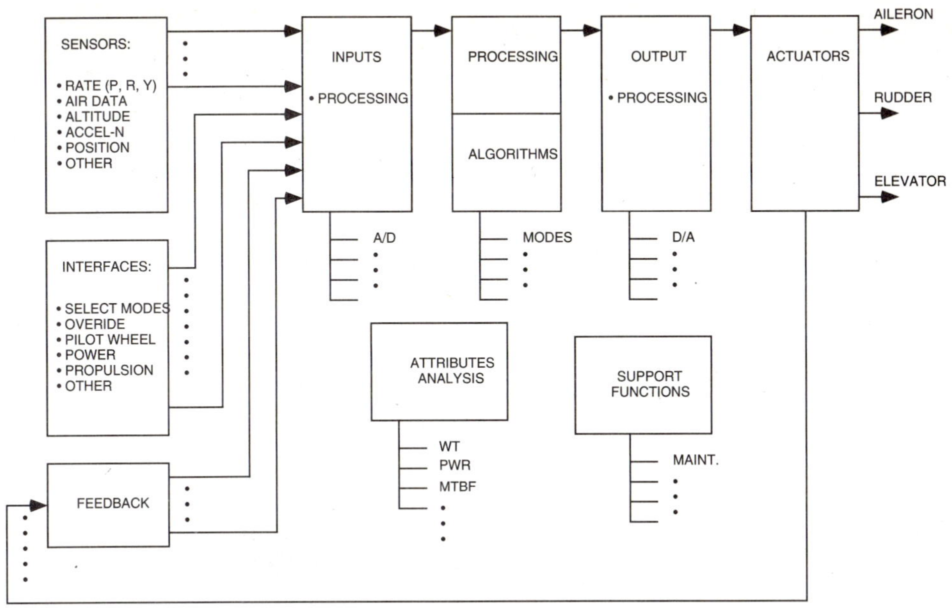

Fig. 3-6. A functional analysis for a flight control system

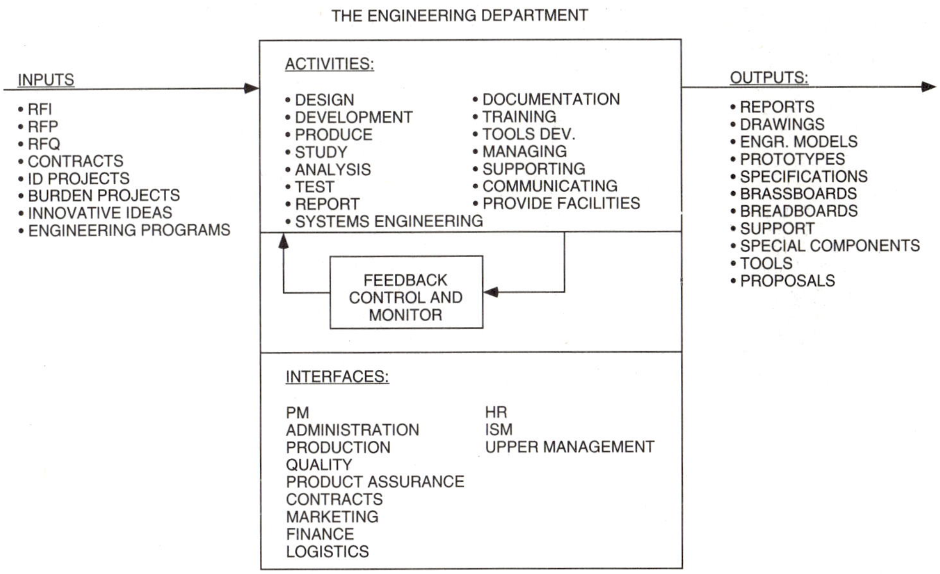

RFI = REQUEST FOR INFORMATION
RFP = REQUEST FOR PROPOSAL
RFQ = REQUEST FOR QUOTE
ID = INDEPENDENT DEVELOPMENT
PM = PROGRAM MANAGEMENT
HR = HUMAN RESOURCES (PERSONNEL DEPARTMENT)
ISM = INFORMATION SYSTEMS MANAGEMENT

Fig. 3-7. Engineering department mission

Step one. Brainstorming (all the activities of the engineering department):

- Brainstorming
- Reviewing
- Selecting
- Proposing
- Improving
- Writing
- Quoting
- Supporting
- Documenting
- Costing
- Interviewing
- Drawing
- Communicating
- Thinking
- Innovating
- Solving
- Remembering
- Building
- Planning
- Deploying
- Mentoring
- Designing
- Leading
- Iterating
- Testing
- Optimizing
- Marketing support
- Developing
- Cost reducing
- Reporting
- Patenting
- Size reducing
- Tool development
- Analyzing
- Weight reducing
- PM support
- Producing
- Directing
- Product support
- Training
- Managing
- Logistics support
- Studying
- Monitoring
- Simulating
- Specifying
- Functionally
- Rewarding
- Developing new technology
- Analyzing
- Out to lunch
- Drafting
- Appraising
- Controlling
- Laboratories
- Feedbacking
- Systems engineering

Step two. Reduce brainstormed functional activities to hierarchial:

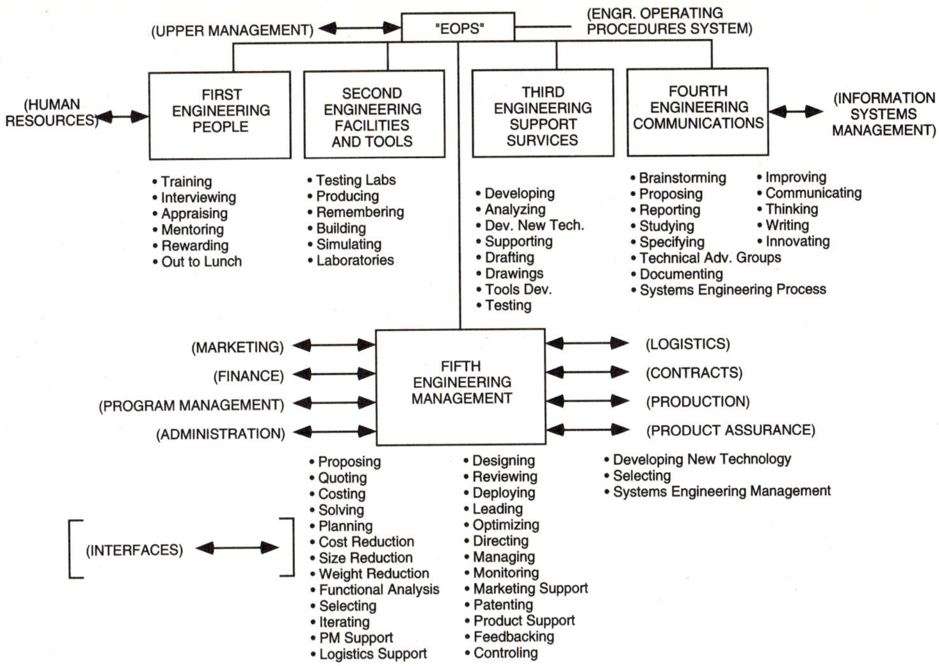

Fig. **3-8.** Functional analysis of the engineering department

Step three. Brainstorming (all the "people" activities of the engineering department):

- Organization
- Supervision
- Surveys
- Job descriptions
- Contract engineering
- Feedback
- Grades
- Work assignments
- Evaluations
- Transfers
- Interviewing
- Defining
- Appraisals
- Management
- Interface to HR department
- Time cards
- Planning
- Education
- Special awards
- Skills matching
- New employee orientation
- Objectives and goals
- Mentoring
- Interface to PM department
- Training tracks
- Improving
- Firing
- Laying off

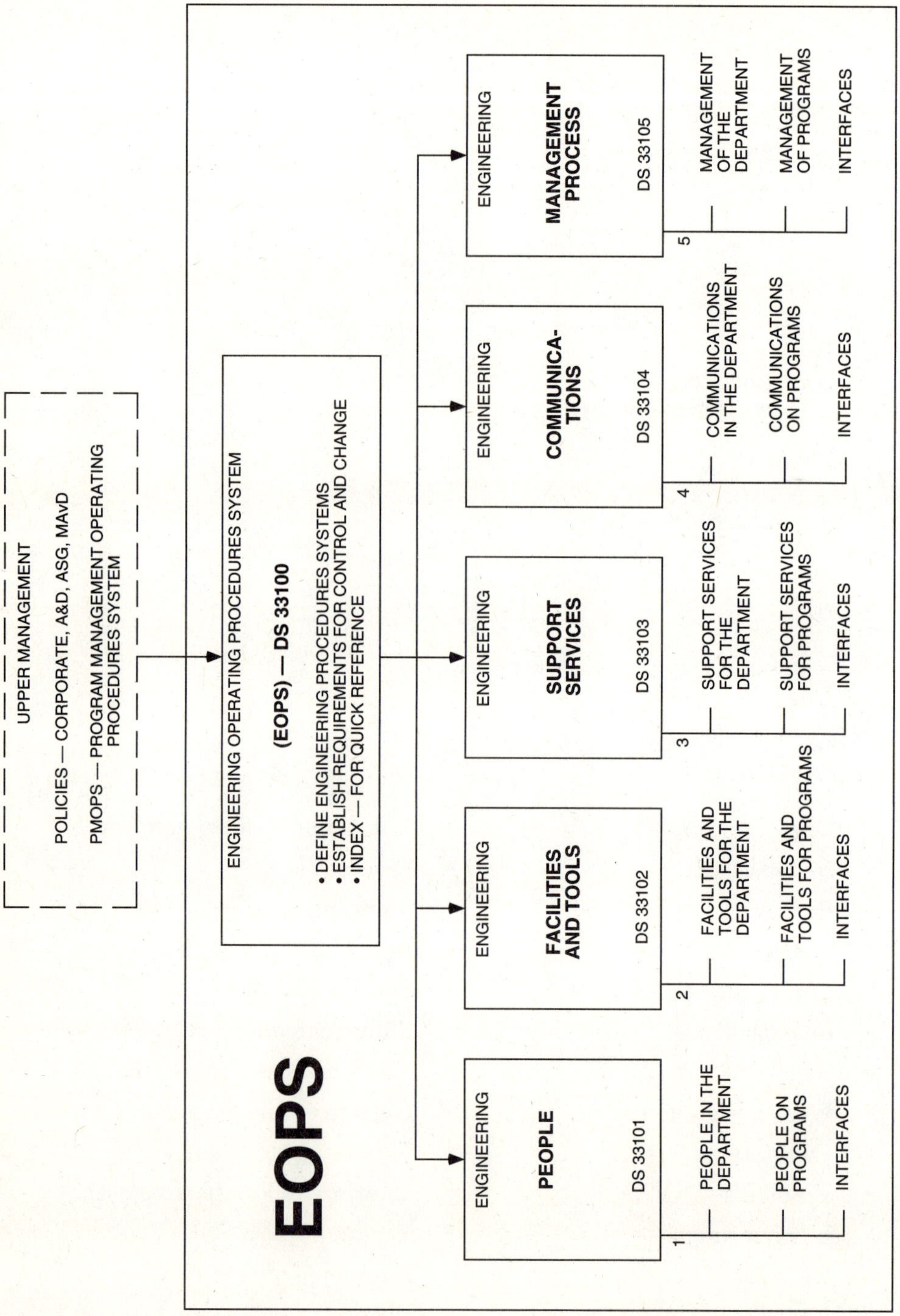

Fig. **3-9.** EOPS—functional analysis leads to a development specification tree hierarchial diagram

- Remembering
- Hiring technical
- Resumes
- Leading
- Skill bank
- Communications
- Optimizing
- System engineering skills
- Manpower/workload transfers in, transfers out
- Directing
- Interating
- Interfacing
- Promoting
- Paying
- Benefits
- Security
- Honesty
- Retirement
- Stability
- Mobility
- Goodwill parties
- Lunches
- Pride
- Day care center
- Sports
- Facilities
- Engineering
- Designing
- Testing
- Analyzing
- Developing
- Training
- Teaching
- Learning
- Cooperating
- Reviewing
- Specification development

Step four. Reduce brainstormed functional people activities to hierarchial:

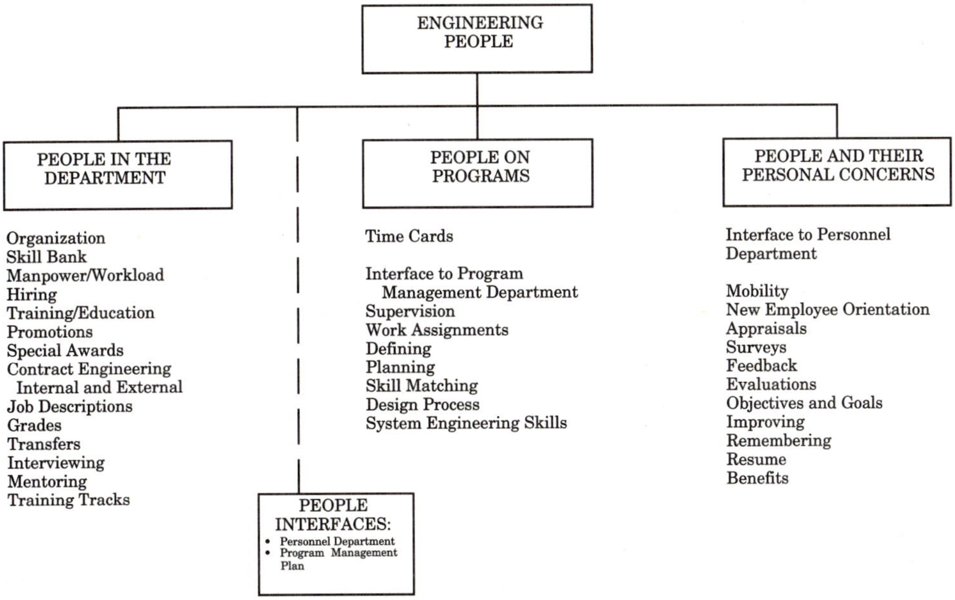

Fig. 3-10. Continuation of functional analysis for the engineering department to lower levels showing, for example, engineering people functions

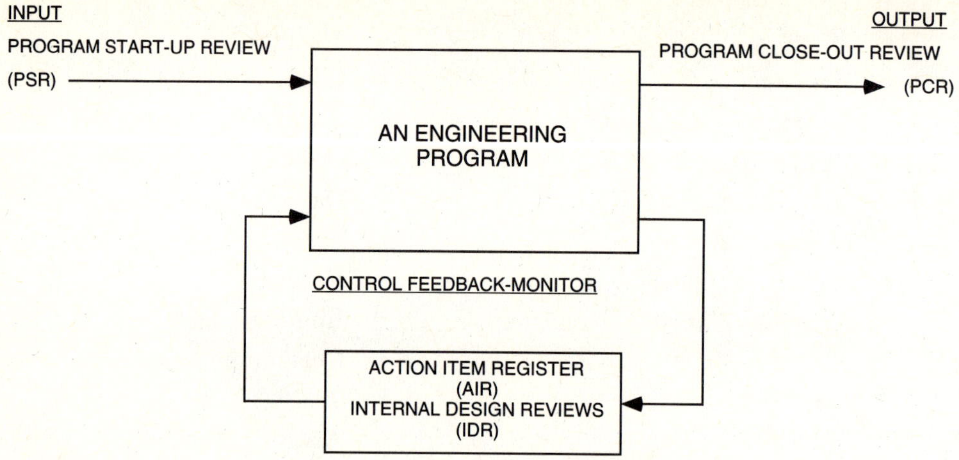

Fig. **3-11.** Engineering department program management control functional diagram

Steps five/six. The continuation of functional analysis for the engineering department to lower levels does lead to a control function. The management function analysis results in a program management control process action function element as shown in figure 3-11.

This example is applicable to a wide range of procedure type systems. If you remove the word "engineering" from figure 3-9 and 3-11, you can replace it with a term appropriate to your specific system, such as your post office, county government, university, or medical clinic.

Example 3 (deriving the total system functional diagram):

This figure shows how the hierarchial and control functional diagrams can be combined to provide the total system functional diagram. This example is generic and is therefore applicable to almost all systems.

Example 4 (flow out of missions):

This figure shows that when you are in the demonstration phase of a system's life mission, you will have to do functional analysis of the deployed system sortie mission, system life mission, and also a functional analysis of the specific demonstration program.

In chapter SEE-6, I will show you how to use the functional analysis process to solve difficult technical problems. In chapter SEE-8 you may use the functional analysis process to do a skill survey for your special system type, which may be in the field of law, civil engineering, government, education, medicine, or social work.

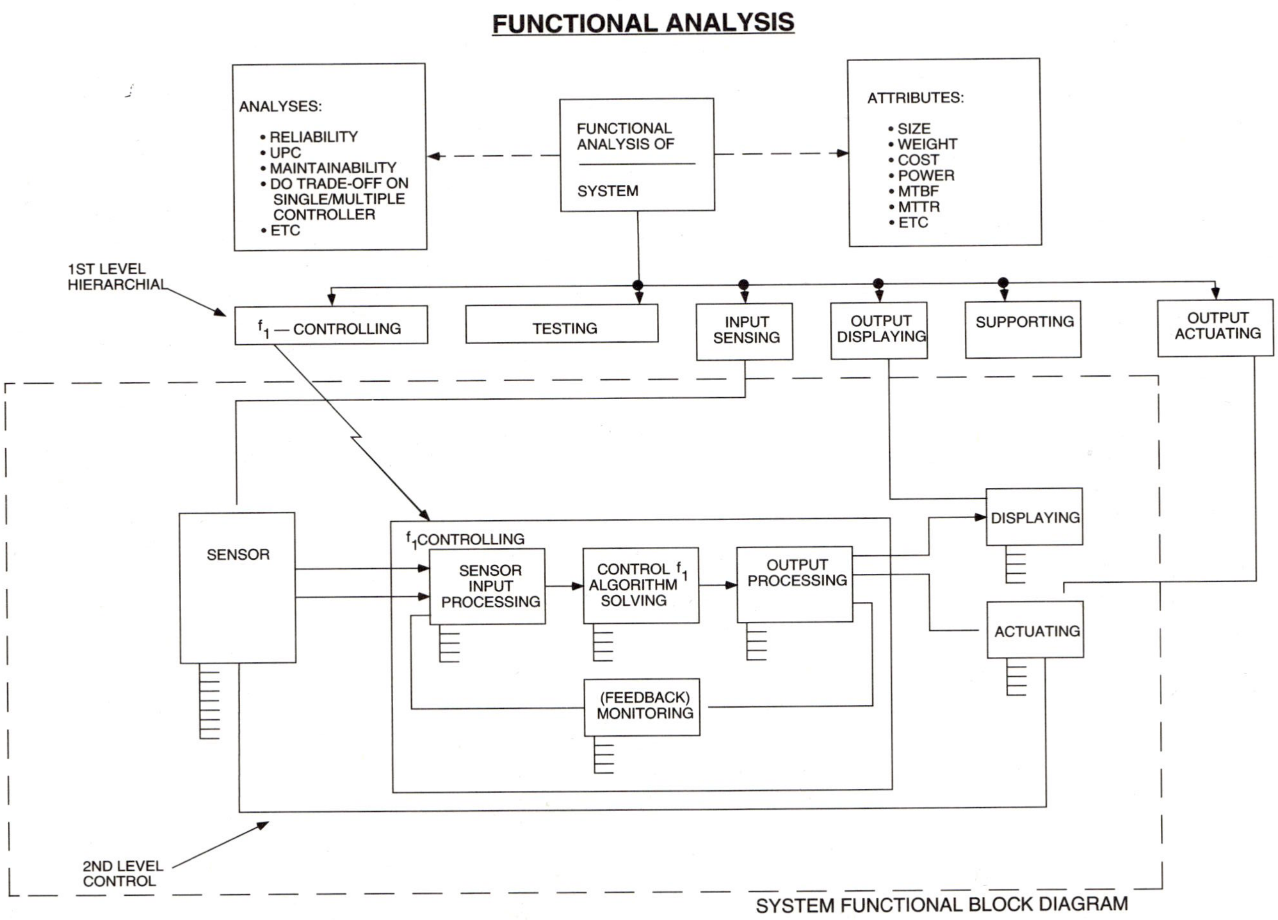

— ALL SYSTEMS MUST HAVE A CONTROL FUNCTION!
— ALL ANALYSES AND TRADE-OFF STUDIES ARE IMPORTANT AND SHOULD BE RECOGNIZED AND STARTED NOW!

Fig. **3-12.** How the total system functional block diagram is derived — through functional analyses

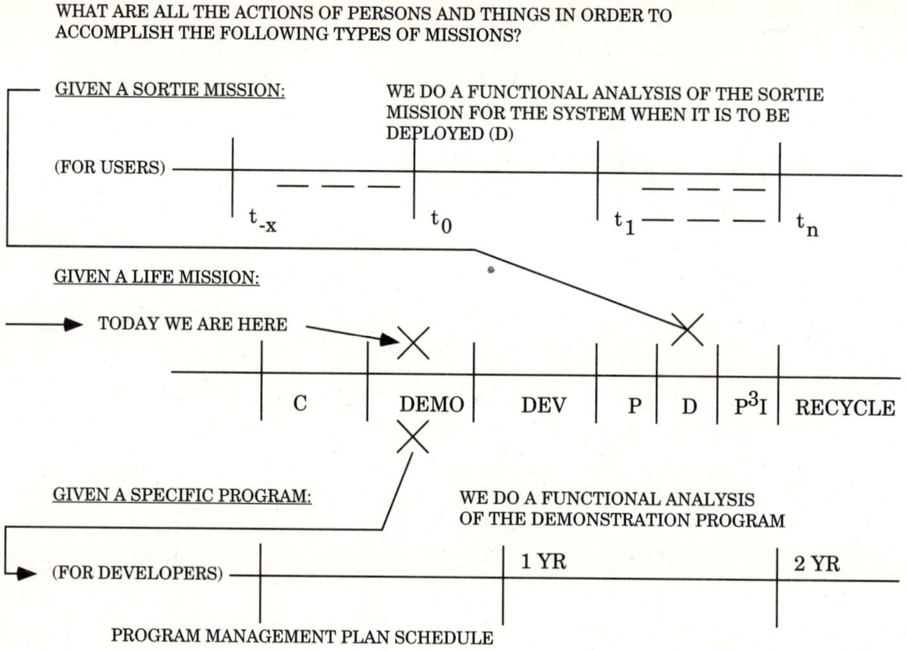

Fig. **3-13.** How functional analyses flow out of multiple mission types: sortie mission, life mission, and a specific program

3.2 Trade-off Studies

A trade-off study is an analysis methodology that systematizes the decision-making process for selecting the optimum solution to a problem. The "problem" in the context of defining the system deals with the transference of mission analysis, functional analysis, and attribute requirements into system implementation—the "physical" side of the system perspective cube. Optimizing the implementation of hardware, software, procedural data, personnel, and facilities may require trade-off studies.

You use trade-off studies to select the optimum design configuration, supplier, proposal, or procedure, and you use them generally for any complex problem solution. Documentation of your trade-off studies will range from a criteria/decision chart entry in a data book to a formal report.

Trade-off studies are done to formalize the decision-making process so that your decisions will reduce risks and optimize your chances of successfully implementing your system. Good-ole-boy, dictatorial, committee consensus, and off-the-top-of-the-head decisions have resulted in poor system implementations in many past systems. By learning and honestly applying the trade-off study methodology, you can reduce the possibility of poor decisions. That's why you do trade-off studies. The methodology forces you to gain knowledge, develop solutions, and do analyses.

Trade-offs are easy to make when the alternative solutions are few or non-complex. In those cases, it probably is not necessary to do a formal trade-off study. When the alternates are obviously even or low-risk, you can expedite an alternate selection without a trade-off.

However, when the alternates are complex, the risks high, the selection criteria contain a large number of variables, and/or the selection criteria are highly subjective, you must do a trade-off study.

3.2.1 Trade-off Study Methodology

Prior to doing the trade-off study, you should establish the trade-off decision criteria (size, weight, cost, power, feasibility, risk, etc.):

1. Set decision criteria priorities (for example, 1 = lowest cost; 2 = lowest risk; 3 = most feasibility; 4 = lowest power).
2. Select feasible solution alternates by brainstorming for alternatives; then, define feasible alternatives in sufficient detail to perform comparisons using models or detailed block diagrams. Note: all alternates must be able to meet the requirements to ensure performing the specific function required for mission success.
3. Assess each alternate per the criteria.
4. Rank alternate solutions under prioritized criteria.
5. Generate a weighting factor score chart or scoring method with a tie-breaking criteria.
6. Fill in a decision chart and rank the final result.

Document your trade-off results and implement the selected solution.

This methodology should be applied by every system engineer for trade-off decision making. The training embodied in the process will help you develop an expeditious systematic approach to doing trade-offs. You will learn how to develop decisions very quickly with confidence that they are optimal.

As you develop this skill, you will discover that the completeness of data collection for criteria and alternates is key to the process. You will learn to ask the right questions, and you also will learn to establish priorities and determine risks.

3.2.2 Trade-off Study Examples

The following examples reveal the wide diversity of trade-off study subjects, from highly technical to highly subjective.

Example 1 (computer control unit—CCU trade-off study)

This example is a technical trade-off of three alternate implementations for the computation function of a system (refer to fig. 3-14). This trade-off was performed in the early 1980s, but it is still pertinent because it shows the depth of technical detail required to give a technical trade-off credibility. The USN AYK-14 standard computer configuration consists of three AYK-14 computers (two general processor module—GPMs—one input-output processor—IOP) with required support assemblies to provide the required computation function for a total avionic system. The AYK-14 is a USN standard avionics computer. Two other configurations also are provided as alternates, one uses four 5301 computers (Company Avionics Standard), plus required support assemblies, while the last alternate uses four 1750 (USAF Standard) computers with required support assemblies. The next two pages of the trade-off study show ranking against many criteria. The last page shows the selection of five top criteria for the final decision chart, ranking of the top five, and the then scoring (AYK-14 = 28, 5301 = 25, 1750 = 37; 5301 is best). In addition, two other decision-scoring criteria are applied: sum of the 1s for all criteria, and total of all rankings for all criteria. The 5301 is best for both of these scoring methods also. This is an example of a highly technical trade-off based on three preliminary system designs. It should be obvious that considerable work has been done to baseline the three alternates for this trade-off. The system engineer has benefited by gaining knowledge and design experience, so trade-off studies can provide excellent low-risk training.

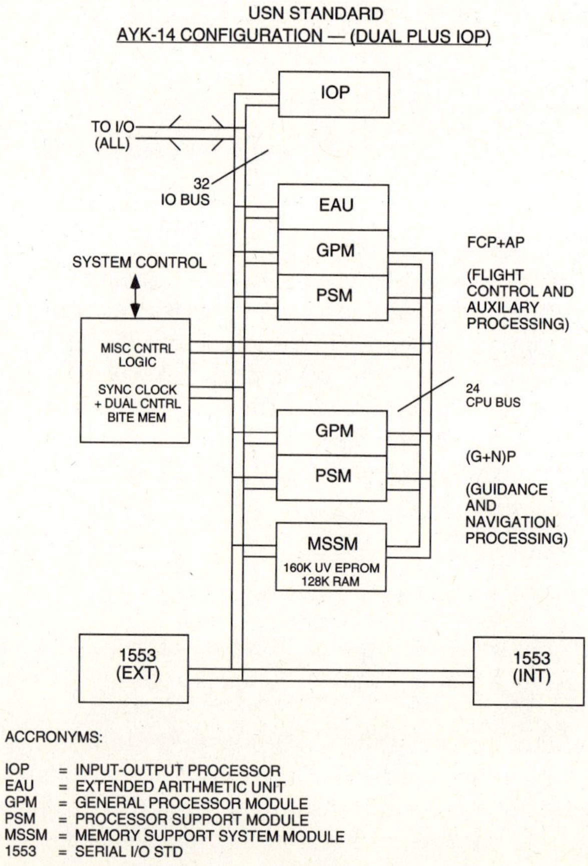

Fig. **3-14a.** CCU—trade-off study

HONEYWELL STANDARD
5301 CONFIGURATION — DUAL-DUAL)

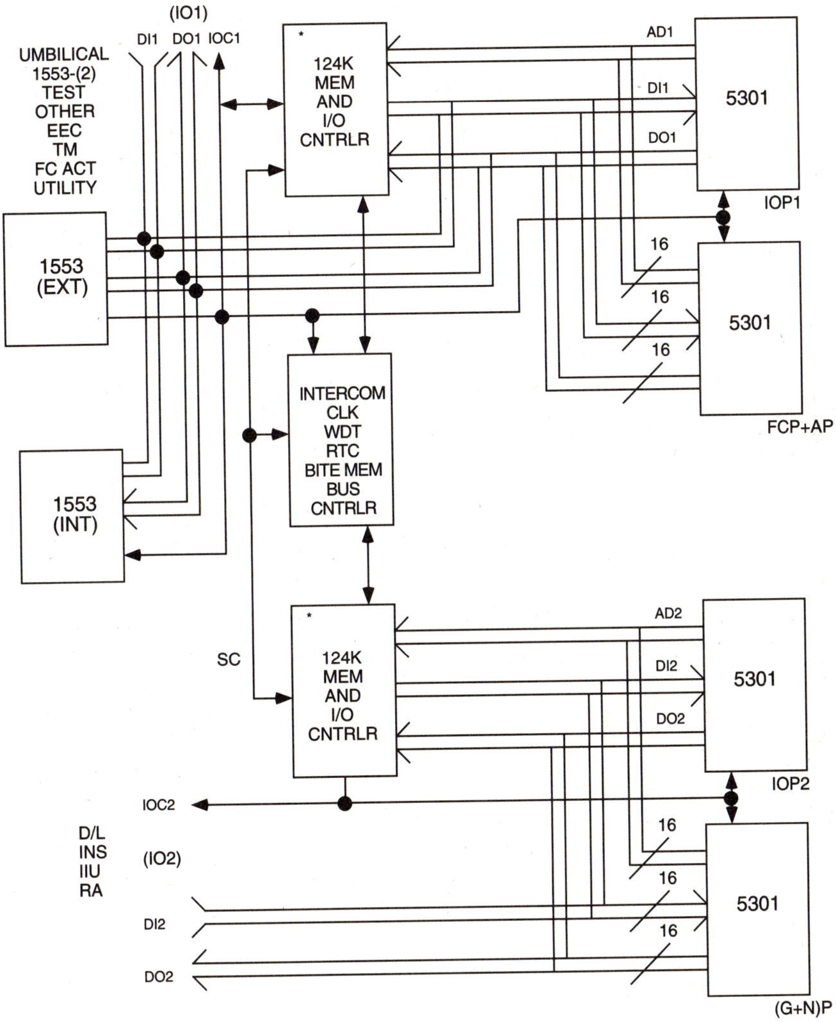

ACCRONYMS:

IOP = IO PROCESSOR
IO = INPUT-OUTPUT
FCP = FLIGHT CONTROL PROCESSING
AP = AUXILARY PROCESSING
AD = ADDRESS (BUS)
DI = DATA IN (BUS)
DO = DATA OUT (BUS)
G+N = GUIDANCE & NAVIGATION
EEC = ELECTRONIC ENGINE CONTROL
TM = TELE METERING

FC ACT = FLIGHT CONTROL ACTUATORS
D/L = DATA LINK
INS = INERTIAL NAVIGATION SENSORS
IIU = INTERTIAL INSTRUMENTS UNIT
RA = RADAR ALTIMETER
CLK = CLOCK
WDT = WATCH DOG TIMER
RTC = REAL TIME CLOCK
BITE = BUILT IN TEST EQUIPMENT
SC = SYSTEM CONTROL

Fig. 3-14b. CCU—trade-off study

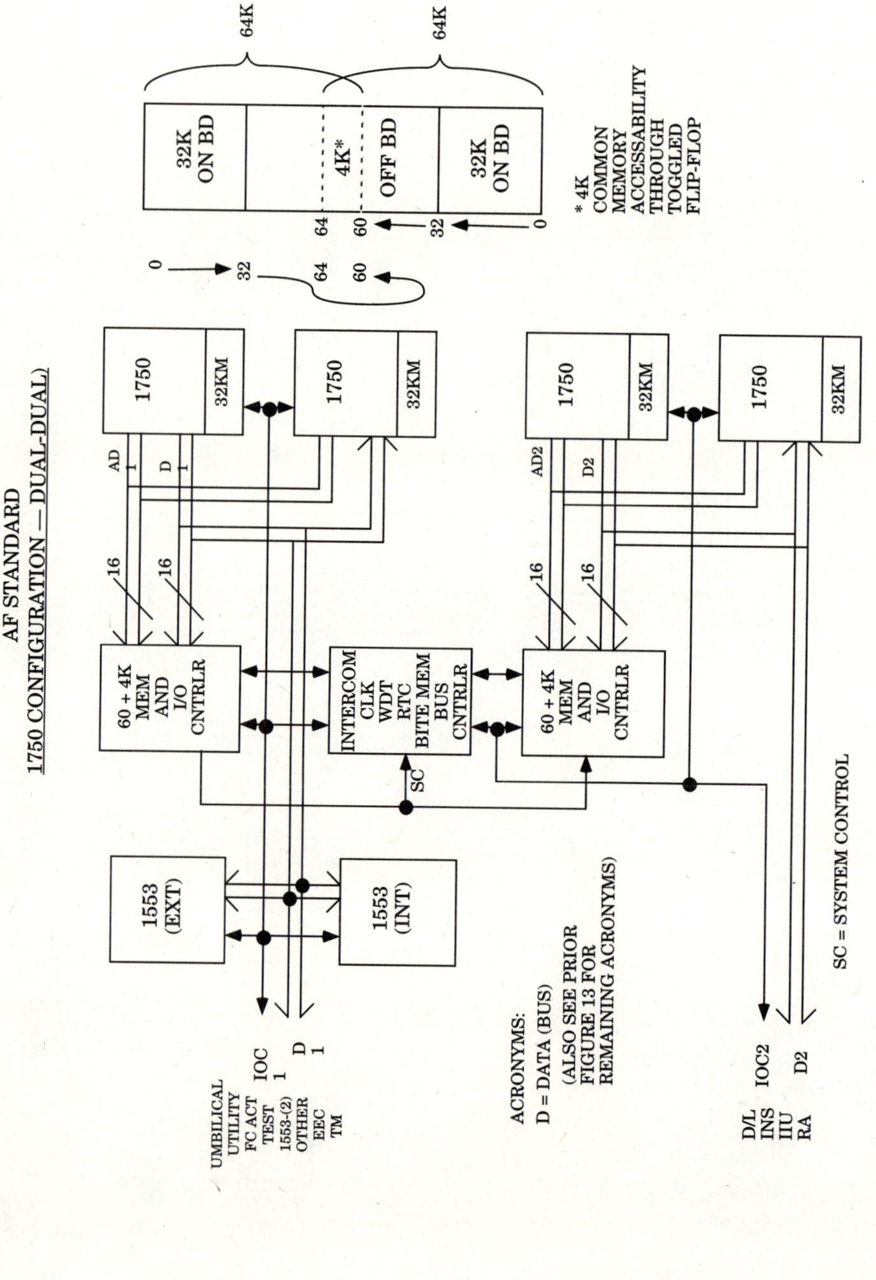

Fig. 3-14c. CCU—trade-off study

TRADE OFF DATA

Criteria	System Architectural Configuration (Alternates)					
		AYK-14 (USN)		5301 (HI)		1750 (AF)
COST	3	70K (50 to 85)	1	27K (20 to 30)	2	38K (30 to 40)
AVAILABILITY	2	CDC prod line	1	HI prod line	3	Prototypes in 9 months
SIZE	3	(6 x 9 x 0.85)/6 cards (6 x 9 x 1.5)/4 cards	2	(6 x 6.25 x 0.7)/7 cards (6 x 6.25 x 1.5)/2 cards	1	(6 x 6.25 x 0.7)/9 cards
WEIGHT	3	18 #	2	9 #	1	8.5 #
POWER	3	217 W nom	1	80 W nom	2	85 W nom
VOLUME	3	356.4 cu. in.	2	296.25 cu. in.	1	236.25 cu. in.
THROUGHPUT	1	760 GPM/EAU + 560 GPM +290 IOP = 1610	3	4 x 350 @ 4.5 MHz clk = 1400	1	4 x 400 (DAIS-mix!) @ 16 MHz = 1600
# CARDS	3	10 cards	1	9 cards	1	9 cards
BUSSES	2	CPU-24 IO-32	3	CPU-48/pair IO-32	1	CPU-32/pair IO-16
I/O	1	IOP (PIO, DMA, DIO)	1	DIO, PIO, DMA	1	DIO, PIO, DMA
POWER TYPE	1	5 VDC	1	5 VDC	3	5 VDC + inj. current
SFTWR DEV COST HRDWR SFTWR	3	VAX facility + CCU 500K CCU VAX	1	SEU (2-Dual 200K SEU SEU	2	RTSS (single) 100K (in development) RTSS RTSS
MEMORY	1	80K/160K/256K UV-EPROM 128K RAM (208 to 394K)	2	248K mixed	2	248K mixed
ADDRESSABILITY	2	525K	3	256K - (paging possible)	1	256K - (paging available (compatible MMU & BPR) (up to 1M)

Fig. **3-14d.** CCU—trade-off study

TRADE OFF DATA

Criteria	System Architectural Configuration (Alternates)		
	AYK-14 (USN)	**5301** (HI)	**1750** (AF)
RISKS	2 Architectural configuration workable, availability, saleable to AF, cost, sftwr dev. equipmt schedule	1 Saleability to USN & AF, memory limit, no HOL	3 Fairchild viable, saleability to USN, chip pkg'g problems, sftwr dev. equipmt, new untested product, I/O sprt untested
PAST HISTORY	1 Good	1 Good	3 In process
PRESENT STATUS	1 Established and works	1 Established and works	3 TBD - In process
FUTURE STATUS	1 P[3]I in process	3 Status quo	2 AF std (& DOD?)
HOL	1 CMS-2 & Ada possible	3 None	2 J73 & Ada possible
TESTABILITY -(A WASH)-	1 Pass	1 Pass	1 Should pass
INSTRUCTION SET -(A WASH)-	1 AYK14 (USN std)	1 5301 (Honeywell)	1 1750 (AF std)
MAINTAINABILITY -(A WASH)-	1 Pass	1 Pass	1 Should pass
RELIABILITY	2 Most # of parts (but proven)	1 Proven (e.g., 111856 hrs/5301)	3 Least # of parts (not proven)
1553 CAPABILITY	1 Available	1 Available	3 In process
BUS DRIVE CAPABILITY	1 Unknown	1 Limited FO of 10 & C < 30pfd	1 Limited FO of 10 & C < 30pfd
EXPR. PERSONNEL AVAILABILITY SYSTEM HARDWARE SOFTWARE	3 (From DSD?) But even then limited	1 Many 5 Many 10 Many 20	2 Limited 2 Limited 4 Limited 4
ACCEPTABILITY	1 High	3 Low	2 Medium

Fig. 3-14e. CCU—trade-off study

DECISION CHART

RECOMMENDATION	CRITERIA WEIGHTING:	1-COST	2-POWER	3-AVAILABILITY	4-RISK	5-ACCEPTABILITY	ESTABLISHED 5 CRITERIA TOTAL	(ALTERNATE) #1's FOR ALL CRITERIA	(ALTERNATE) TOTAL OF ALL RANKINGS FOR ALL CRITERIA
		COMPUTER CONTROL UNIT FUNCTION TRADE OFF					SELECTION BASIS		
	ALTERNATE	(RANK x WEIGHT = SCORE)							
	AYK-14	3 x1 = 3	3 X2 = 6	2 x3 = 6	2 x4 = 8	1 x5 = 5	28	14 1'S	48
X	5301	1 x1 = 1	1 x2 = 2	1 x3 = 3	1 x4 = 4	3 x5 = 15	(25)	(17 1'S)	(43)
	1750	2 x1 = 2	2 x2 = 4	3 x3 = 9	3 x4 = 12	2 x5 = 10	37	13 1'S	48

Fig. 3-14f. CCU—trade-off study

Example 2 (technical advisory groups (TAGs)—a study)

This is an example of a completely different type of trade-off; it is a highly subjective trade-off of a procedural system. This study was performed in the 1980s but is still pertinent because it demonstrates how to develop credibility for a subjective trade-off study. The study is preplanned by establishing a chronological methodology that, once established, is adhered to. The methodology established follows the fairly common problem-solving process: study, problem definition, brainstorming, viable alternates, trade-off criteria, decision analysis (chart), and, finally, a recommendation.

The criteria for this trade-off are very subjective. Therefore, each criteria comparison must be done with care to establish valid comparisons. The scoring for "acceptance" requires consideration of *only* "acceptance" and also requires putting oneself in the other person's shoes: i.e., I am a design engineer (DE); how do I accept alternates 1, 2, 3, 4 and 5? Which one is the best for me? Then ask the same question from the program manager's (PM) perspective, likewise, the chief engineer's (C), and then from the director of engineering's (D) point of view. You may consider actually surveying "users" to get their acceptance rating.

This particular example also shows a different scoring methodology in step 6, which is intended to show "bang for the buck"; that is, how much acceptance and effectivity will we get for the funds provided to implement each alternate? In addition, some very subjective gates are established to eliminate alternates 1, 3, and 5.

The presentation of a highly subjective trade-off study such as this, in this way, is intended to establish credibility. The trade-off is based on a thorough study and honest comparisons. It also makes sense, uses data from named experts, and results in a logical acceptable conclusion. I have abbreviated the names, titles, and credentials of the people named in this study.

I have provided only key portions of the TAGS trade-off study here, but I will point out the important aspects of a totally subjective trade-off. 1) *You do your homework to gain knowledge and study what presently exists. 2) You talk to and study "experts" and then mention them in your report to give the trade-off credibility. 3) You establish a selection criteria that allows for surveying "users." 4) You make your decision based on acceptance factors and effectivity factors. 5) Your decision is also based on a "bang-for-the-buck" comparison of the alternate solutions.*

Technical Advisory Groups (TAGs)—A Study by Jack H.

Methodology

Chronologically:

1. **Knowledge gaining study** existing Technical Advisory Groups (TAGs).
2. **Problem Definition**—What are the basic problems TAGs are to correct?

3. **Brainstorming**—What are the alternate TAG implementations that can solve the problem? Also, what are the non-TAG solutions to the problem that may supplement or replace TAGs?
4. Reduce brainstorming to a list of **Viable Alternates**.
5. Establish **Trade-off Criteria**.
6. Perform **Decision Analysis**.
7. Consolidate results into a **Recommendation**.

Step 1: knowledge gaining study. I have read Jim H.'s METAG (Mechanical Engineering TAG) file, talked to Wayne P. and Bob G., as well as attended a number of their PSTAG (Power Supply TAG) meetings. I've also read Ivert A.'s CIRTAG (Circuits TAG) file that included data and reports of the Florida CIRTAG (that sort of combines our METAG, CIRTAG, and PSTAG). Bj. B. gave me a proposed charter for a SOFTAG (Software TAG), and Dick R. recently instituted an ATAG (Accelerometer TAG), so I have studied these charters. In addition, I have had a number of discussions with people involved with various TAGs. I need to finish this off with a rereading of the Engineering Directives that establish the TAGs, and a review of the design problems that instigated the various TAGs, and I'm ready to make some comments.

(Comments left out in the interest of brevity.)

Knowledge gaining study conclusions: TAGs have generally had a positive impact but need improved charters with required training, external communications, and effectivity criteria. In addition, a more systematic structure should be established for technical advisory groups to replace the haphazard TAG establishment response to problems. Effectivity measurements such as number of successful designs, number of difficult problems solved, number of new technologies, number of training sessions, etc., should be established. Acceptance and support for TAGs must be developed throughout engineering and other divisional organizations.

Step 2: problem definition. We have had problem programs that show a lack of coordination and communication. Many of the problems show a lack of technical expertise, of system engineering expertise (SEE), of discipline, of optimization of technical expertise (right people in the right place at the right time), of appropriate team building, of marketing foresight, of good specification development process, and of good design reviews; the problems also reveal a lack of technical synergism within the division, of technical synergism with the customer, of engineering efficiency, of technical competency in our managers, and a lack of synergism with other groups in the company (production, quality, marketing, etc.), implying that we are:

A lot of lackadaisical lackeys with lots of lacks?

No! I don't think so. Along with the problem programs, we have successful programs, successful money-making production products, excellent teams, teams with great esprit de corps. Therefore, when we work on fixing our problem, we should be careful that we don't throw out the baby with the dirty bath water.

The system engineering skills survey has shown that we have engineers with self-perceived unbalanced skill profiles and also a lack of system engineering expertise (SEE). Therefore, it follows that we have some teams with unbalanced skill profiles and lack of SEE. The challenge is to recognize these skill deficiencies, and then create the desire to do something about them. Skills can be improved by mentoring and training, but we also need to better understand the skills available and skills required for specific programs to ensure that our teams have the proper skill mix to perform well. Thus, we say, "know your people." Therefore, in summary, the statement of the problem and desired solution is:

Statement of Problem: We have had serious problems that have impacted some of our programs and proposals. These have indicated that some of our teams have skill deficiencies in assigning people, system engineering, mechanical design, software design, customer liaison, power supply design, technical competitiveness, analysis, and, especially, communication. Therefore, we need to provide a means that will improve our engineering department to eliminate these problems. In addition, these means will be developed in such a way that they raise the esprit de corp of the engineering department (our mission).

Step 3: brainstorming. (Reader Note): The actual brainstorming has been left out of this book because it was too lengthy; however, the brainstorming resulted in discovering over fourteen other groups, committees, projects, training courses, and review processes that existed to solve the stated problem. We recommended that we must *consolidate and eliminate, not proliferate,* committees and groups if we wish to solve the problem.

(Resuming the trade-off study): My brainstorming has diverged. The reorganization of all these services, functions, groups, committees, etc. is a new mission to work on at a later date. My immediate study is only to consider technical advisory group (TAG) implementation. What are the viable TAG implementation alternates that will solve the stated problem? I hope some of the alternates may help consolidate some of the multiplicity of disjointed groups and services as well as provide an improved TAG system.

Step 4: list viable alternate implementations for TAGs.

1. None — **Eliminate TAGs** altogether (let each company operation make its own decision).

2. Weak — **Academic-Only TAGs**, similar to the Mil-Std-1750 computer users group methodology. This group was formed by informal surveys, questionnaires, and newsletters that established communication between interested people. Those that consistently

responded were placed on the user group permanent mailing list. Some travel and coordination meetings were held. This 1750 users group covered *all* interested engineering departments in the company. This methodology established excellent communications between interested people and could serve as an example for TAGs.

3. Semi-Weak **Leave TAGs As They Are**—based on the concept, "if it works, don't fix it."

4. Normal Process **Expand and Improve the TAGs We Have**—(normal development process. Enhance what works; eliminate what doesn't work.) Add systems engineering TAG (SETAG) and software TAG (SOFTAG) with charters similar to METAG and PSTAG. Improve all charters to add training, effectivity measurements, and improved external communication (using TAG wheel and user group methodology). Establish a TAG chairperson's committee that will be chaired by the chairperson of the SETAG. Also, provide more funding support to show we really believe in TAGs.

The recommended TAG wheel is shown in figure 3-16.

5. Strong Consolidate all the funds now being spent to solve the "problem" into a reformed TAG structure similar to the previous TAG wheel. This should consolidate the multiple number of committees, groups, organizations, support services, tool developers, new technology teams, research programs, and training programs that have been established to solve parts of the "stated problem."

This restructuring should be simple with an expansion of the TAG wheel as a baseline. Establish permanent full-time chairpersons for each TAG. TAGs would be given authority for design review process, engineering operating procedures, career development, all design support functions, tools development, and new technology department. This is a matrix type approach, with more than just advisory function. In fact, instead of calling it TAG—technical advisory groups, it should probably be renamed technical control group (TCG).

Step 5: establish trade-off criteria.

1. Acceptance (A): Will the engineers and managers wholeheartedly accept and dedicate themselves to support the implementation? This criterion will be established as a requirement for all four levels: 1) design/development engineers, 2) managers/supervisors/work directors, 3) chiefs/section heads, 4) directors of engineer-

ing. It must then be accepted by all levels. If it is unaccepted and unsupported, it won't work.

(Score in four levels 0 to 25)

2. Effectivity (E): Will the implementation be effective in solving the stated problem? Five measures of effectivity are:

1) PDB—Prevent design problems
2) MT—Maintain technology current with state of the art
3) PC—Provide communication
4) PT—Provide training
5) DR—Provide design reviews

(Score in five levels 0 to 20)

3. Cost: (Number of people to fund and $ budget?) What resources ($ and people) need to be diverted from mainstream product development activity?

Note: Look at cost as comparison, not actuals. Also note that the high cost of solution no. 5 is offset by reducing other expenditures. Cost does not appear to be as big a concern as acceptance and effectivity for this decision analysis.

The trade-off criterion for this decision is basically all-subjective. Usually, we have some numerics in decision charts (weight, power, and real cost) that provide a more comfortable decision analysis. This analysis does have nine separate trade-off criteria (four levels of acceptance + five measures of effectivity) however, and when we put on the blinders and impartially compare the solutions on each of these nine criteria one at a time, the analysis has validity. An alternate scoring method might be doing a survey of engineers and managers to get their scoring of acceptance and effectivity of the solution alternates.

Step 6: perform decision analysis (see fig. 3-15). This decision chart provides a considerable amount of data regarding the selection criteria scoring and the final results. The results are:

"NONE"—is unacceptable because of poor effectivity. We need some TAGs to carry us into autonomy, and we still will need cross operation and division communication.

"LEAVE AS IS"—is unacceptable because what we have needs improvement and expansion. Both acceptance and effectivity are low.

"TOTAL CONTROL GROUPS"—are marginal. Acceptance at the working level will be poor if disciplined watchdog approach is used. (If we intend to go this way, we should work into it over a couple years to establish acceptance.) Cost of AE is poorest, but this should be offset by eliminating or reducing other areas.

"ACADEMIC"—gives the most bang for the buck, but only gives a communications forum.

"IMPROVED"—improves what we have, which can improve our effectivity. Also, if we incorporate the academic methodology (at no cost), we can improve our AE to 5.05. This will be our recommendation because it is the best AE and no. 2 in cost effectiveness.

Step 7: recommendation (see fig. 3-16). We should expand and improve the TAGs we have. Add systems engineering TAG (SETAG) and software (SOFTAG). Provide improved charters for all TAGs (SETAG, SOFTAG, METAG, PSTAG, CIRTAG) to include effectivity measurement criteria, training advisory for track/skills, tools, and communication within the TAG wheel. Also establish an interdivisional SETAG that ties our TAG wheel to Florida's, to Seattle's, to West Covina's, and to Boston's. The interdivisional SETAGs will be loosely coupled in a manner similar to that in the solution 2 methodology.

In addition, it is recommended that a study be conducted regarding the multitude of groups, committees, tool developments, support services, etc. that divert mainstream resources. The objective of this study will be to eliminate nonproductive and duplicative activity. We will consolidate, eliminate, and not proliferate.

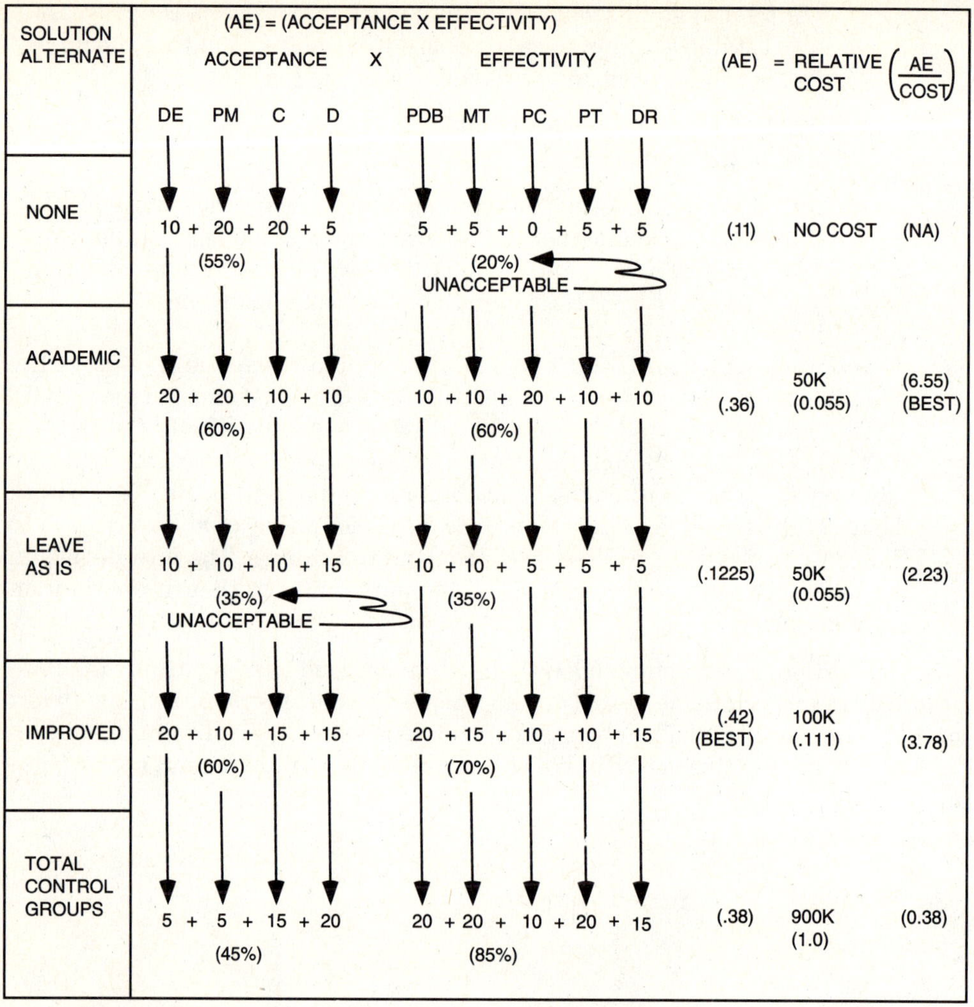

EFFECTIVITY:
PDB = PREVENT DESIGN PROBLEMS
MT = MAINTAIN TECHNOLOGY
PC = PROVIDE COMMUNICATION
PT = PROVIDE TRAINING
DR = DESIGN REVIEWS

ACCEPTANCE:
DE = DESIGN/DEVELOPMENT ENGINEERS
PM = PROGRAM MANAGERS
C = CHIEF ENGINEERS
D = DIRECTORS OF ENGINEERING

Fig. **3-15.** Decision analysis structure

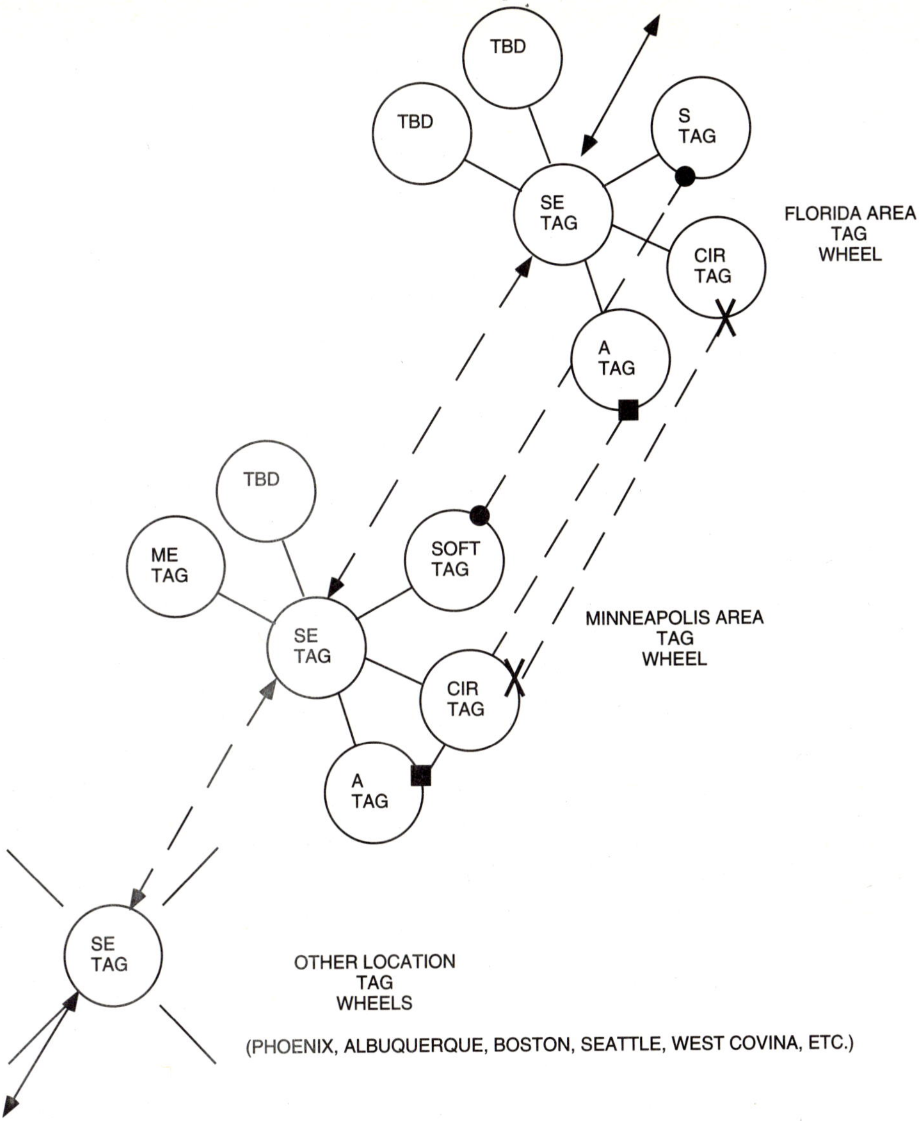

Fig. **3-16.** Recommended TAG wheel structure

Example 3 space shuttle communications bus trade-off

This example, done in the 1970s, shows how to generate a weighting factor score chart. A scoring system that ranks the alternates by percentages, where 100 percent would be perfect, helps people understand the trade-off results.

This trade-off was done to select the optimum functional implementation for a data bus to be part of the Space Shuttle System.

Step 1: set priorities for the categories of decision.

- Category 1—Optimum System (mission effectiveness)
- Category 2—Least Weight
- Category 3—Least Power
- Category 4—Optimum GSE (ground support equipment)
- Category 5—Optimum Software

CAUTION: These categories (or criteria) must be top system level attributes. If you use lower-level categories such as High Order Language (HOL) used, testability, or bus structure, you may get a system optimized on these categories but completely unacceptable in top system levels such as weight, power, and cost. Also, never make a decision on only one, two, or three categories.

Note that Category 1 above is the highest priority and, therefore, will have the highest weighting factor. The number of categories selected is directly proportional to time-to-decision and decision-validity. The time-to-decision is proportional to matrix size because the more categories and alternates, the larger the matrix and the more time it will take to properly access rank per category. Decision-validity is also proportional to the matrix size; thus, the larger the matrix the more valid the decision.

Step 2: select feasible solution alternates. As shown on the left side of figure 3-20, these are:

a. Present system

b. Load memory through 30-foot extension

b1. B plus a new memory load design

c. Load memory through a new 250-foot Manchester-encoded bus

c1. C Plus verify using data bus or D

d. Load through existing 250-foot Manchester data bus

Step 3: assess each solution alternate per category separately. Numerically estimate items such as cost, weight, time, power, etc., and then rank them appropriately. Subjectively analyze those items such as optimum system, optimum ground support equipment, optimum software, customer acceptance, management acceptance, risks, etc. Writing a sentence or two to describe each item can be helpful. Rank these subjective items. This data collection and ranking function requires the most time and effort in decision making. To perform it accurately and objectively is very important to the final ranking.

Ranking based on numerical estimates of data is quite easy; however, ranking of nonnumerical judgment type considerations can be difficult. To obtain an objective ranking, it will be necessary to put on blinders to all other categories so the ranking will not be distorted. It also may be helpful to have three, five, or seven people independently provide ranking, and then sum these to reach a final ranking.

Figure 3-17 is a chart of the Step 3 process for ranking optimum GSE (Category 4). Note that seven additional subcategory ranking criteria are used. Establishing additional criteria for subjective items is highly recommended.

Step 4: tabulate rankings on a composite ranking chart (fig. 3-18). Note that "ties" are reasonable in ranking. The judgment on optimum software, for example, shows a four-way tie for first place. You may wish to establish a "tie-breaker" category as shown in the column called "parts-delta." This tie breaker is used in case of a first-place tie in the final decision, further followed by a first-place tie in the Category 1 column, etc. Usually, if you have that many ties, the final two tied alternates must be so close that you can probably just flip a coin to decide).

Step 5: generate a weighting factor (WF) score chart (see WF score chart, fig. 3-19).

A = No. of alternates

C = No. of categories

RW = Rank weight

RW = From (A-1) to (0)

WF = Weight factor

$$\text{MWF} = \frac{100}{(A-1)\,(1 + 2 + 3 + 4 + \ldots C)} = \text{Minimum WF (MWF)}$$

C1WF = MWF

C2WF = 2C1WF

C3WF = 3C1WF

etc.

If A = 6 and C = 5, then:

$$\text{MWF} = \frac{100}{5(1 + 2 + 3 + 4 + 5)} = \frac{100}{5(15)} = \frac{100}{75} = 1.333$$

Step 6: fill in the decision chart. With the scores from RW—Step 4 and the WF—Step 5 score chart, total and rank for final result. Then record on the decision chart as shown in figure 3-20. In case of ties, the highest score in category 1 wins, or, if another tie occurs in category 1, 2, 3, 4, 5 then the tie-breaker column from figure 3-18 is used.

In addition to the tie-breaker problem, note that, logically, a "don't

ALTERNATE =	A PRESENT SYSTEM	B LOAD 30-FOOT EXTENSION	B1 B PLUS NEW LOADER	C LOAD 250-FOOT MANCHESTER NEW BUS	C1 PLUS VERIFY THROUGH DATA BUS	D LOAD 250-FOOT MANCHESTER DATA BUS
ONE SET GSE OR REQUIRES DIFFERENT GSE	5	5	5	4	3	2
EASE OF USE ON PAD	1	2	3	4	4	5
EASE OF USE ON BENCH	5	5	5	4	4	3 SEPARATE GSE
TOTAL GSE SYSTEM COSTS	5	4	4	3	2	5
FLEXIBILITY FOR PROGRAM CHANGES	4	3	3	4	4	5
FASTER LOADER (LOGISTICS ON PAD)	0	0	0	4	3	5
RELIABILITY OF MEMORY LOAD SYSTEM	5	4	3	2	1	5
TOTAL RANK	25	23	23	25	21	30
FINAL RANK	2	3	3	2	4	1
RANK WEIGHT	4	3	3	4	2	5

Fig. **3-17.** Step 3—(example 4 of 5 categories) ranking process for optimum ground support equipment (GSE)

TIE BREAKER (PARTS DELTA)	ALTERNATE	CATEGORY 1 JUDGEMENT OPTIMUM SYSTEM	CATEGORY 2 NUMERICAL WEIGHT/SYSTEM	CATEGORY 3 NUMERICAL POWER/SYSTEM	CATEGORY 4 JUDGEMENT OPTIMUM GSE	CATEGORY 5 JUDGEMENT OPTIMUM SOFTWARE
+88 (2)	A	5 (RW = 1)	6 54.3 LB	4 9.6W	2	2
+56 (1)	B	4 (RW = 2)	5 37.5 LB	2 8.4W	3	1
+128 (4)	B1	3 (RW = 3)	3 20.5 LB	1 6.3W	3	1
+227 (5)	C	6 (RW = 0)	4 21 LB	5 12.8W	2	1
+115 (3)	C1	2 (RW = 4)	2 6.94 LB	3 8.9W	4	1
+190 (6)	D	1 (RW = 5)	1 4 LB	6 16.8W	1	3
		RANK	RANK	RANK	RANK FROM FIGURE 3-17 EXAMPLE	

Fig. 3-18. Step 4—ranking chart for memory load study showing ranking and RW determination

care" is possible. If a "don't care" situation should occur, the "don't care" will be flagged with a "DC" and will be given a rank weight (RW) that is the average of all other RWs in that category. This will be done so that the "don't care" does not inappropriately gain or lose points in the evaluation, but receives the average score. The number on the decision chart should also carry a subscript "DC" indicating "don't care."

EXAMPLE:

Assume six alternates, with RW = 5 for best, RW = 4 for 2nd, RW = 0 for poorest. Now if RW of A = 5, B = 4, C = 3, D = 2, and E = DC, we calculate RW for E to be:

5

4

3 14/4 = 3.5

2

———

14)

So E = 3.5/DC.

The "don't care" possibility can occur when there is a basic difference between alternates such that the selection criteria are not applicable to all alternates. If, for example, you wish to compare control system implementations, you may compare an analog computer implementation with a number of digital implementation alternates. If one of the criteria is "software complexity," the analog system is decidedly ranked as the best. Likewise, this is so for "software support equipment required." By contrast, the analog system is poorest when you consider "ease of control algorithm modification." A criterion of "amount of spare memory," however, provides a problem because it does not apply to the analog implementation. It would not be fair to penalize the analog implementation for not having spare memory; likewise, it would not be fair to give it unfair credit; therefore, we use the "don't care."

The results shown in the figure 3-20 decision chart have no DCs or tie-breaker situations. The Alternate "D" solution of loading a 250-foot Manchester data bus was ranked number 1, with a final total score of 77.34% out of 100%.

		SCORE						
	WF	RW=0	1	2	3	4	5(A-1)	NO. of Δ's
C1WF	1.33	0	1.33	2.67	4.00	5.33	6.67	5
C2WF	2.67	0	2.67	5.34	8.01	10.68	13.34	10
C3WF	4.00	0	4.00	8.00	12.00	16.00	20.00	15
C4WF	5.33	0	5.33	10.66	15.99	21.32	26.65	20
C5WF	6.67	0	6.67	13.34	20.01	26.68	33.35	25
		0	WF = WEIGHTING FACTOR				-100.01	75 UNITS

C1WF = CATEGORY 1 WF

RW = RANK WEIGHT

Fig. 3-19. WF score chart (for six alternates). Given six alternates, establish a rank weight of (0, 1, 2, 3, 4, 5), then calculate WF to give a max score of 100.

ALTERNATE	CATEGORY 1	2	3	4	5	FINAL TOTAL SCORE	FINAL RANK
	OPTIMUM SYSTEM (WF = 6.67)	WEIGHT FLIGHT SYSTEM (WF = 5.33)	POWER FLIGHT SYSTEM (WF = 4.00)	OPTIMUM GSE (WF = 2.67)	OPTIMUM SOFTWARE (WF - 1.33)		
A PRESENT SYSTEM	RW = 1 6.67	RW = 0 0.00	RW = 2 8.00	RW = 1 10.68	RW = 4 5.33	30.68	6
B LOAD 30-FOOT EXTENSION	RW = 2 13.34	RW = 1 5.33	RW = 4 16.00	RW = 3 8.01	RW = 5 6.67	49.35	4
B1 B PLUS NEW LOADER	RW = 3 20.01	RW = 3 15.99	RW = 5 20.00	RW = 3 8.01	RW = 5 6.67	70.68	3
C LOAD 250-FOOT MANCHESTER NEW BUS	RW = 0 0.00	RW = 2 10.66	RW = 1 4.00	RW = 4 10.68	RW = 5 6.67	32.01	5
C1 C PLUS VERIFY as in D	RW = 4 26.68	RW = 4 21.34	RW = 3 12.00	RW = 2 5.34	RW = 5 6.67	72.01	2
D LOAD 250-FOOT MANCHESTER DATA BUS	RW = 5 33.35	RW = 5 26.65	RW = 0 0.00	RW = 5 13.34	RW = 3 4.00	77.34	1

NOTES:
RW = RANK WEIGHT (BEST = 5, 2ND = 4, - - - , POOREST = 1)
WF = WEIGHING FACTOR
SCORE = (RW) (WF)
TOTAL = SUM OF SOURCES (HIGHEST SCORE IS BEST)

Fig. **3-20.** Decision chart for memory loading for space shuttle (weighted to give 100 score max) showing final total score and final rank

3.3 Summary of SEE-1, -2, and -3

Our objectives in the first three chapters of this book have been to develop:

1. **Your ability to quickly:** Do a skills analysis, team build, and:

   SEE 1 — See the big picture regarding your subsystem (SE perspective cube).

   SEE 2 — Visualize and understand the mission of your subsystem (sortie and life mission analyses).

   SEE 3 — Reduce the mission to actions required to perform the mission (functional analyses).

   SEE 3 — Make optimal decisions regarding areas of implementation trade-offs (trade-off studies).

2. **Your ability to slowly:** perform in great detail and precision the skills taught in SEE-1, -2, and -3.

3. **Your ability to know when to go fast and when to go slow!**

   Slow: The SEE-1, -2, and -3 skills were applied by the Apollo system developers to the high-risk Apollo Moon Lander program slowly and with great detail and precision.

   Fast: The SEE-1, -2, and -3 skills were applied by a single part-time engineering consultant to a low-risk advertising display system development quickly and adequately.

3.4 Problems and Questions

1. Match the Following:

a) A function	A) Usually occurs at the end of the concept definition phase
b) System definition function	B) Functional block diagram
c) Functional decomposition	C) Must be done prior to design
d) System requirements review	D) Ultimate incorporation in specifications
e) Input to functional analysis	E) Accuracy apportionment
f) Output from functional analysis	F) An action to be done
g) Output from functional analysis	G) Trade-off studies
h) Output from functional analysis	I) Mission analysis
i) Evaluation of alternatives	H) Hierarchial description
j) Requirements allocation	J) Uses system engineering process

2. True or False?

    a) A computer is a function.
    b) Functional analysis is used to help define a system.
    c) Functional analysis is used to help develop the requirements for a system.
    d) Structured analysis is not functional analysis methodology.
    e) Mission analysis and functional analysis are both done so we don't forget any of the elements of our system.
    f) Functional actions may be accomplished by equipment, people, facilities, software, data, or a combination thereof.
    g) A detailed procedure is a function.
    h) Computing is a function.
    i) Functional analysis asks: What are the actions of persons or things required to perform the mission?
    j) System definition function is basically just functional analysis.

3. Match the following:

a) System perspective	A. Functional block diagram/specification tree
b) Inputs to functional analysis	B. Time line
c) Mission analysis	C. Mission analysis
d) Functional analysis	D. Time/physical/actions/ hierarchial/conditions/specific costs and dates
e) Output of Functional Analysis	E. Actions of persons and things

4. True or False?

    a) Functional analysis is represented on the system perspective cube by the "actions to perform" side.
    b) Implementation of the functions developed results in the items listed on the "physical" side of the system perspective cube.
    c) Functional analysis and mission analysis are skills that can be applied to all three levels of the hierarchial side of the system perspective cube.
    d) Functional analysis and mission analysis are skills that can be applied to all system time phases of the time side of the system perspective cube.
    e) Functional analysis is unrelated to the system perspective cube.
    f) You can do functional analysis without knowing the mission of the system.
    g) A hierarchial chart showing a tree of detailed specifications is an adequate output of functional analysis for some systems.
    h) Functional analysis deals with "actions"; therefore, it does not completely define a system or establish all the requirements for a system. System attributes and mission analysis time line requirements are also required.
    i) Brainstorming requires that no foolish or incorrect remarks will be recorded.
    j) Brainstorming is a catalytic and innovation developing activity.

PROBLEMS REQUIRING FUNCTIONAL ANALYSIS SKILLS

5. Do action function elements for: An automobile, a home, a house, a washing machine, a library, an educational system.
6. Do a functional analysis of the mission sortie in SEE-2, helicopter example.
7. Do a functional analysis of a personal mission. Provide a hierarchial chart with attributes and analyses blocks included. Also, provide any of the following if applicable: a specification tree, state diagrams, functional flow diagram, and a functional block diagram.
8. Using the engineering department example as a format, do a functional analysis of a procedure system for one of the following: post office, county government, a university, medical clinic, a professional sports organization, your church, police department, a grocery store, or a similar process/procedure system.

QUESTIONS ABOUT TRADE-OFF STUDIES

9. Match the following:

a) Gates	A. The alternate will meet the specification
b) A model	B. Establish the trade-off criteria
c) Feasible	C. A methodology to select the best implementation
d) Brainstorming	D. A method of describing an alternate
e) Step 1	E. Spare memory criteria in a manual (has no memory) test equipment alternate
f) "Don't care" condition	F. Subjective criteria comparison
g) AYK-14 (for USN)	G. USN standard computer
h) Blinders on	H. Used to find all suitable alternates
i) Prior to trade-off study	I. Set priorities and gates
j) Trade-off study	J. All alternatives must fulfill these requirements

10. True or False?
    a) Trade-off studies are performed during all phases of a system: concept, demonstration, development, production, deployment, and recycle.
    b) Trade-off studies are performed for all system development function decisions.
    c) When you have years of experience, you don't have to use trade-off study methodology.
    d) Selecting feasible solution alternates (step 2) is the most important part of the trade-off study methodology.

e) Decision chart methodology is a subset of the trade-off study methodology.

f) A serial trade-off study methodology is where you first select the optimum alternate for the highest priority criterion (Level 1); then you determine all the possible variances of the Level 1 alternate that are then compared on the next priority criterion to select the optimum variance of the Level 1 alternate for the Level 2 alternate—working your way through all the priority levels to a final selected alternate. This is an acceptable methodology.

g) Only known and accepted "experts" can make acceptable subjective trade-offs.

h) "Putting the blinders on" means to rank the alternates based on a single specific criterion, without reference or interaction to other criteria.

i) Making a poor trade-off decision can cause a system to fail.

j) Trade-off studies do not require much work.

PROBLEMS REQUIRING TRADE-OFF STUDIES

11. Brainstorm all the ways you can perform your job. Do a trade-off study on the top three viable alternates.

12. Do a trade-off study of alternates for implementation of one of the functions from your functional analysis of the helicopter mission sorties of chapter SEE-3.

13. Do a trade-off study of alternates for implementation of one of the functions from your functional analysis of your present project.

14. List at least three possible ways to solve each of the following problems:

    a) You don't have enough money
    b) You have too much money
    c) People are starving
    d) People are overeating
    e) The national debt
    f) Congestion on our highways
    g) Not enough jobs
    h) Poor jury selection process

15. Make a list of questions you would ask to better define one of the problems listed in problem 3.14.

The Proposal

(Baselining and Management)
(Tell me what it is
and how we are going to get it)

In chapter SEE-1, you learned some system engineering terminology, established your skill profile, did some team building, and related the ***system engineering perspective*** cube to your present job, project, or mission.

In chapter SEE-2, you expanded upon the perspective cube "time" axis by being exposued to ***mission analysis*** for system sorties and lives. You related mission analysis to your present job and developed skill in applying mission analysis. You recognized that the purpose of mission analysis is to help develop a baseline definition of the system and develop the requirements for that system.

In chapter SEE-3, you learned how to use your mission analysis datum to perform ***functional analysis***, which leads to the ultimate goal: a system functional design which will perform the system mission. You also learned how to do trade-off studies, which are used for optimization of system implementation for specific functions of the system.

In chapter SEE-4, you will take the material generated in chapters SEE-1, -2, and -3 and use it to help you define a ***baseline system*** (fig. 4-1). You will study an example of what a systems engineer does to kick off a proposal effort. In conjunction with this, you will also be exposed to development specification process, system engineering management, the system engineering management plan, the program management plan, and program start-up review process.

You have accumulated a lot of data at this point in our system engineering process: systems perspective worksheet, sortie/life mission analysis time line diagrams, functional analysis diagrams, trade-off studies. You also have collected customer data such as the statement of work and procurement specification. These, however, do not make a system, so how do you organize and make use of all this data? How do you work to lower levels? Use your system perspective

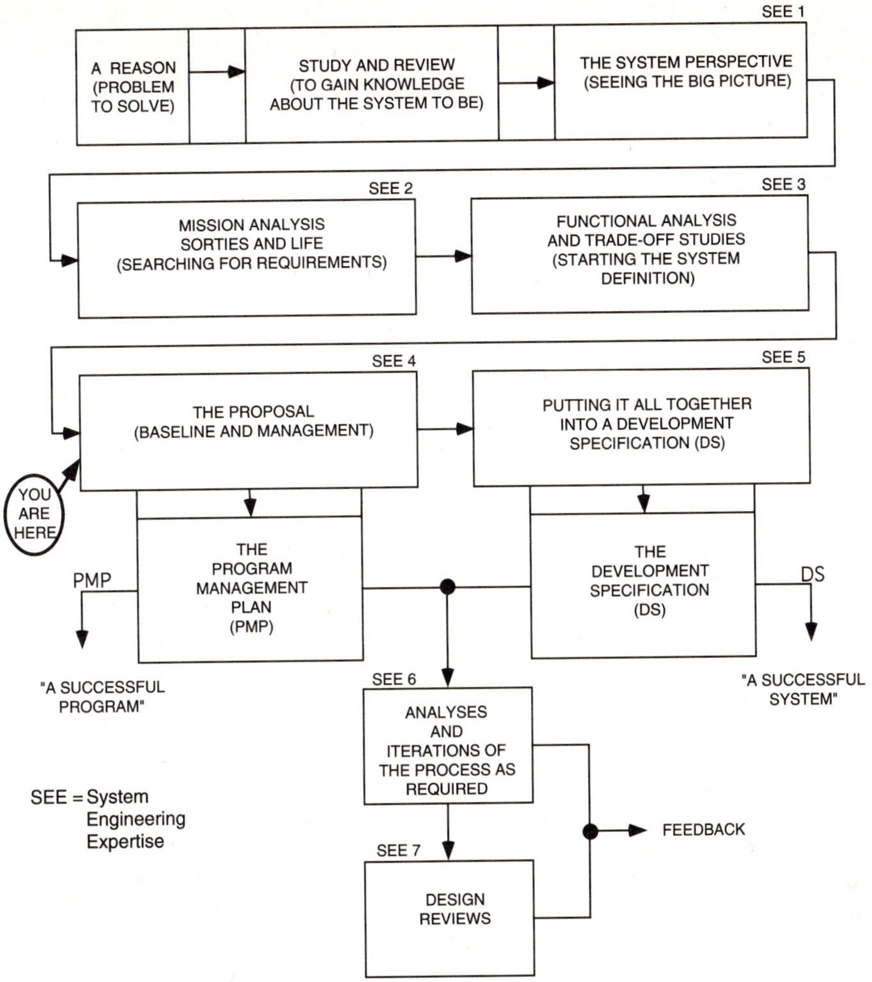

Fig. **4-1.** The system engineering process

skills. Your problem is to convert this data to a system definition that can perform your system-to-be's missions:

1. My definition must be functionally complete: It will define the physical attributes of the system (hardware, software, documentation, people, tools/facilities), the actions to perform, its place in a hierarchy, its hierarchy, constraining conditions to apply, and where it is in the time phases plus specific costs and dates.

2. I have to expose my system to all the people. Because of this, I must define it sufficiently so it can be developed, defined quickly, and especially defined so all the people who are developing the system and interface to it will know enough about it to get started.

3. I will put myself in the shoes of *all the people* who will interface to this system. What do they need to know? What questions must be answered?

4. I must not design the system details; that is to be done later. My mission analysis, functional analysis, and trade-off studies will be sufficient to ensure that my baseline implementation is functionally complete. Contract-driven, design-constraining conditions must be transmitted to all the system designers.

5. My definition must cover both technical and business requirements.

How do you systematize and organize to control the system development process? First you will baseline define your system in technical terms. Then you will show how you will present the system to the developers (the proposal effort). Finally you will organize the technical and business aspects of the system development into two key documents: the development specification and the program management plan. I will show you how you can tie the technical and the business aspects of the program together. The development specification and program management plan will be seen as controlling documents to regulate the development of the system. I will also show how to organize the people to develop an appropriately skilled team to develop your system.

4.1 Defining a Baseline System

You now must take your trade-off study results, your functional analysis and mission analysis diagrams and convert them to a baseline system definition. You know that your functional diagrams provide a description of all the actions of persons and things required to successfully perform your sortie mission. The action function element for developing the baseline system definition is shown in figure 4-2. A picture is worth a thousand words, so draw pictures to define your baseline system. The outputs of figure 4-2 are pictures for the most part.

I will show you examples of the types of pictures that initially launch a system development project. The baseline I have selected for the example is the Total Avionics Control System (TACS). The customer had done a major system mission analysis and functional analysis and established a baseline that included a TACS as a subsystem of a missile. I am responsible for the TACS, which has been top-level baselined, in working with the customer, into a Command and Control Unit (CCU) and a Sensor System (SS) that I am to develop as subsystems of my TACS. You will note the use of acronyms in the following examples. The basis for their use is to shorten long titles or much used terms to improve communication. For example, the acronym RADAR stands for radio detection and ranging. Common use of this acronym has resulted in its acceptance as a word. You must, however, avoid overusing acronyms because too many of them can make your writing incomprehensible. The following pictures, sketches, figures, tables and diagrams for TACS were originally done by very quick hand sketches

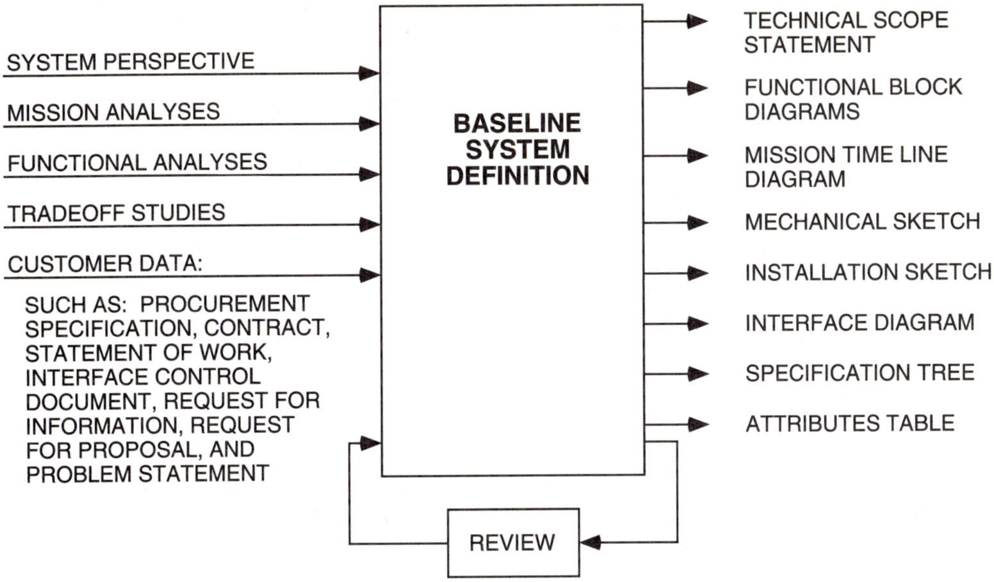

Fig. **4-2.** Baselining

by a single systems engineer in one day. These rough sketches were then modified, updated, and improved as the system was developed. I have converted the original rough sketches in this book to formal drawings to ensure readability. The first page of the baseline must be a title page that provides a method for keeping a highly visible revision record for the baseline. Figure 4-3 is a title page example.

The next part of the baseline definition will be a SCOPE statement that reads as follows:

SCOPE:

Scope of TACS: Control systems for missiles and remotely piloted vehicles (RPVs) are too costly because they have evolved from a multiple of stand alone subsystems. These subsystems have separate boxes, mounting, cooling, power supplies, interfaces, control algorithms, and suppliers. The TACS will integrate these systems by providing a single box, mounting, cooling method, power distribution, interface method, consolidated control algorithms from a single supplier, with selection and development of a second geographically different noncompetitive team member supplier.

4.1.1 Time Line of System Mission

You need to baseline the basic sortie mission your system is being developed to perform. This is derived from your prior mission analysis. Figure 4-4 is a sortie mission time line analysis.

PROPOSAL

SPECIFICATION NO. [ ]

PART I OF II

TITLE: TOTAL AVIONICS CONTROL SYSTEM-608 (TACS-608)
(BASELINE SYSTEM DEFINITION)

	SIGNATURES	TITLE	DATE
PREPARED BY	**Jack Hunger**	System Engineer	
CHECKED BY			
APPROVED BY			

☐ CUSTOMER FUNDED — CONTRACT NUMBER

☒ COMPANY — PROJECT NUMBER

REVISIONS							
LTR	DESCRIPTION	DATE	APPROVAL	LTR	DESCRIPTION	DATE	APPROVAL
—	Proposal Effort						

PAGE	
REVISION	
PAGE	PAGE 1 OF
REVISION	

Fig. 4-3. Title page

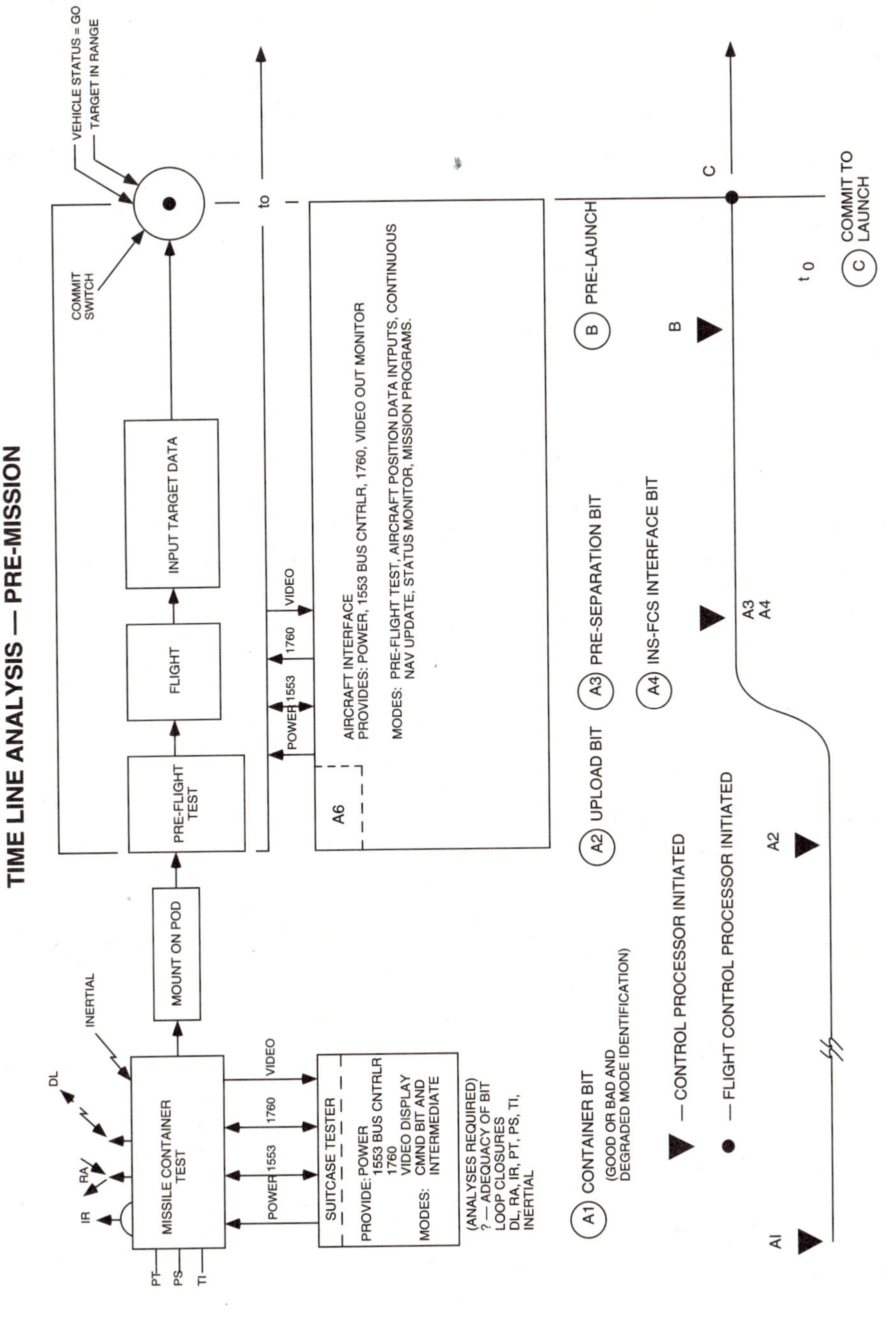

Fig. 4-4a. Time line analysis (pre-mission)

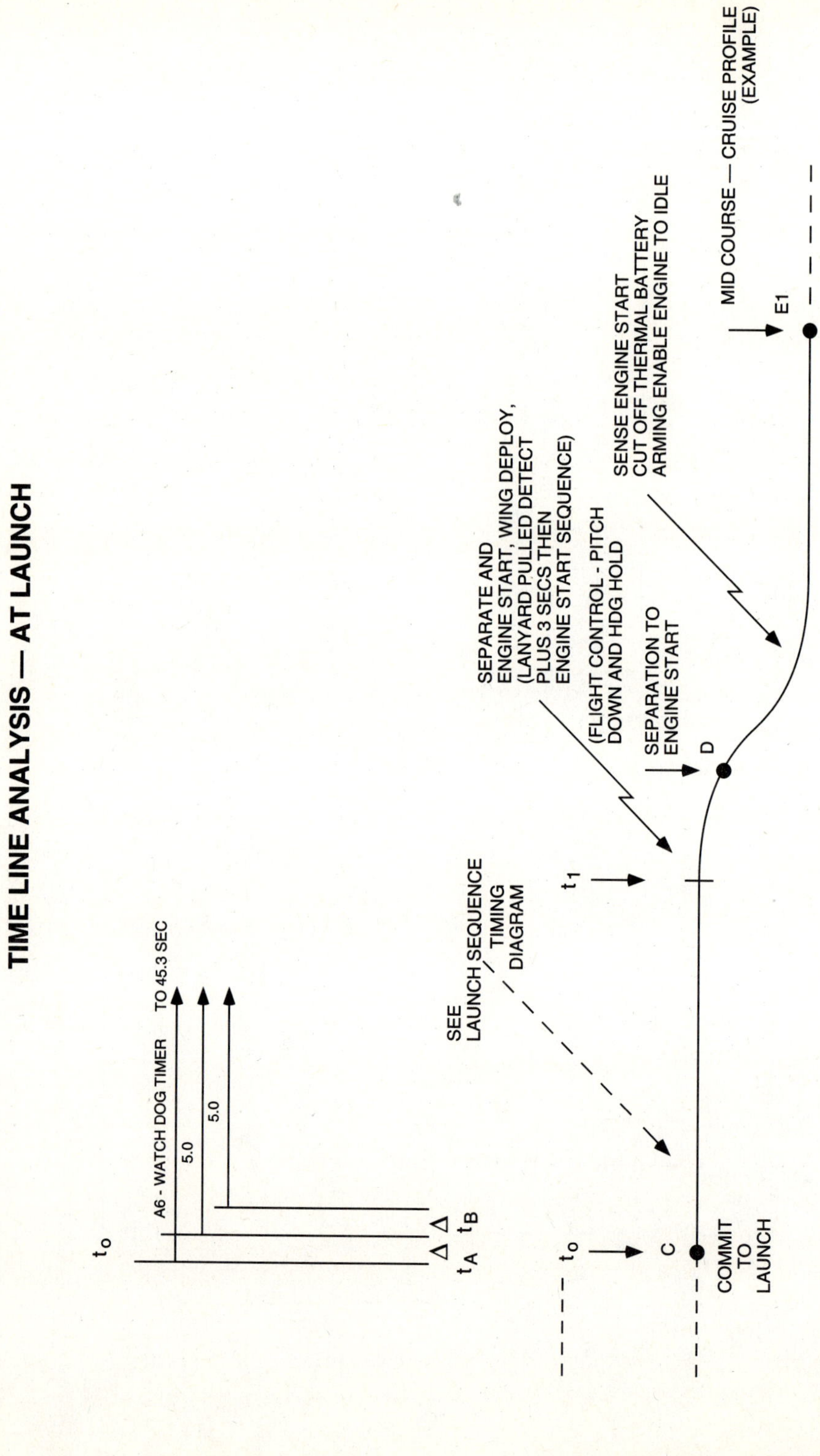

Fig. 4-4b. Time line analysis (at launch)

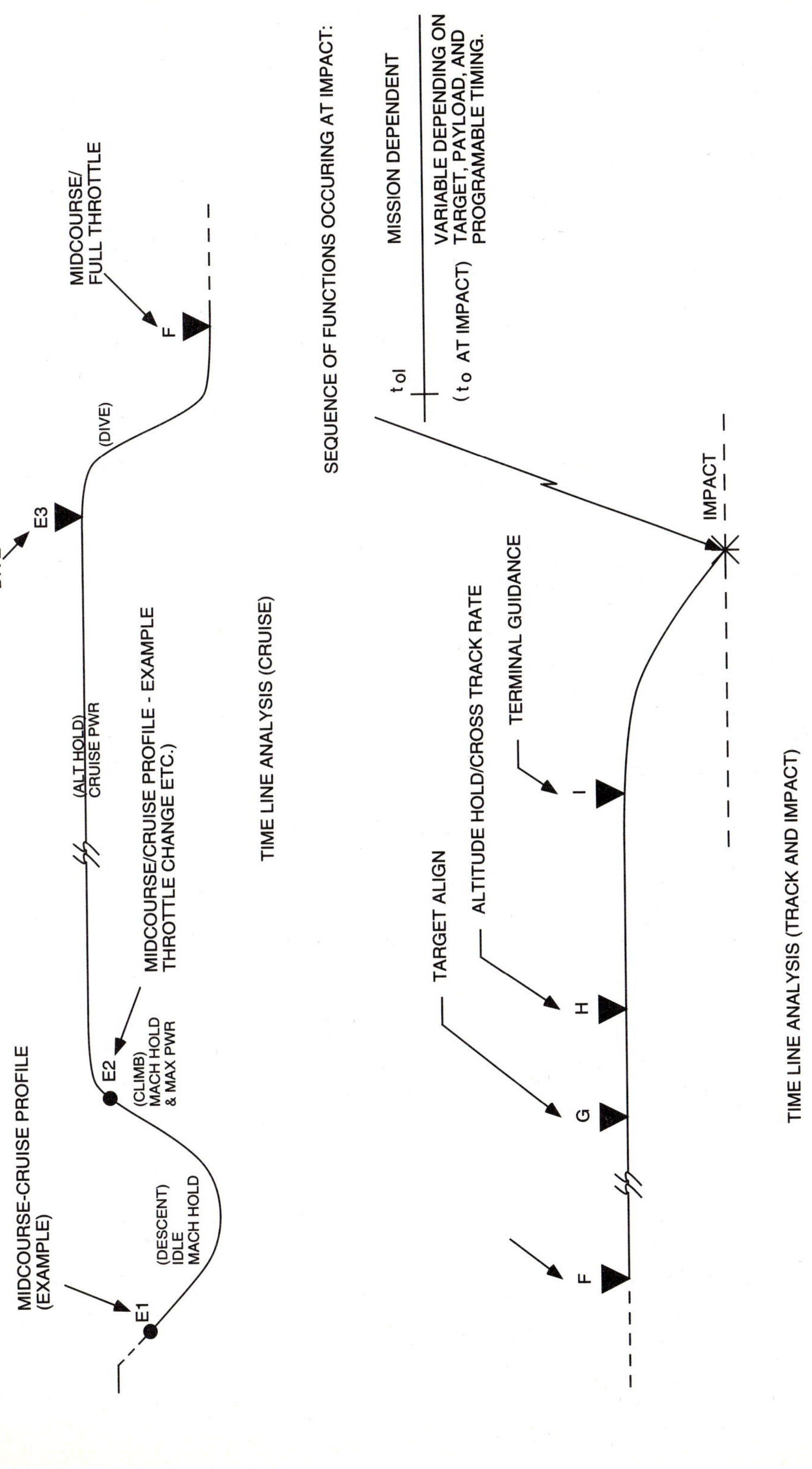

Fig. **4-4c.** Time line analysis (cruise, track and impact)

4.1.2 Functional Block Diagrams

Figure 4-5 is a functional block diagram of the command and control unit (CCU) for our TACS system. This diagram is the result of a functional analysis of the CCU. The sequence or analysis that leads to this diagram is: 1) mission analysis; 2) functional analysis of the mission to identify the top-level functions required to perform the mission (functions of sensing, commanding, controlling, and navigating were then synthesized into a TACS); 3) lower-level functional analysis of the TACS avionics subfunction (the resulting subfunctions were then synthesized into a sensor system and a command and control unit, or CCU); and 4) lower-level functional analysis of the CCU that resulted in the functional hardware elements, functional software elements, and functional test elements.

This diagram also shows interfaces and therefore serves as a device for individual designers to start the design of their specific functional elements. The analog to digital converter designer can now start functional analysis of the specific element, "input conversion." In like manner, the engine control software designer can start lower-level functional analysis of the software-specific element, "engine control." The complete functional block diagrams for the TACS would include diagrams for sensor system (SS) and customer vehicle tester (CVT) as well as the CCU.

4.1.3 The Specification Tree

You have functional block diagrams, and now you want to provide a diagram that establishes the documentation that leads and controls the system development process. Figure 4-6 shows how the system definition and requirements will be organized into a hierarchy of development specifications. The top system development specification TACS must be initiated *immediately,* and how to do this will be the subject of chapter SEE-5, "The Development Specification."

The specification tree should be kept simple, reflecting the selected top-level partitioning. Carrying it too low can be restrictive and may limit the designer's innovation. Let the development process carry the specification tree to lower levels as the system implementation evolves. If you have a problem with "selected top-level partitioning," just go with the basics—system hardware, software, and test equipment.

Also, view the specification tree as associated with deliverable products (specific hardware, software, and tester elements) and, therefore, associated with specific development work tasks that are performed by specifically skilled people. The lead total system systems engineer does the system development specification immediately to *lead* the development process. Next, the lead hardware systems engineer, the lead software systems engineer, and the lead test systems engineer do their development specifications. They then each individually lead or personally do the associated development tasks in their respective areas.

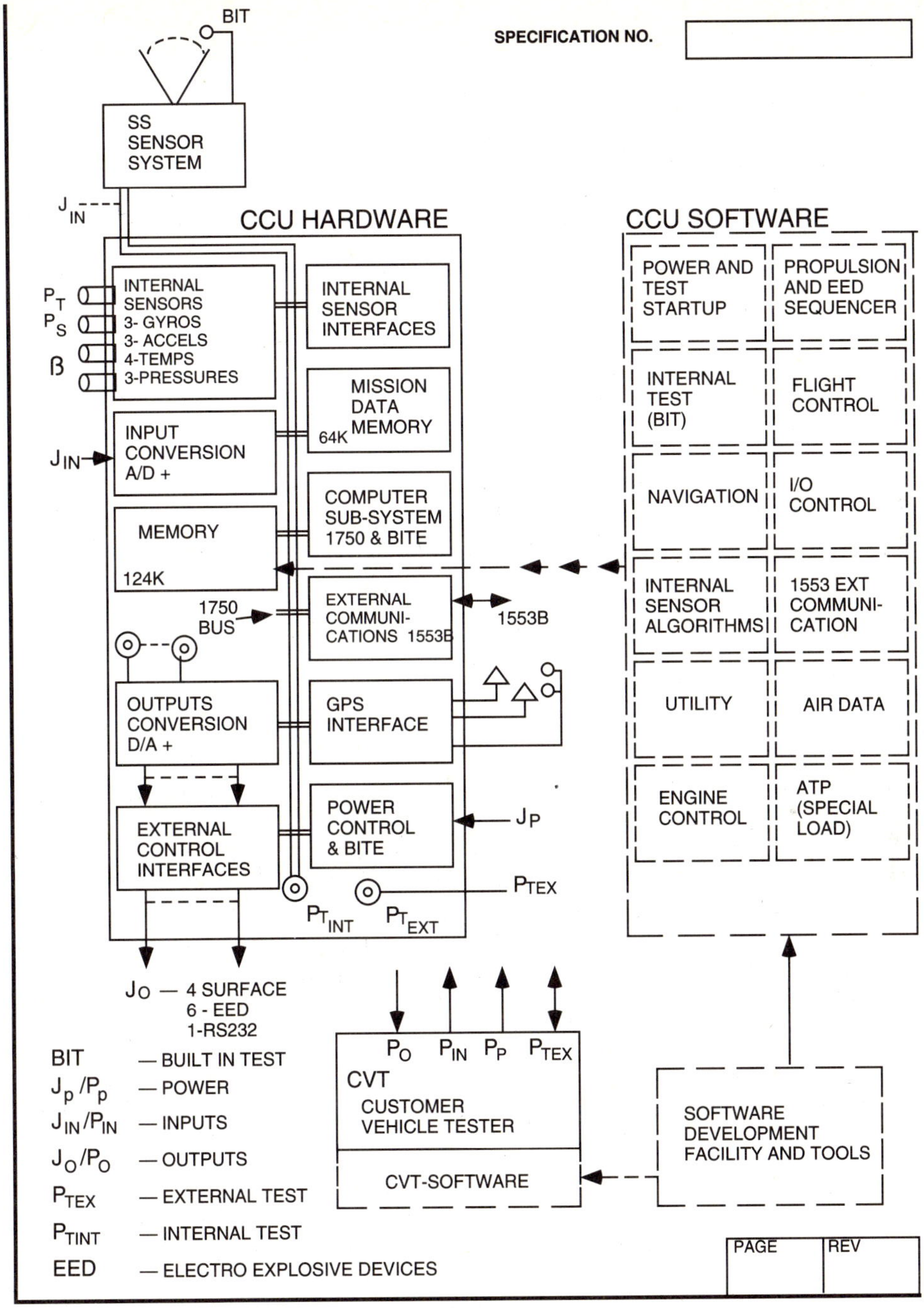

Fig. **4-5.** Functional block diagram of the CCU

4.1.4 The Interface Diagram

Figure 4-7 provides a simple diagram of the interfaces for the TACS. This figure represents the first step in the definition of the interfaces. Interfaces must be established as part of the system definition. They must be backed up later as the system evolves by wire lists and detailed specifications on each specific interface.

Interface diagrams often lack a well-defined boundary that bounds the "system" so it is very clear what is in it and what is merely descriptive of the inter-

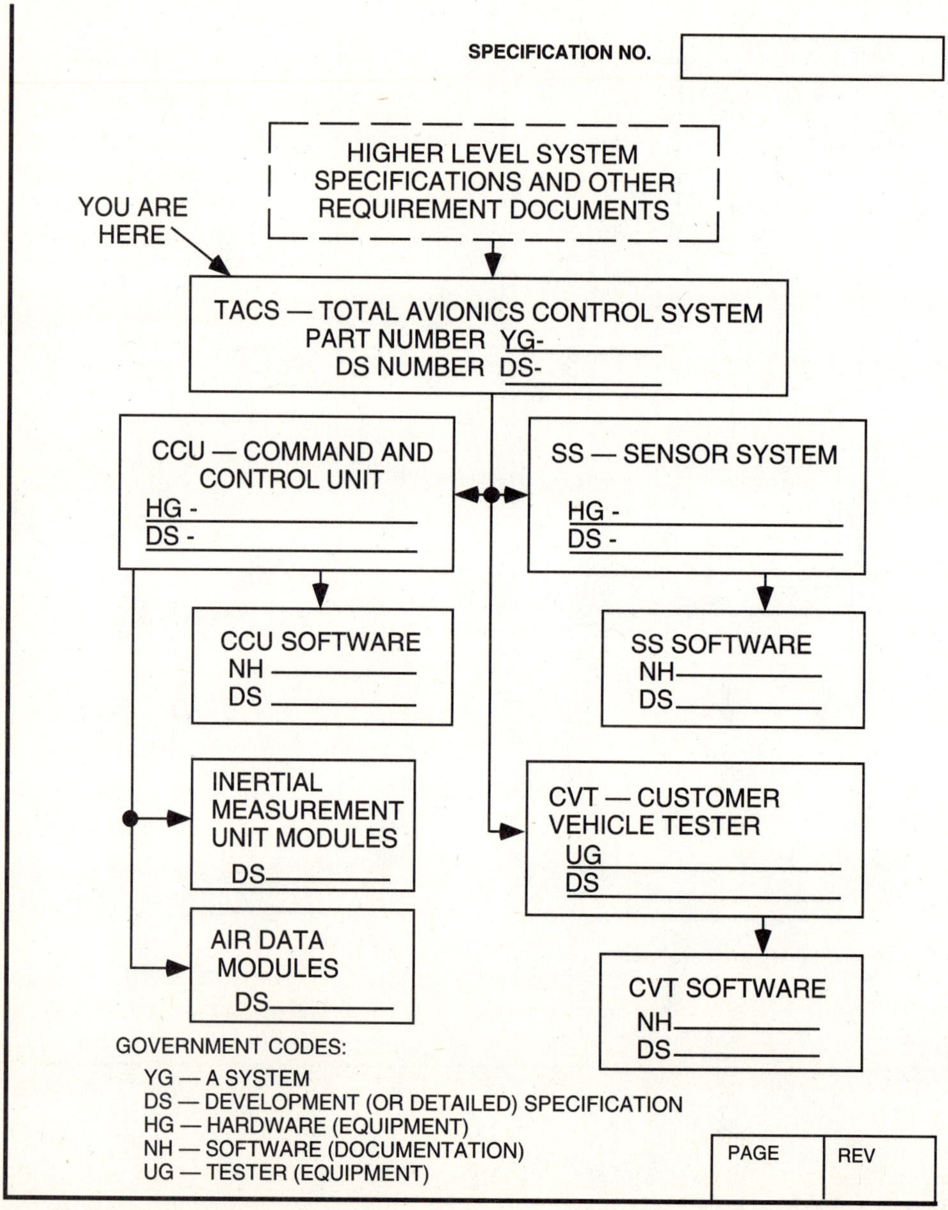

Fig. 4-6. Specification tree

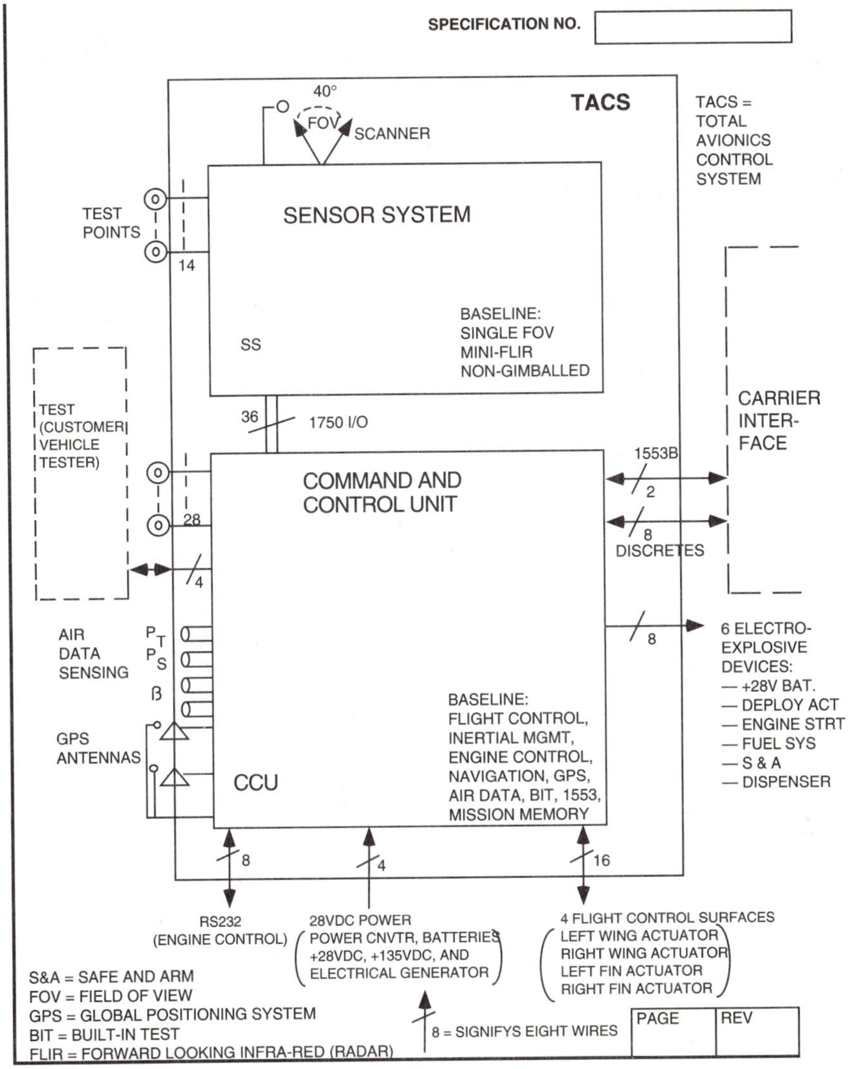

Fig. **4-7.** TACS and its interfaces

face. Draw a complete box around "the system." Peripheral interface definitions can be bracketed or placed in dashed line boxes.

Identify all the inputs, outputs, and externally routed feedbacks to your system to make sure you have not forgotten any interfaces. The human operator control, power, environmental sensors, and test interfaces are many times forgotten, so use your system perspective cube analysis and mission analysis studies as prompters when you sketch out the interfaces.

This diagram shows not only the TACS external interfaces but also some of the internal interfaces. In addition, it reflects the mission analysis because it provides system features that show capability for performing the desired mission.

4.1.5 Mechanical Sketches and Installation Drawings

The baseline system definition must include a physical mechanical sketch describing the system's hardware configuration. Figure 4-8 illustrates the TACS assembly and installation details. Baseline attributes are provided to show estimated volume, weight, power, and center-of-gravity location. These estimates are iterated upon as the system design evolves. They will be discussed later in this

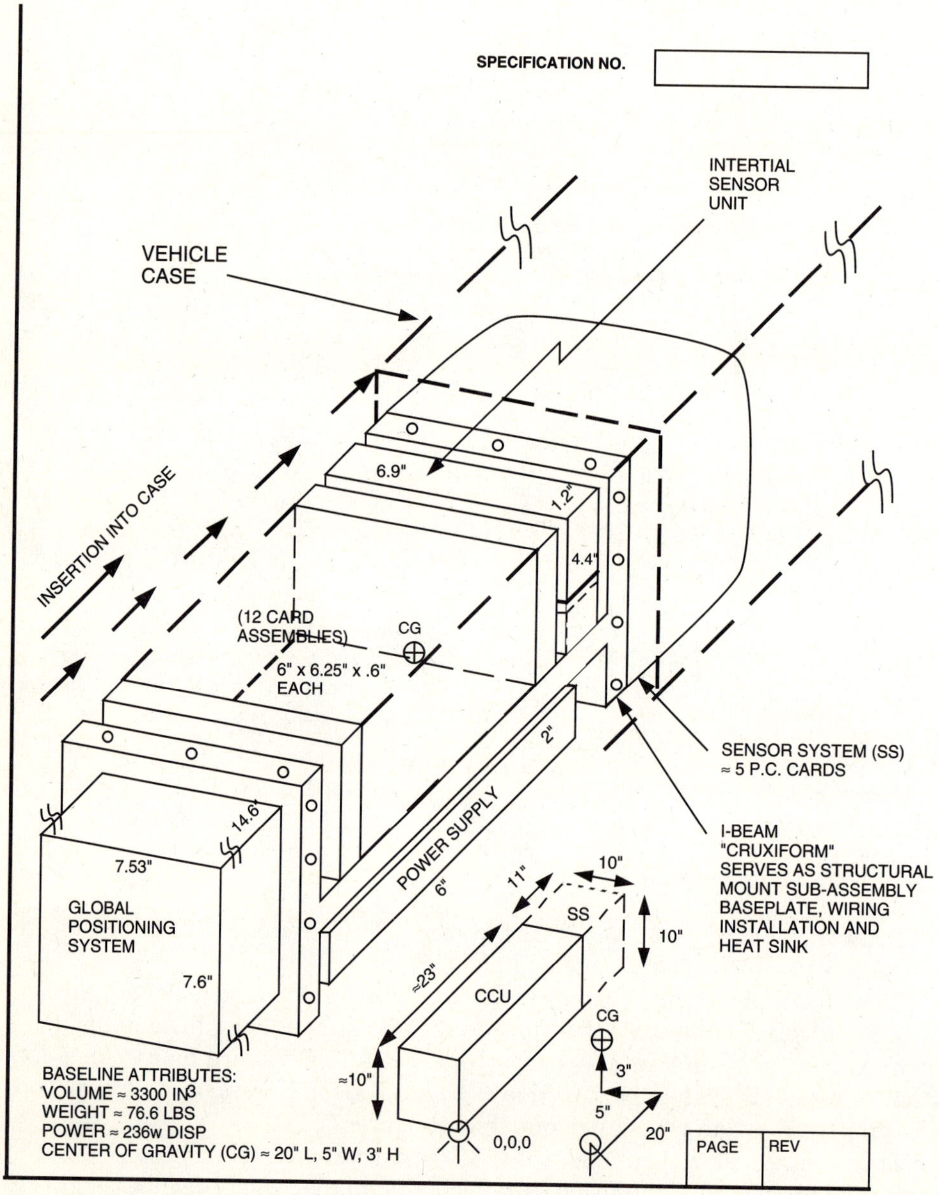

Fig. **4-8.** TACS assembly and installation

book as some of the system's "technical performance measures" (TPMs). This initial sketch also serves as a starting point for thermal profile analysis, structural analysis, mounting analysis, and subassembly packaging design requirements.

4.1.6 Detailed Block Diagram

The process of developing a baseline system requires working the design down to lower and lower levels. The detailed block diagram takes the interface drawing, the functional diagrams, the top-level partitioning, the mechanical configuration with installation and includes these with trade-off study results to provide greater detail. This detail definition is best led by a single highly skilled system engineer receiving support from a cadre of specialty experts. Leading this detailed development process requires a balanced set of skills. The system engineering skills survey profile analysis, which you were given in SEE-1, is intended to point out the need for balanced, good overall skills in hardware, mechanical, software, analysis, and program management areas. It should be your goal and objective to develop your skills sufficiently in all these areas to have the ability to develop the total system down to the detail that will provide sufficient data for design specialists to start work. You will select existing designs that are applicable, access risks, and do trade-offs of key functional areas to optimize the implementation selected. You will not always be right, but you will lead with confidence because you realize that the system development process is iterative. You will correct your mistakes and do it over until you create a system adequate to perform the mission. Obviously, you desire to reduce the number and seriousness of iterations by being mostly right with your initial baseline system.

Figure 4-9 is an example of a detailed block diagram developed by a single system engineer with good overall skills and support by a cadre of experts. This TACS diagram includes working notes as the system is being baselined. Interfaces are shown, risky design areas are indicated by "R" and system data rates of data by "Hz" (electronic control signal frequencies will be considerably higher in mega Hz, for example), circuitry is partitioned to cards, power dissipation of all the assemblies are estimated, throughput is estimated, and test features are developed. This diagram is a hardware detailed block diagram. It is usually three by five feet in size and has considerably more information than that shown in this book. I have simplified my diagram for clarity. The objective of this diagram should be to provide a *very detailed single piece of paper* that shows your complete system so the designers develop a total big picture of the baseline, its interfaces, and characteristics. A similar diagram can also be developed for the software system. Note: As the system development specification evolves, these diagrams will be simplified by moving the working notes to the requirements section of the development specification and also translating some of the data down to other development specifications in the specification tree.

4.1.7 Bill of Material

The baseline will include a top-level bill of material as shown in figure 4-10. This list of major system components is needed so system cost, weight, power,

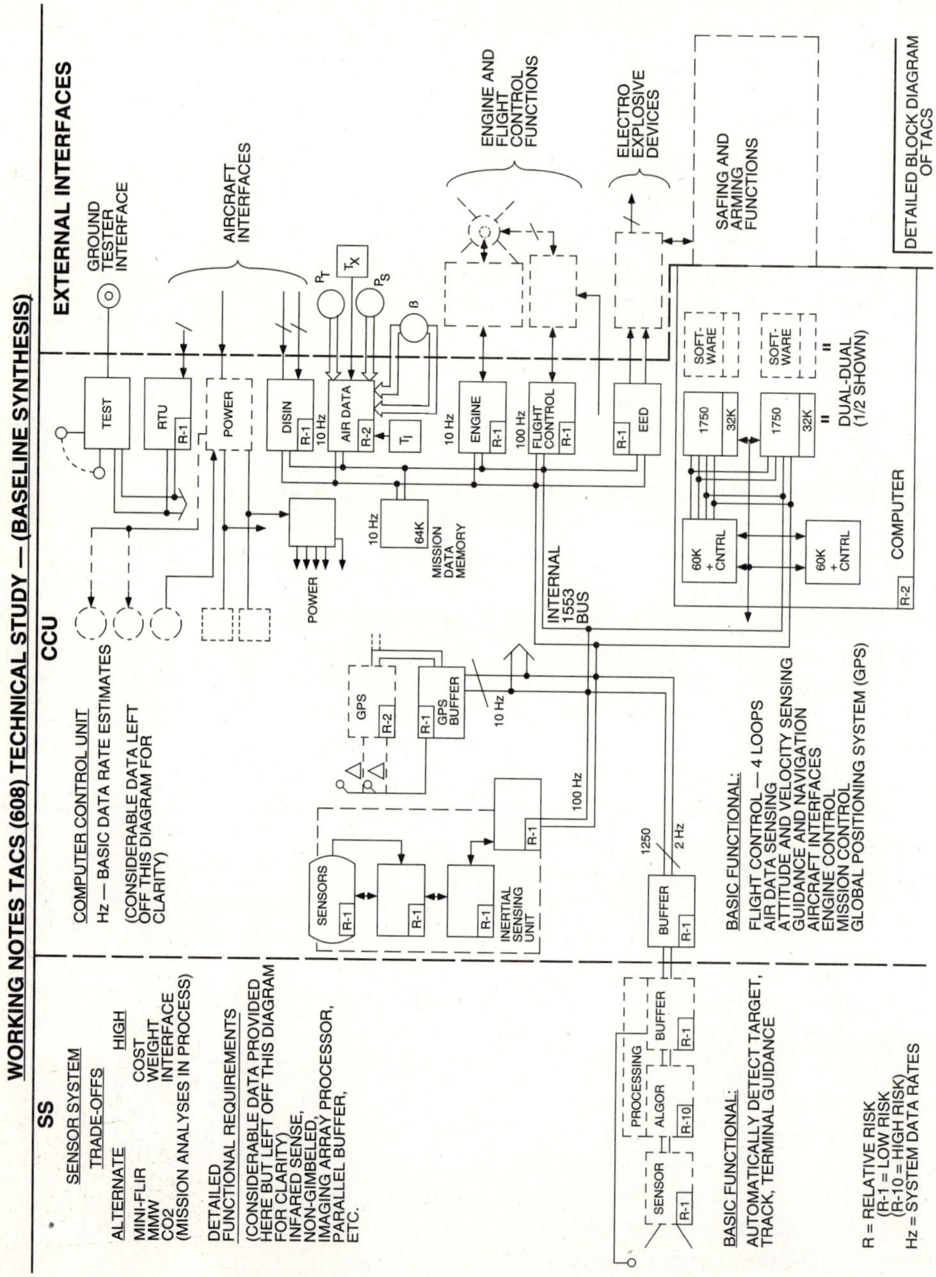

Fig. 4-9. Detailed block diagram of TACS—simplified for clarity (normally this diagram is 5 x 3 feet and has considerably more information)

volume, reliability, producibility, testing required, logistics, and other similar types of analyses can be started. The baseline bill of material is a very important communication instigator. Those who will manufacture, test, and service the system need this data.

4.1.8 Attributes Table

The attributes table of figure 4-11 is an example of one of the most important parts of our baseline system because it provides the basis for trade-off studies, technical performance measures (TPMs), tracking and the development of a

SPECIFICATION NO.

CCU — BILL OF MATERIAL — TABLE

QUANTITY	DESCRIPTION
4 —	HDP 1750A/SBC COMPUTER BOARDS AS DEFINED BY DS28991-01 (REF-1) 6" X 6.25" X 0.6"
1 —	1553B RTU (COMPATIBLE TO 1750 SBC) REF-1 6" X 6.25" X 0.6"
1 —	MISSION DATA MEMORY BOARD (NEW DESIGN) 6" X 6.25" X 0.6"
1 —	DIGITAL BUFFER - AIR DATA (RE-PACKAGE) 6" X 6.25" X 0.6" (CAVD SUPPLIED)
1 —	DISCRETE I/O, A/D, D/A, RS232, FOR ENGINE AND FLIGHT CONTROL INTERFACE (NEW DESIGN) 6" X 6.25" X 0.6"
3 —	IMU — INERTIAL MEASUREMENT UNIT-CARDS PURCHASED (NEW PACKAGE FROM 478 SYSTEM) 6" X 6.25" X 0.6" (MAVD-SO SUPPLIED)
1 —	MEMORY EXPANSION & I/O CONTROL (NEW DESIGN) 6" X 6.25" X 0.6"
1 —	CONTROL LOGIC (NEW DESIGN) 6" X 6.25" X 0.6"
1 —	DIGITAL BUFFER CARD (NEW DESIGN) 6" X 6.25" X 0.6"
1 —	POWER SUPPLY ASSEMBLY - 100W (ADOCS AND FSW EQUIVALENT)
1 —	ISA (MAVD-SO H710)
1 —	GPS (CONTACT CUSTOMER)
1 —	I-BEAM CRUX/FORM ASSEMBLY

PAGE	REV

Fig. **4-10.** CCU—bill of material table

SPECIFICATION NO.

SUB-ASSEMBLY ATTRIBUTES OF TACS

ATTRIBUTE		WEIGHT	POWER DISP	COST	VOLUME	MTBF
TACS TOTALS =		76.875#	236W	183K$	2011 IN^3	333 HRS (λ =.003)
CCU	MISSION COMPUTER	8.6 CARDS 12.0 MTG'G 5.0 P.S.	65.3 (80% EFF)	41K	12 CARDS 1 PS ASSY 393	2000 HRS (.0005)
CCU	INERTIAL MEASUREMENT UNIT	7.7	35	12K	232	1000 HRS (.001)
CCU	GLOBAL POSITIONING SYSTEM	23.1	85.7	45K	837	2000 HRS (.0005)
SENSOR SYSTEM (SS) (EOD-BASELINE)		3.37 CARDS 17.1 FLIR	50	85K	549	1000 HRS (.001)
SS TRADE-OFF (REFERENCES)		83-C02 50-MMW	400-C02 500-MMW	35K-C02 30K-MMW	678-C02 864-MMW	500 HRS (.002)

INTERFACE ASSUMPTIONS FOR TACS

ENGINE CONTROL	RS232 COMMANDS ONLY
FLIGHT CONTROL	LOW LEVEL/POWER DC, DC FEEDBACK
SAFE & ARMING DISPENSER	6 DUAL HI/LO SQUIB DRIVERS
POWER CONVERTOR	UN-INTERRUPTABLE +28 & +135 VDC
VEHICLE UMBILICAL CNTR	1553B RTU, PRESENT-AWAY-ADD. DISCRETES
TEST	CVT — TEST CONNECTOR (61 PIN)
BATTERIES	MOUNTED IN VEHICLE +28, +135
INTERFACE	ASSUMPTIONS — BASELINE

PAGE	REV

Fig. **4-11.** Attributes table and interface assumptions

balanced system design. The system engineer will trade-off weights and volumes against cost to obtain the optimal balance between these attributes. In section 4.1.9, "Attributes Balancing," I will show the methodology used for the trading off of system attributes to evolve a balanced system.

In addition, graphs may be provided that may show environmental requirements, system response characteristics, memory maps, and other required system attributes. Tables may be used to list special characteristics and assumptions of interfaces and subassemblies. Tables and graphs help to define the

SPECIFICATION NO. [ ]

6.0 NOTES

6.1 PRIORITY: 1-MISSION EFFECTIVENESS, 2-RELIABILITY, 3-LOW COST, 4-LOW WEIGHT

6.2 COST ESTIMATE: -PROPOSAL/PHASE I, USE COMPUTER AIDED-COST ESTIMATING SYSTEM FOR ESTIMATES (REF DS32615-01)

6.3 INITIAL OPERATIONAL CAPABILITY DATE: IOC 1989/1990 (FIRST MISSION AVAILABILITY)

6.4 VELOCITY 0.7 MACH

6.5 ALTITUDE h SPEC REQUIRED

6.6 SOFTWARE ESTIMATE BASELINED AT 60K LINES OF CODE
30K TO 120K RANGE (SOURCE T. LAHN AND M. WARD)

6.7 USE: MIL-1535 REGISTERED COMPONENT (CUSTOMER REQUIREMENT)
MIL-570-1750 PER FAIRCHILD 9450
JAS-1553B AFS-J35,J37 AS PROPOSED (SOURCE: J. KIMBALL OR MODEL 360 SOURCE R. GAABO)

6.8 CENTROID ERROR PROBABILITY (CEP)
18 METERS IN 1989 IOC
3 METERS IN 1990 IOC

6.9 ACRONYM DICTIONARY

6.10 MISSION ANALYSES DETAILED TO LOWER LEVEL

6.11 AYK14 IMPLEMENTATION (REFERENCE-ONLY)

PAGE	REV

Fig. **4-12.** Miscellaneous baseline notes

baseline system. In the notes of the baseline you can provide additional important data as shown in figure 4-12.

4.1.9 Attributes Balancing

When systems are developed with only one or two attributes as development requirements, you lose. Systems designed only to provide the attribute of lowest cost can have poor reliability. Many times people select presidents, buy cars, get married, and develop systems on the basis of a single attribute and, consequently, have imbalances for which they have to compensate. You need to develop a sensitivity to all the attributes of people, things, and systems in order to make decisions based on a balance of these attributes. The objective of attributes balancing is to develop a system that is "only adequate" to perform the required missions for the system. It will perform the mission, but is not overdesigned.

You must establish very early in the development of your system a complete table of all the system's key attributes. Figure 4-13 shows such a table for a major system. You can trade off these attributes against each other to evolve a balanced system. You may, for example, select a low-cost, low-reliability component that will lower your system cost (good) and, at the same time, reduce your reliability (bad). That is OK, however, because your reliability is still adequate for the mission of the system. A method for doing an attributes vs. requirements optimization to get a balanced system is shown in figure 4-14. You are searching for the common requirements that provide for low cost and also will provide reliability improvement. The arrow up in the equation signifies you need to raise the value of that system functional requirement to get the desired impact on the dependent system attribute, while the arrow down signifies lowering should be done. Arrows that point both up and down indicate that the impacting requirement could make the dependent system attribute go up or down depending on how the functional requirement is implemented. If maintainability is accomplished by designing a simple system, the system cost will go down. If, however, the maintainability is accomplished by using complicated internal test and throw away of high-cost assemblies, you can cause the system cost to go up. Functions such as maintainability, testability, and computer-aided design may require special trade-off studies to optimize their implementation to achieve a balanced system.

At the top of figure 4-14 you write an equation of those functional requirements that will reduce the cost of the system. The terms in this equation are based on a brainstorming process where you ask yourselves, "What are all the possible actions of persons or things that may reduce the cost of our system?" Note that you are using your generic functional analysis skills. You also write an equation of those functional requirements that will improve the reliability of the system. If you combine all the common terms of these two equations, you have an equation listing those terms that will both lower the cost of your system and raise its reliability. This methodology can be used to help develop a balanced system.

One of the attributes you may establish on your attributes table might be "System Quality Rating (SQR)." This attribute might be defined by an equation such as: SQR = f1(MTBF-up) + f2(LCC-down) + f3(USER ACCEPTANCE RATING-up) + f4(MISSION EFFECTIVENESS-up) + f5(MISSION READINESS-up) + f6(WAIVERS-down) + f7(DEVIATIONS-down) + . . . fn(SPECIFIC SYSTEM CRITICAL ATTRIBUTE-like weight-down) ÷ (DIVIDED BY) THE TOTAL NUMBER OF FUNCTIONS(fn). In this equation each function is a percentage of the desired quality goal for that function with a max limit of 100 percent for any function. This equation shows how we can establish an equation that can act as a continuous monitor of the quality attribute of your system. SQR can become a tool to provide incentive to your customer support people, your production people and your system design people to constantly improve the system quality. Function 1 (f1) deals with reliability Mean Time Between Failures (MTBF) that is based on actual customer usage. Function 2 (f2) deals with the cost to the customer to own and use your system life cycle cost (LCC). The other functions are self-explanatory.

MISSION EFFECTIVENESS	TARGET HIT WITHIN ±3 FEET WITH 95% CONFIDENCE
MISSION READINESS	98% MIN
LCC (LIFE CYCLE COST)	$600K MAX
MTBF (MEAN TIME BETWEEN FAILURES)	2000 HRS MIN
WEIGHT	250# MAX
SIZE	3000 IN. CU. MAX
POWER	500 W MAX
MTTR (MEAN TIME TO REPAIR)	1.0 HOUR MAX
SYSTEM QUALITY RATING	80% RATING MIN
THROUGHPUT	6 MIPS (DAIS-MIX) MIN
CAPACITY	4 MEGABYTE MIN

Fig. **4-13.** Example: attributes of a major system

DEPENDENT SYSTEM ATTRIBUTE	FUNCTIONAL REQUIREMENTS
COST UNIT PRODUCT COST (UPC) (GOES DOWN ↓ WHEN:)	MEAN TIME BETWEEN FAILURES (MTBF) ↓ + MAINTAINABILITY ↕ + ENVIRONMENTAL STRESS SCREENING ↓ + TESTABILITY ↕ + ACCURACY ↓ + SIZE ↑ + WEIGHT ↑ + STANDARDS ↓ + ENVIRONMENTAL ↓ + SPEED ↓ + DESIGN MARGIN ↓ + COMPLEXITY ↓ + NEW TECHNOLOGY ↕ + DESIGNERS CAPABILITY ↑ + POWER DISSIPATION ↓ + PRODUCTION TESTING ↓ + PRODUCTION CALIBRATION ↓ + RELIABILITY DEVELOPMENT TESTING ↓ + PRODUCTION WORKMANSHIP ↑ + SPARE-PINS, CARDS, THROUGHPUT, MEMORY, ETC. ↓ + KNOWLEDGE ↑ + REQUIRED COMMONALITY ↓ + REQUIRED 2ND SOURCES ↑ + SPECIFICATION CLARITY ↑ + REQUIRED PROCESS ↓ + MISSION EFFECTIVENESS ↓ + REQUIRED COMPUTER AIDED DESIGN (CAD) ↓ + ACTUAL CAD ↕ + REQUIRED PARTS QUALITY ↓ + ACTUAL PARTS QUALITY ↑
RELIABILITY MEAN TIME BETWEEN FAILURES (MTBF) (GOES UP ↑ WHEN:)	ENVIRONMENTAL ↓ + PARTS QUALITY ↑ + COMPLEXITY ↓ + DESIGN MARGIN ↑ + ACCURACY ↓ + MISSION EFFECTIVENESS ↓ + RELIABILITY DEVELOPMENT TESTS ↑ + ENVIRONMENTAL STRESS SCREENING ↑ + TESTABILITY ↕ + SIZE ↑ +WEIGHT ↑ + STANDARDS ↑ + COMMONALITY ↑ + 2ND SOURCES ↑ + SPECIFICATION CLARITY ↑ + PROCESS IMPROVEMENT ↑ + KNOWLEDGE ↑ + SPEED ↓ + POWER DISSIPATION ↓ + WORKMANSHIP ↑ +SPARE-PINS, CARDS, THRUPUT, MEM ↓ + PRODUCTION CALIBRATION ↓ + REQUIRED CAD ↓ + ACTUAL CAD ↕ + NEW TECHNOLOGY ↕ + DESIGNERS CAPABILITY ↑ + MAINTAINABILITY ↕
COST AND RELIABILITY UPC ↓ AND MTBF ↑ WHEN:	ENVIRONMENTAL ↓ + COMPLEXITY ↓ + WEIGHT ↑ + ACCURACY ↓ + SPEED ↓ + MISSION EFFECTIVENESS ↓ + SIZE ↑ + SPECIFICATION CLARITY ↑ + KNOWLEDGE ↑ + POWER DISSIPATION ↓ + PRODUCTION WORKMANSHIP ↑ + PRODUCTION CALIBRATION ↓ + DESIGNERS CAPABILITY ↑ + SPARE-PINS, CARDS, THRUPUT, MEMORY ↓ + PROCESS IMPROVEMENT ↑ + WORKMANSHIP ↑

When you deliberately design your system with reduced mission effectiveness, lower accuracy, lower speed, and reduced complexity, you can improve both system cost and reliability. When you allow the system to operate in a benign environment, to have high weight, size, and power dissipation, and also to reduce the requirements for spare (pins, cards, computer speed (throughput), and memory), you also can improve system cost and reliability. When you increase knowledge, specification clarity, workpersonship, designers capability and process improvement and also reduce production calibration requirements, you will improve system cost and reliability.

Fig. 4-14. System attributes relationships study—an example of the methodology for a developing a balanced system

4.1.10 Caution! The Baseline Is Not a Development Specification!

Defining a baseline system is only part of the system development specification process. The establishment of requirements is equally important. You will study in SEE-5 the development specification *process* where you take the baseline, add requirements from the mission analysis, and then start the iterative process of taking the infant development specification to maturity.

The baseline *must* provide ten items:

1. A title page for revision record
2. A scope statement
3. A time line of system mission
4. Top-level block diagrams
5. A specification tree
6. An interface diagram
7. Mechanical sketches and installation drawings
8. Detailed block diagram
9. Bill of material
10. An attributes table

Leaving any of these out results in an incomplete system baseline definition. Additional graphs, tables, flow diagrams, mission analysis diagrams, timing diagrams, and memory maps can be added to further define the baseline, but the prior ten items must be provided.

4.2 The Proposal Process and System Engineering Management (SEM)

4.2.1 System Engineering and "The Proposal"

Every system has a customer to whom you must present a proposal of some sort. For example, the customer for an engineering department could be the marketing department, who decides what products to develop. Your customer could be a country, a government, a department, or the general public. You may propose a new system, a cost reduction of a present system (a value engineering change proposal—VECP), the improvement of an old system, or reestablishment of an old system (such as reestablish more stay-at-home cottage industries).

The systems engineer usually develops the baseline system initially for a proposal to the customer but first exposes the baseline system in a preliminary development specification to the proposal team at a proposal kickoff meeting. Figure 4-15 shows the system engineer's mission to technically lead a proposal team by using system engineering process (SEP) and system engineering man-

Time	Role	Activity
t_{-1} ———	SEP	• CUSTOMER INTERFACE
	SEP	• ESTABLISH STATEMENT OF PROBLEM
	SEP	• ESTABLISH CONCISE OVERALL OBJECTIVES
	SEP	• DEFINE THE MISSION, DO ANALYSIS
	SEP	• DEVELOP FUNCTIONAL REQUIREMENTS
	SEP	• DEVELOP ENVIRONMENTAL REQUIREMENTS
	SEP	• ESTABLISH INTERFACE REQUIREMENTS
	SEP	• DEVELOP SPECIAL REQUIREMENTS
t_0 ———	RFP	→ — — — (REQUEST FOR PROPOSAL) ———
t_1 ———	SEP	• PARTITION AND DEFINE THE SYSTEM
	SEP	• PROVIDE DEVELOPMENT SPECIFICATIONS (DS) TREE
	SEM	• PROVIDE WORK BREAKDOWN STRUCTURE (WBS) TREE
	SEM	• PROVIDE TEAM TREE
t_2 ———	SEP	• PROVIDE BASELINE SYSTEM DEFINITION BY PRELIMINARY DS
t_3 ———	SEM	• PROVIDE PROGRAM MANAGEMENT PLAN (PMP)
t_4 ———	SEM	• PROPOSAL KICKOFF MEETING—PRESENT BASELINE DS, PMP, ALONG WITH CUSTOMERS RFP DATA TO THE PROPOSAL TEAM
t_n ———	SEM	• SUPPORT THE TEAM
CONTINUE SEP/SEM ITERATIVE PROCESS AS REQUIRED		

Fig. 4-15. The system engineer and the proposal—mission time line

agement (SEM). In your mission at t_4 you must be prepared to answer questions at the proposal kickoff meeting.

Who is the customer? What is the problem? What are your objectives? How are you organized? Who does what? How is the system specified? Who can answer my questions? What should I do? What are my interfaces? What is the system? How is it positioned? What does the request for proposal request? What is the statement of work? Do you have a work breakdown structure, a development specification, an interface control document, a program management plan? What is the cost-estimating process? Schedules? Quantities? When is the proposal to be done? How should you present your proposal? To whom will you present the proposal?

The proposal manager and systems engineer must provide answers to these questions by some prekickoff meeting work. They need to work with the customer, study, do analysis, and expedite a preliminary system development specification and program management plan.

The following hypothetical example shows a typical proposal process:

• OUR CUSTOMER IS:	"Xerxes"—A large, long-standing customer of ours; over the years we have done some very good work for them.
• STATEMENT OF PROBLEM:	Control system for remote piloted vehicles (RPVs) are too costly.
• CONCISE OVERALL OBJECTIVE (Scope):	Produce a modern repackaged single box, integrated avionics controller that will be called "total avionics control system" (TACS). It will control the link, flight, propulsion, navigation, itself, and mission functions. It will provide flexible, expandable input/output, test itself, communicate its health, be safe to the user, and be producible at a cost not to exceed a unit product cost of approximately 150K $ for a quantity of 1000 systems.
• DEFINE THE MISSION (Mission Analysis):	The various missions will be defined by scenarios. The controller will provide control for various RPVs in a multitude of missions as shown in the baseline definition.
• DETERMINE FUNCTIONAL REQUIREMENTS (Functional Analysis):	The missions will be analyzed to determine the functions required. Flight maneuverability equivalent to an F-16 fighter, navigation 0.1 NM (nautical mile) max error proportional over a 1 NM to 100 NM flight, propulsion sequence controller, power and test startup control, internal communication, internal test, external communication, flexible input/output, personnel training, power system—electrical, maintenance, logistics support, output control interfaces, input sensor interface being typical requirements that will result from the mission and functional requirements analyses.
• DETERMINE ENVIRONMENTAL REQUIREMENTS:	(Based on mission analysis, the environmental requirements will be determined.) Altitude: sea level to 40,000 ft; temp: minus 65° to plus 160°F; vibrations per provided curve 5 to 2000 Hz; shock: 25 ± 1

G; Electromagnetic Interference (EMI) levels, Electromagnetic Pulse (EMP) levels, and nuclear per battlefield 2000 mission scenarios.

- DETERMINE SPECIAL REQUIREMENTS:

(Based on mission analysis and objective, some special requirements will be determined.) Installation: to fit within 2-ft long by 1-ft diameter tube with a wire raceway similar to the MK-46 torpedo structure.

Cooling: Convection to aluminum casting and external environment.

Mounting: Hard mount.

Reliability: Parts selection—(Mil-ER) and non-standard parts are allowed; do failure mode and effects analysis (FMEA), safety analysis, 10-year shelf life analysis + 10-hour mission analysis.

Maintainability: logistics, and testability. Features will be developed and proposed by the assigned development engineers.

Production design-to requirements: unit product cost (UPC) = 150K$ (baseline) and compatible to present production process, Life Cycle Cost (LCC = 200K$ (baseline).

Weight: 150 pounds max.

Size: Within installation volume per baseline.

Power: Mil-Std-704 28Vdc with battery backup controller max power of 500W for 10ms; nominal power is 350W.

- PARTITIONING SYSTEM DEFINITION:

Based on all requirements: mission, functional environmental and special; now baseline define the system and provide a preliminary system development specification (DS) and program management plan (PMP).

It will be our design aim that the DS spec tree and the PMP's work breakdown structure WBS task tree will tell our designers what they are to do.

This means that the first version of the system DS and the PMP must be adequately developed now.

Figure 4-16 shows the relationships between the DS spec tree, the PMP-WBS task tree, and the PMP staffing tree. We first by SEP functional analyses establish the required functions and their requirements (1—functional elements). Then we establish the work to be done to acquire the functional elements (2—work elements). Lastly, we select specific skilled team members with the skills to do the work to acquire the desired functional elements (3—skill elements).

- ORGANIZATION AND TIMING: The preceding is presented to the proposal team by providing a system DS (including a complete baseline system definition) and a PMP (with WBS) in a kickoff meeting. Prior to this kickoff presentation, it would be appropriate for other proposal team members to be reading, studying, and generating "concern" questions from the customer-supplied RFP, SOW, and ICD. Note: It should be understood by team members that they are engaged in a "study only" subsequent to the kickoff meeting.
- THE PROPOSAL KICKOFF MEETING: The structures of the DS spec tree, the WBS task tree, and the team tree are now presented as part of the preliminary DS and PMP. Team members are instructed to expand their specific areas of WBS tasks by developing preliminary design definition, proposal writeup material, and the cost estimating data required for the WBS tasks assigned to them. The proposal team asks questions, and the lead systems engineer program manager and/or proposal manager provide answers.

Conclusions:

The baseline questions to be answered are done so at the proposal kickoff meeting. This is accomplished by some hard up-front disciplined work by the proposal manager and the system engineer. The key aim for a successful proposal is the early development of a preliminary PMP with WBS and a system DS before starting the whole team. The matching of team skill capabilities to WBS work element skill needs is another key function the proposal manager and/or the system engineer must perform. The use of engineering skill bank based on self-assessment skill surveys similar to the survey in SEE-1 is recommended as an aid to this key function.

Cost estimating for the proposal can be expedited by use of a computer-aided cost-estimating system. This type of system can be based on parametric models, which are validated against actual data from prior projects by experts. In SEE-5 examples from the computer-aided cost estimating system (CA-CES) development specification number DS32615, I provide some details for this proposal tool.

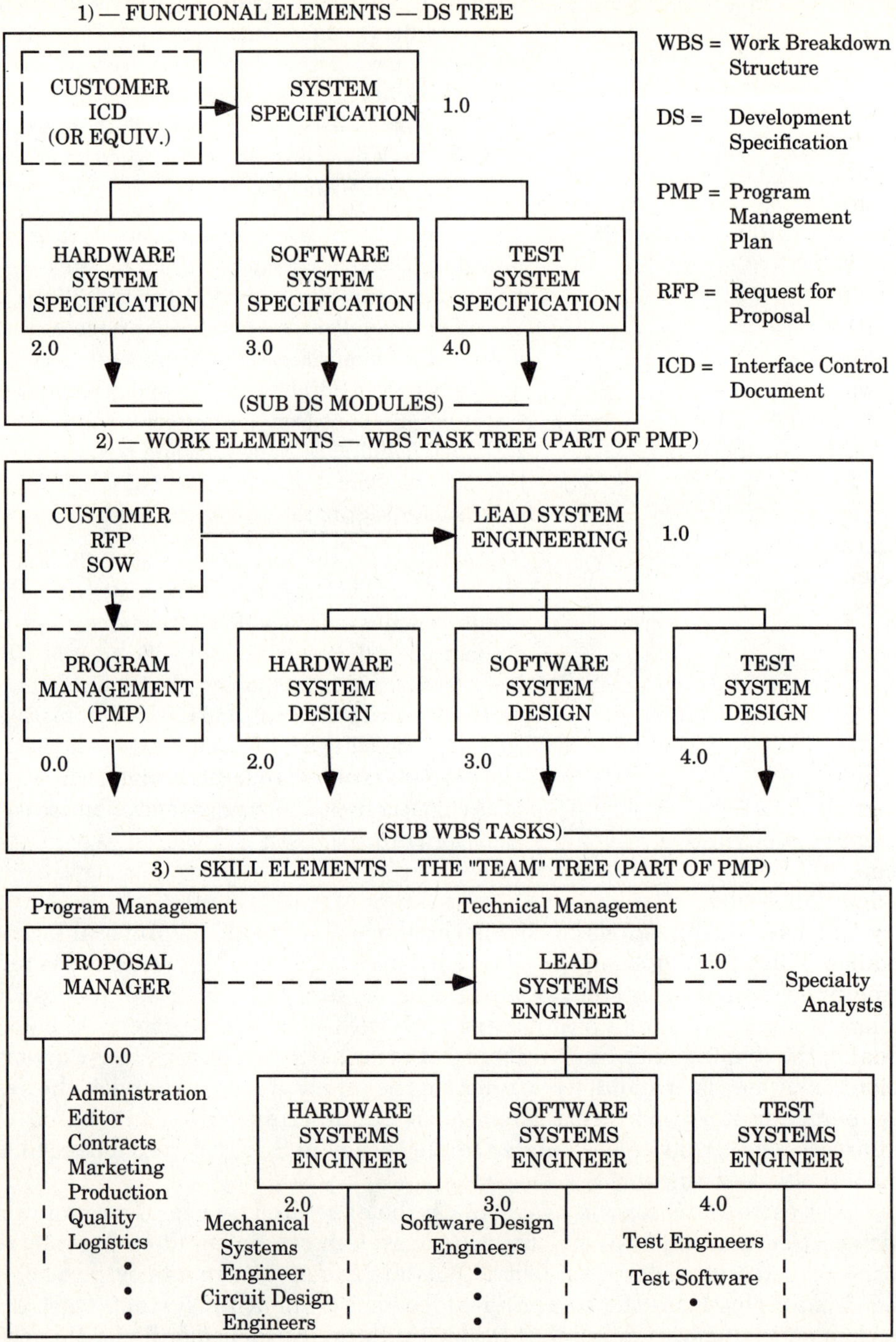

Fig. **4-16.** The three elementary trees (we three tree)

4.2.2 Partitioned Elements

Functional elements:

The generic functional analysis-derived elements shown on the DS specification tree in figure 4-16 (system, hardware, software, test equipment) are the result of the functional analysis and baseline of the system. These reflect the functions required to perform the desired mission. The specification shows the partitioning down of the technical requirements.

Work elements:

The elements depicted on the WBS task tree in figure 4-16 are the work elements relating to the work that has to be done to develop the functional elements required to develop a system that can perform the desired mission. The WBS shows the partitioning down of the work to be done.

Skill elements:

Skills are embodied in the team members in the "team" tree (fig. 4-16). The lead systems engineer (1.0), for example, will have the skills to perform the system engineering task (1.0) in the WBS to establish the requirements within the system DS (1.0) in the DS spec tree. The team tree shows the partitioning down of responsibility.

4.2.3 Development Specification Process

System development specifications—living documents

Initially, the purpose of a development specification is to *guide the design of the system*. Part I of the system development specification is provided early in the program; a system definition (your baseline) and requirements for the system are the main objectives at this time. This specification, at the beginning of the program, will be *adequate* (not perfect) to define the requirements for the detailed design. The system engineer must immediately at the start of a program *work very hard and quickly* in liaison with the customer to establish all the necessary detailed data and requirements to ensure that when detailed design is started, sufficient definition is provided to the various designers (analog hardware, digital hardware, software, mechanical and analysis design engineers). System functional description, algorithms, physical definition, installation requirements, partitioning, block diagram, a specification tree, accuracy requirements, environmental requirements, other required documentation (customer technical requirement data), power requirements, I/O interface specifications, and any other special requirements must be provided.

This early on, first-cut, preliminary specification then must be reviewed over and over in an iterative fashion to bring it to maturity, thus controlling and aiding the development process. At the beginning of this process, the number of changes will be large, but as the system develops, fewer and fewer changes will be required to establish completeness and understanding agreement.

Figure 4-17 illustrates four perceived levels of communications. The development specification should be viewed as an example of Level-1 type communications, which are of the highest criticality. If there are misunderstandings, they can be costly or maybe even impossible to recover from. The work level to establish this Level 1 of communication is the highest, but the methodology of utilization of an early on, expedient, iterative specification process makes it possible to be successful at establishing complete understanding.

COMMUNICATION LEVEL	UNDERSTANDING	WORK LEVEL TO OBTAIN	CRITICALITY OF MISUNDERSTANDING
1. Specifications (concise and written by specialists)	Complete	Highest	Costly or impossible to recover
2. Educational (study data collection open discussion to functional understanding)	General	Medium	Somewhat costly, but can usually recover
3. Social	Loose	Low	Normally we can recover
4. Talking to yourself	Yours	Constant	Stress

Fig. **4-17.** Levels of communication

Prior to system testing, a part II of our specification, will be generated. This development specification part II will provide test and analysis procedures to be performed proving system compliance to the requirements of the development specification part I.

Subsequent to qualification of the system, the development specification (parts I and II) may be updated and enhanced to fulfill new objectives: aid to production, deployment of the system, marketing aid, and guideline for subsequent similar system developments. While the initial specification development specification part I was provided for the design engineers and development specification part II was for the test evaluation and data analysis engineers, this update is now to be provided with additional data as required for production engineers, field service engineers, marketing engineers and trainee engineers.

Military standard 490

(Mil-Std-490) is the recommended guide for writing a specification. The general outline for a Mil-Std-490 type specification is as follows. Note the comments and subparagraph numbering are only recommendations based on my thirty-five years experience with development specifications as a personal tool. Your successful experiences may have a different format, numbering, etc. That's ok; it is *your* tool, so use what works for you to *lead* the development.

(1.0) ***Scope.*** Translation of system perspective to condensed statement of what this system and this specification is all about.

(2.0) ***Applicable documents.*** Listing of all the pertinent documents that provide inputs and/or control the activities to be performed. List customer, government, company, and other technical documents associated with this system. You may list any document that is pertinent to the development of the system, a magazine article, a report, or any other document you want to use to lead your system designers.

(3.0) ***Requirements.***

3.1 System definition: This is your baseline figures with a brief write-up describing them (interface drawing, block diagrams, specification tree, mechanical sketches, and tables). Remember, this is just a baseline definition—it is not the design.

(3.2) Specific requirements

- (3.2.1) Performance requirements
- (3.2.2) Interface requirements
- (3.2.3) Mechanical requirements
- (3.2.4) Environmental requirements
- (3.2.5) Hardware design requirements
- (3.2.6) Software requirements
- (3.2.7) Reliability, maintainability, and safety requirements
- (3.2.8) Testability requirements
- (3.2.9) Documentation requirements
- (3.2.10) Producibility (production) requirements
- (3.2.11) Installation and logistics requirements
- (3.2.12) Other requirements

(4.0) ***Quality assurance.*** How this system is to be inspected, tested, audited, and analyzed to ensure conformance to the requirements of this specification.

(5.0) ***Preparation for delivery.*** How this system is to be delivered for deployment

(6.0) ***NOTES.*** Acronym dictionary, ordering data, intended use, supporting technical data, user's manual, development specification revision history, important trade-off study reports, marketing studies, etc.

You will in SEE-5 use this outline, your baseline system definition, and your mission analysis data to prepare a system development specification. I will show how, through use of the development specification process, you can lead and control the system development of any system.

4.3 System Engineering Management (SEM)

The system engineering program manager is Ralph. Ralph understands mission analysis, time, timing, and money. Figure 4-18 shows how Ralph establishes incentives using time and money to enhance a mission to ship a system on schedule. The customer would have been satisfied to receive the system on or before August 21. By setting three dates, August 1, August 15, and August 21, he establishes not only a "win" scenario, but also a potential "win-win" scenario. You, as system engineering program managers, get the SEM timing right by using your mission analysis skills. You develop success oriented scenarios, play acting, walkthroughs, mission analyses, schedules. You also track and follow these schedules by listening and watching, by "getting a feel" and especially by being sensitive to all the people involved (customer, designers, etc.).

I will expose you to four SEM components: 1) the program brief, 2) the program management plan (PMP), 3) the system engineering management plan (SEMP), and 4) the program start-up review (PSR). The system engineer will not be successful unless SEM is used in planning and controlling the system development process.

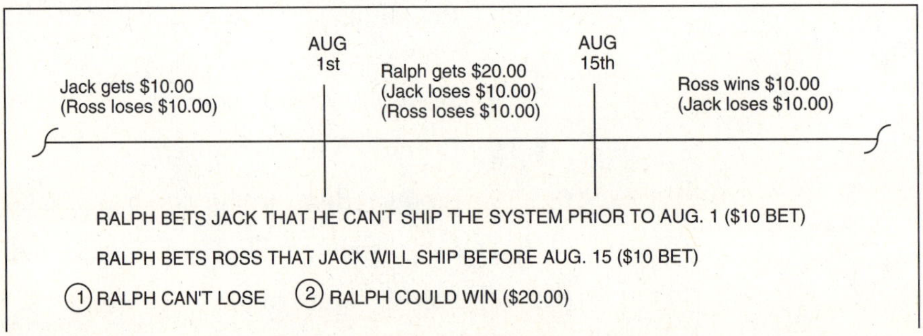

Fig. **4-18.** SEM mission analysis example "time and money"

System engineering management (SEM) is defined in Military Standard MIL-STD-499A as "the planning and control of a totally integrated technical effort of system engineering, design engineering and engineering specialties to meet technical performance, cost, and schedule objectives." You cannot do system engineering process (SEP) successfully without doing system engineering management. It is very important for all systems engineers to develop skills and apply the disciplines of system engineering management.

In large programs the specific system engineering responsibilities and tasks for a program are defined by the program manager in the program's system engineering plan (SEMP) that is a key part of the program manager's program management plan (PMP). However, on small programs it is common that the lead systems engineer is totally responsible for the entire program. This means that the total program PMP as well as the top system development specification (DS) are the lead system engineer's responsibility.

Program managers must be "systems engineering program managers" and systems engineers must be "program management systems engineer." This is so because if small program leaders are to become large program managers, they need the experience of successfully managing (SEM) and technically leading (SEP) a small program.

A key issue is skills; the program manager, the lead systems engineer, the system engineer program manager, and the program manager system engineer must take a self-assessment skill survey like the one in chapter SEE-1 to honestly determine what skills they are lacking. It is the skills and knowledge you don't have that will get you into trouble. You then must establish a plan of support (your team with experts) to compensate for your skill deficiencies.

What are the functions that system engineering management are to manage?

To help you continue your study of system engineering management (SEM) I have paraphrased material from Honeywell's System Engineering Guidelines,

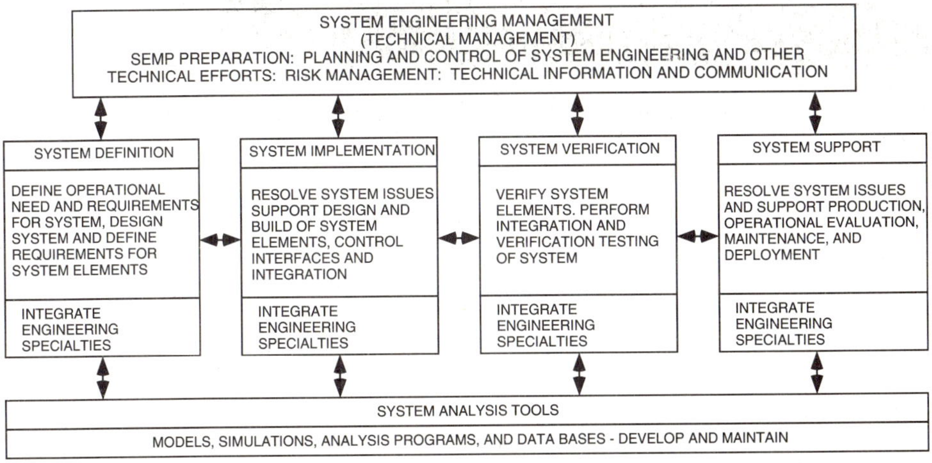

Fig. **4-19.** System engineering methodology

Volume 1, SEM, and Honeywell's System Engineering Guidelines, Volume 2, Functions.[1, 2] The Honeywell System Engineering Guidelines consist of three volumes. Volume 1 provides an introduction to system engineering. Volume 2 is a what-to-do, and Volume 3 is a how-to-do manual. This text is based in part on the concepts developed by Volumes 1 and 2 of these guidelines. The system perspective cube, for example, is an expansion of the cube given in the guidelines, which, in turn, was based on original works by Art Hall of Bell Labs.

"What-to-do" system engineering functions (excerpted with modification from Honeywell System Engineering Guidelines, Volume 2). The system engineering methodology requires six system engineering functions. Figures 4-19 and 4-20 summarize these functions and show their interrelationships. The relative level of system engineering effort for these functions throughout the life of a system are shown in figure 4-21. Typical system phases and events are shown in figure 4-22. The first two functions discussed are those that apply to all of system engineering-system engineering management (SEM) and system analysis tools (SAT).

Function	Task
SEM System Engineering Management (Function)	SEM-1 System Engineering Management Plan Preparation SEM-2 System Engineering Planning and Control SEM-3 Risk Management SEM-4 Technical Information and Communication
SAT System Analysis Tools (Function)	SAT-1 Tool Development SAT-2 Data Bases SAT-3 System Analysis Tools Maintenance
SDF System Definition (Function)	SDF-1 Mission and System Requirements Analysis SDF-2 System Functional Definition and Allocation SDF-3 Requirements Allocation and System Synthesis SDF-4 System Evaluation SDF-5 Specifications SDF-6 System Integration
SIF System Implementation (Function)	SIF-1 Interface Control SIF-2 Resolving System Issues SIF-3 Checkout and Integration Control
SVF System Verification (Function)	SVF-1 Verification Methodology and Plan SVF-2 Verification Requirements Definition and Analysis SVF-3 Verification Procedure Validation SVF-4 Verification Data Analysis SVF-5 System Verification Supporting Analyses Simulations
SSF System Support (Function)	SSF-1 Production Support SSF-2 Customer Support

Fig. **4-20.** System engineering functions and tasks

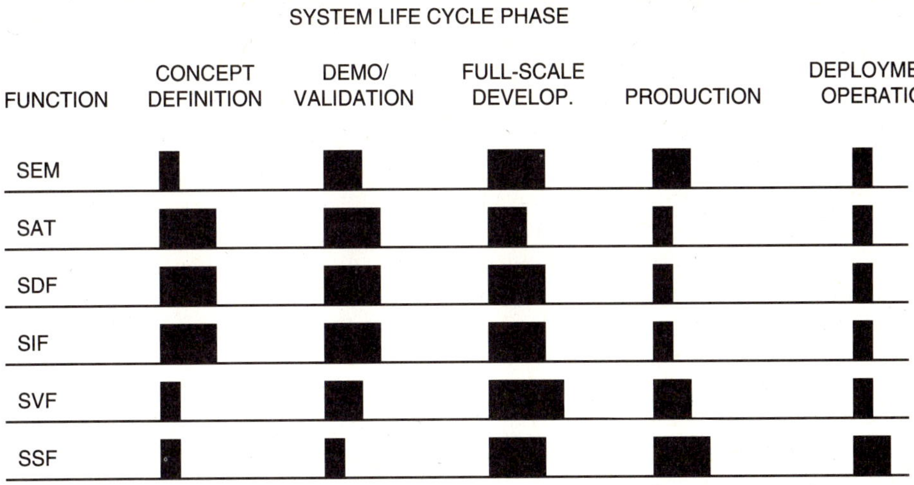

Fig. **4-21.** Relative level of system engineering effort

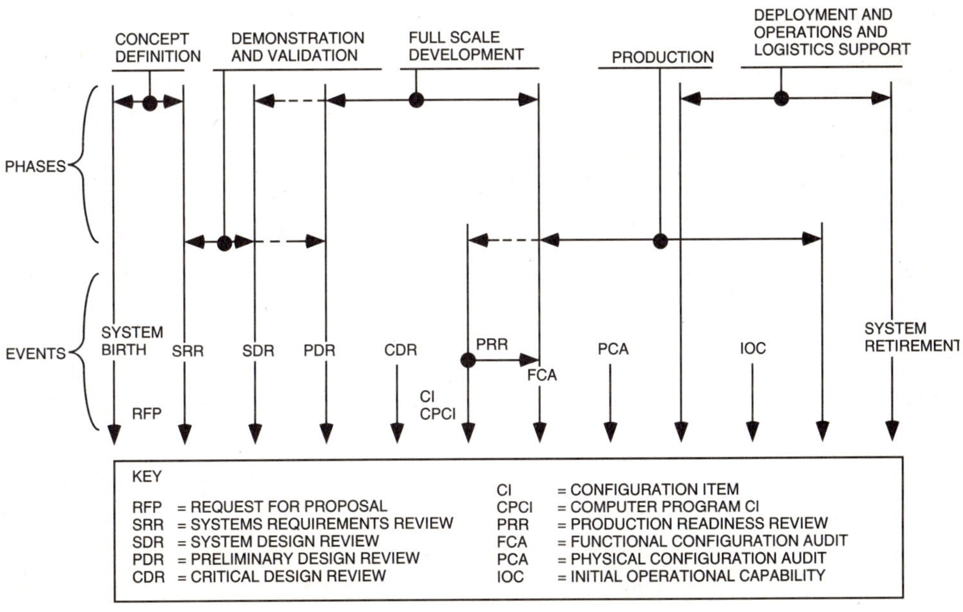

Fig. **4-22.** System phases and events

The management of the system engineering methodology, functions, and tasks throughout all the system phases and events requires planning.

The document that describes in detail how the above activities are managed and conducted on a program is the system engineering management plan (SEMP). This may be as small and simple as a one-page memo or as large as a few hundred pages. The size and complexity of the SEMP depends on the size and complexity of the program to which it is to be applied. The SEMP for a government contract will typically be larger than that for an equivalent-sized commercial program because of various government requirements (reference MIL-STD-499A or B as appropriate). The following SEM-1, 2, 3, 4 are the four specific system engineering tasks for accomplishing the system engineering management functions:

SEM-1: System engineering management plan preparation

SEM-2: System engineering planning and control (SEE-4.3.1)

SEM-3: Risk management

SEM-4: Technical information and communication

SEM-1: system engineering management plan. A system engineering management plan (SEMP) is essential to a well-managed program. It will be part of the program management plan (SEE-4.3.1). The SEMP describes how the system engineering activities will be managed and conducted. The SEMP must be tailored to match the size and complexity of the associated program. A small, low-risk program should have a brief plan, while large or high-risk programs require more detail. The SEMP must reflect an effort that is adequate but no more. A small, low-risk program has a smaller program team and a simpler product. Therefore, communications and control are easier. Indeed, one person may have several responsibilities on such a program. Thus the SEMP can be very brief. On the other hand, a large or high-risk program has a large program team and a large or complex product. Communication and control are more difficult and more critical. Consequently, the SEMP for such a program must be very well thought out and very detailed. The SEMP (and/or the PMP depending on specific contract requirements) must document the plans and other SEM data developed under the SEM-2, SEM-3 and SEM-4 functions.

SEM-2: system engineering planning and control. The program management plan (PMP) provides the requirements for the system engineering planning and control. See Section 4.3.1 of this chapter for a definition of the PMP and how to prepare one. Note that SEM-2 is part of the PMP, but I wish to develop your skills to do the bigger system above the PMP because your team must include a program management systems engineer.

SEM-3: risk management (excerpted with modification from Honeywell System Engineering Guidelines, Volume 2). One of the tasks of system engineering management is SEM-3 risk management that is to identify and quantify technical risks. Program management then authorizes the resources deemed necessary and appropriate to avoid or ameliorate the risks. Risk occurs when there is a probability that a planned event will not be attained within constraints (cost, schedule, performance) by following a specified course of action. Risk assessment is performed by system engineering to identify and quantify those elements of a program or system that have high risk. Note that a high technical risk can occur because of any one or more of the constraints of cost, schedule, or performance. Inadequate resources, insufficient time, or marginal performance can each cause an element to have a high risk.

Management identifies three types of unusual conditions that tend to create high risk: technical risk conditions, programmatic risk conditions, and program implementation risks. All programs, big or little, will have risk management techniques applied.

System engineering should coordinate the risk assessment and a risk reduction plan as part of the PMP/SEMP on every program. The technical progress of a program should be closely watched for developing risk areas. Not all risks can be readily identified at the beginning of a program. Moreover, circumstances may occur that change the risks. Revisions to the risk reduction plan should be made as necessary. Program management must be intimately involved in the development of the risk reduction plan and is responsible for authorizing resources to carry out the plan.

SEM-4: technical information and communication (excerpted with modification from Honeywell System Engineering Guidelines, Volume 2). The SEM-4 function should ensure that all system engineering activities are documented in a timely manner and communicated to the entire program team. Emphasis should be placed on making these data readily available in a format and detail that facilitates ready assimilation. These data must be maintained current through regular updating. Configuration control, data management, and nomenclature management procedures should be appropriately documented in the PMP/SEMP and used in accordance with your company policies and procedures for data dissemination and control. System engineering should ensure that all required customer and company standards, specifications, documents, guidelines, etc., pertaining to the technical activities are appropriately delegated to the various program organizations, either through the appropriate system/subsystem specifications or the PMP/SEMP.

4.3.1 Program Management Plan (PMP)

The PMP will be tailored to the size of the development program. A small, simple, low-cost, low-risk, single engineer development program can be run without a PMP merely by indicating "PMP not required" on a single page program brief. Most programs, however, will require a PMP whose size and depth will be proportional to program size, cost, risk, and number of work elements.

Purpose—PMP. It is important to put the PMP purpose in perspective with regard to the development specification (DS) and the Work Breakdown Structure (WBS).

The PMP. The PMP is just a plan. After the program is over you can file the PMP for future reference or even throw it away. These plans are yours to use to manage your program. You must expend a lot of concentrated effort to establish plans at the start-up of a program. These plans must be discussed and reviewed. Even more important, these plans must be rereviewed every week. They can then be modified, but the contractually established schedule must always be met.

The DS. In contrast, the DS is a living document. Your goal is to continually improve the product/system definition and the precision of the product/system requirements provided by the DS. The DS is to lead the initial development and then to support the subsequent production and deployment of the product/system. It lives throughout the life of the product/system.

The WBS. The WBS is a subset of the PMP. It defines the work to be done in work modules of inputs/activity to perform/outputs. Figure 4-23 summarizes the PMP function and its relationship to the DS and the WBS.

The SEMP. The SEMP is also a subset of the PMP. It defines the plans for the management of the system engineering tasks to be done.

DS = DEVELOPMENT SPECIFICATION

PMP = PROGRAM MANAGEMENT PLAN

WBS = WORK BREAKDOWN STRUCTURE

SEMP = SYSTEM ENGINEERING MANAGEMENT PLAN

THE DS AND PMP: THE DS IS A "LIVING DOCUMENT." IT LEADS THE TECHNICAL DEVELOPMENT, THEN ULTIMATELY SERVES AS THE CONCISE SOURCE OF TECHNICAL DEFINITION OF THE SYSTEM AND THE REQUIREMENTS THEREOF FOR THE *LIFE* OF THE SYSTEM. THE PMP IS ENTIRELY DIFFERENT. IT DIES AT THE END OF A SPECIFIC DEVELOPMENT *PROGRAM* PHASE. IT IS A *PROGRAM* MANAGEMENT PLAN, EACH PHASE OF THE DEVELOPMENT LIFE, AND PRODUCTION AND LOGISTICS SUPPORT. IT HAS A NONRECURRING SPECIFICALLY FUNDED PROGRAM THAT HAS TO BE MANAGED, REQUIRING A PLAN—A PMP. THE WBS AND THE SEMP ARE JUST A PART OF THE PMP.

Fig. **4-23.** DS, PMP, and WBS definition

Numbering. The table of contents and numbering for the DS, PMP and the WBS follow a consistent common pattern. The Scope is Section 1.0 and provides a statement of what the scope is for the specific document. The DS 1.0 Scope is basically to lead the development of the specific product/system. It tells the designers about the product or system mission. The PMP 1.0 Scope tells the work team about the specific program schedule and development costs, and the WBS 1.0 Scope talks about the specific work to be done. Figure 4-24 shows the basic outlines and numbering of the DS, PMP and WBS.

DS		PMP		WBS	
1.0	SCOPE	1.0	SCOPE	1.0	SCOPE
2.0	APPLICABLE DOCUMENTS	2.0	APPLICABLE DOCUMENTS	2.0	APPLICABLE DOCUMENTS
3.0	REQUIREMENTS (TECHNICAL)	3.0	REQUIREMENTS (MANAGEMENT)	3.0	REQUIREMENTS (WORK TO DO)
3.1	SYSTEM DEFINITION	3.1	PROGRAM DEFINITION	3.1	WBS DEFINITION
3.2	SPECIFIC REQUIREMENTS	3.2	SPECIFIC REQUIREMENTS	3.2	SPECIFIC REQUIREMENTS
	SPECIFIC TECHNICAL REQUIREMENTS		SPECIFIC MANAGEMENT REQUIREMENTS		SPECIFIC WORK TO DO REQUIREMENTS
	(CONTENT DEPENDS ON SYSTEM TYPE)		(CONTENT DEPENDS ON PROGRAM TYPE)		(CONTENT DEPENDS ON PROGRAM WORK REQUIREMENTS)
4.0	QUALITY ASSURANCE	4.0	QUALITY ASSURANCE	4.0	QUALITY ASSURANCE
5.0	PREP FOR DELIVERY	5.0	PREP FOR DELIVERY	5.0	PREP FOR DELIVERY
6.0	NOTES	6.0	NOTES	6.0	NOTES
	DICTIONARY		DICTIONARY		DICTIONARY

Fig. **4-24.** Systematic numbering for SEM process

Preparing a PMP. The following provides instructions and examples for preparing a program management plan. The PMP will be viewed as a consolidation of management documentation. If you are required on your contract to provide a system engineering management plan (SEMP)-per MIL-STD-499 or equivalent, make it a part of your PMP, not a separate document. If a special software development process is required such as DOD-STD-2167 or equivalent, make the software management plans part of your PMP. If software technical specifications are also required on your contract, these will become part of your technical development specifications, which are identified in your system baseline development specification tree. All plans belong in your program management plan and all your technical definition and requirements belong in your development specifications. Thou will consolidate, eliminate duplication and not proliferate!

PMP title page. The PMP will have a title page similar to that in the example shown in figure 4-25, which includes sign-off and a revision record. The PMP will be assigned a number such as the program financial charge account number W7777.

PMP outline. The PMP will be outlined and numbered consistent with an outline as shown in figure 4-26. Specific people will be assigned responsibility for the various parts of the plan on this outline. These assignments will help expedite communication between team members.

PMP 1.0 scope. The "program brief" will be used as part of the PMP scope. An introductory paragraph describing the program in a conversational way will also be provided.

Figure 4-27 is a program brief that is designed to act as a monthly program status report form. This form is typical of the simple functionally complete type of program overview statement required to initially define a system development program. The program brief may be used alone to manage a small program or may be backed up by a detailed PMP with integral system engineering management plan (SEMP), software development plan, test plan, producibility plan, and other plans.

The program brief provides the mission analysis of the program; both the life mission and the sortie mission of the present program are included. A functional analysis of the program sortie is shown by the action function element: inputs-activities-outputs. The program startup review (PSR), weekly/monthly program reviews, the design reviews, and the action item register (AIR) establish the monitor feedback required for control of the program. The program brief is a condensed and effective tool for system engineering management. My example includes only first names, but you must put the full name and telephone numbers on the brief when you use it for an actual program. This is to ensure you will expedite communication between team members.

PMP 2.0 applicable documents. The PMP in section 2.0 will provide a bibliography of referenced documents by the title/date/number/source-author. These include management documents such as the contract, request for proposal (RFP), statement of work (SOW), problem statement, and your company management procedure documents. The work breakdown structure (WBS), which is part of the PMP, also has a section 2.0 applicable document section where you list the contract number, the statement of work (SOW) and the PMP. Also, note that the WBS and the PMP may be assigned the same top program number such as W7777 that has a PMP-W7777 and an associated WBS-W7777, and all the people working on the program charge their time and material to charge numbers derived from W7777, such as W7777-0001.

Remember that these program management plan applicable documents relate to the sortie mission of your specific program, so they apply only to management plans for the program. Do not include technical documents in your PMP!

Applicable documents in the development specification section 2.0 provide a bibliography of referenced technical documents such as the customer's technical procurement specification, interface control document, prior technical study report, technical standards, technical regulations, safety regulations, and even a technical article from a magazine. Anything that applies to the design and technical development to lead and control the development of the product/system technically should be listed as an applicable document in the section 2.0 of the development specification.

Remember that the development specification applicable documents relate to the sortie mission of the product/system, so they apply only to technical requirements for the life of the system. Do not include program management documents in the development specification.

PMP 3.0 program requirements. The detailed program definition and specific requirements will be reviewed and be understood by all those assigned to work on the program. The work to accomplish program requirements will be done, but will be limited to the following, unless contract renegotiation takes place or the program manager deems additional work is appropriate.

PMP 3.1 program definition. Define your program. You might say, "We have two years and two million dollars of contract funds to perform a development program to meet the requirements of contract XGCE-12, statement of work (SOW) XGCE-49, interface control document (ICD) XGCE-57, and procurement specification (PS) XGCE-124957."

PMP 3.1.1 schedules. An appropriate top-level program schedule will be provided. This is your program sortie mission, of which we will also do a functional analysis. What are all the actions of persons and things required to perform our program mission? Schedules will be worked to lower levels in the specific plans, such as schedule of system requirements development as part of your system engineering plan, software development schedule as part of the software development plan, and test schedule as part of the test plan.

PMP 3.1.2 work breakdown structure. A WBS will be provided and each WBS element will have established all the required inputs, the work to be performed, and method of work performance monitor feedback, and the required outputs. The WBS will also provide the funding allocation and the person responsible for each element. A WBS functional block diagram is shown in figure 4-28. Note that the outputs of elements become inputs of other elements. The work to be done is a sortie mission; therefore, we are applying our mission analysis and our functional analysis skills to generate our WBS functional block diagram.

PMP staffing. Assignments will be provided as part of the WBS. This will reflect appropriate skills and experience to the assigned WBS element work to be done. It is important that these assigned element responsibilities are for those who are specifically doing the work, not just for top-level managers. Lower-level staffing will be identified in the lower-level plans. For example, the software engineering development may be led by a single responsible software systems engineer identified on the WBS who is supported by a team of software development engineers. This software team will be identified in the PMP section 3.2.3 software development plan. The software development plan is provided as an output of the WBS software development element by the responsible software systems engineer.

PMP program costs. In conjunction with the WBS, top-level program funds allocations will be established. These also will be allocated down to lower levels in each individual specific WBS element plan as appropriate.

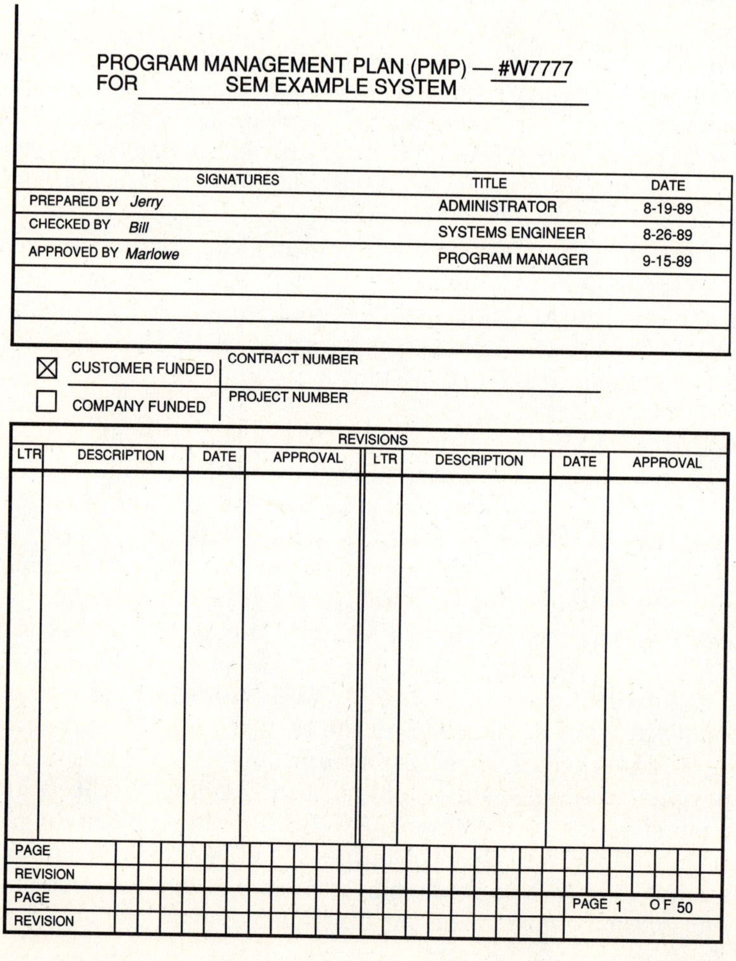

PROGRAM MANAGEMENT PLAN (PMP) — #W7777
FOR SEM EXAMPLE SYSTEM

SIGNATURES	TITLE	DATE
PREPARED BY *Jerry*	ADMINISTRATOR	8-19-89
CHECKED BY *Bill*	SYSTEMS ENGINEER	8-26-89
APPROVED BY *Marlowe*	PROGRAM MANAGER	9-15-89

☒ CUSTOMER FUNDED | CONTRACT NUMBER
☐ COMPANY FUNDED | PROJECT NUMBER

REVISIONS							
LTR	DESCRIPTION	DATE	APPROVAL	LTR	DESCRIPTION	DATE	APPROVAL

PAGE
REVISION
PAGE
REVISION
PAGE 1 OF 50

Fig. 4-25. PMP title page

PROGRAM MANAGEMENT PLAN TABLE OF CONTENTS

Program ______________________________

Responsible Persons

________ 1.0 Scope of PMP and Program Brief

________ 2.0 Applicable Documents

________ 3.0 Program Requirements

________ 3.1 Program Definition

3.1.1 Program schedules

3.1.2 Work Breakdown Structure (WBS), staffing assignments, and program funds allocation

3.2 Specific Program Requirements

3.2.1 Program management process—monitor and control

3.2.1.1 Team building, internal and external

3.2.1.2 Communications, internal and external

3.2.1.3 Risk analysis and plans

3.2.1.4 Monitoring: Schedule performance monitor (SPM), cost performance monitor (CPM), technical performance monitor (TPM), and variances

3.2.1.5 Special contract requirements

______ 3.2.2 System engineering management

3.2.2.1 System engineering management plan (SEMP) per contract

3.2.2.2 Specifications

3.2.2.3 Standards

3.2.2.4 Reliability, maintainability, safety

3.2.2.5 Support specialties integration

3.2.2.6 Configuration management

3.2.2.7 Design reviews

3.2.2.8 Engineering parts management

3.2.2.9 Analyses required and technical performance monitors

3.2.2.10 Logistics

______ 3.2.3 Software Development Plan

______ 3.2.4 Test Plans

______ 3.2.5 Producibility Plans

______ 3.2.6 Subcontract Management Plan

______ 3.2.7 Deliverables and Program Closeout Review

__________ 4.0 Quality Control Plan

__________ 5.0 Preparation for Delivery

__________ 6.0 Notes

__________ 10.0 Appendices (as required)

|

|

|

Responsible Persons:

Initials	Name	Mail Stop	Phone

Fig. 4-26. PMP outline example

PROGRAM BRIEF

PMP # W7777

SCOPE

PROGRAM NAME: SEM EXAMPLE SYSTEM — TYPE: CPFF

CUSTOMER: CAPTAIN ELECTRIC — CONTRACT #: CE-JWH-1

LIFE CYCLE PHASE (FILL IN BAR: [hatched] — PRIOR HI ACTIVITY, [solid] — THIS PROGRAM, [cross-hatched] — FUTURE GOAL)

CONCEPTUAL	DEMO/VAL	FSD	PRODUCTION	DEPLOYMENT	P3I	DISPOSAL

SCHEDULE: START DATE 8-5-89 (INDICATE YEARS AND FILL IN BOTTOM)

STATUS 2 MONTHS AHEAD — 1989 | 1990 | 1991

PLAN — NOW / PROGRAM

ACTUAL

COST 4 M $ — SHARING NONE

STATUS 100K$ UNDER — 1 | 2 | 3 | 4

PLAN — NOW / PROGRAM

ACTUAL

SCALE — (CHECK ONE) 1K ___ 10K ___ 100K X 1M ___ 10M ___

PROGRAM INPUTS	ACTIVITIES	OUTPUTS
CONTRACT CE-JWH-1	SEP/SEM/SEMP/DS/PMP/SRR/	DS
SOW CE-JWH-1	PDR/CDR/PROTOTYPE/TESTS/	PROTOTYPE DESIGN
PROCUREMENT SPEC XCE-12	ANALYSES/DOCUMENTATION/	DOCUMENTATION
PROBLEM STATEMENT	REPORTS/DEMONSTRATIONS	PROTO-HRDWR
	WBS # W7777	3 DEMO SYSTEM-SWS
		REPORTS

PEOPLE

PROGRAM MANAGER: RALPH — ADMINISTRATOR: JERRY

CONTRACTS: MERLE — MARKETING: ELMER

ENGINEERING MANAGER: CHARLIE — SYSTEMS ENGINEER: JOHN

PRODUCTION: GENE — QUALITY: ART

CUSTOMER: CHRIS

PEER CONTRACTORS: AJAX CONSULTING

SUB CONTRACTORS: DALE

DESIGN-MECHANICAL: JIM —HARDWARE: BOB —SOFTWARE: TOM

DRAFTING: TOM — D&E LAB: LEON

RELIABILITY: HAROLD — LOGISTICS: BILL

OTHER: EXPERTS: MEDICAL - DR. BILL, CONTROL - RUSS

P3I = PRE-PLANNED PRODUCT IMPROVEMENT

D&E = DEVELOPMENT AND EVALUATION

Fig. **4-27a.** PMP program brief example

PROGRAM NAME SEM EXAMPLE SYSTEM (PMP # W7777)

COMMUNICATIONS

INTERNAL WEEKLY STATUS MEETINGS, DS PROCESS, PMP PROCESS, PSR, SRR, PDR, CDR, AIR, PCR

EXTERNAL WEEKLY/MONTHLY STATUS REPORTS, PEER REVIEWS, COORDINATION MEMOS, REVIEW REPORTS, CELs.

FACILITIES AND TOOLS

TEST AUTOMOBILE, TEST AIRCRAFT, TEST HUMAN.

RISKS: (SHADE IN RISK IN APPROPRIATE SECTION)

RISK DIAL

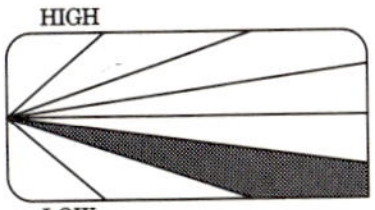

VERY LOW RISK PROGRAM, TECHNICALLY FEASABLE, EXCELLENTLY SKILLED AND EXPERIENCED TEAM, SCHEDULE AND FUNDING ARE MORE THEN ADEQUATE.

GAINS: STAND ALONE SYSTEM APPLICATIONS ARE FAR RANGING, POTENTIAL PATENTS TO ESTABLISH ROYALTIES FOR INTEGRATED SYSTEM APPLICATIONS.

STATUS: (SHADE IN APPROPRIATELY)

TECHNICAL		SCHEDULE		COST		
NO PROBLEMS	○	AHEAD	○	UNDER	○	RED
	○		○		○	YELLOW
GREEN	●	GREEN	●	GREEN	●	GREEN

ADDITIONAL COMMENTS THIS SYSTEM IS A GENERIC EXPERT SYSTEM TO BE USED IN THE DETECTION OF INTERMITTENT FAILURES. THIS PROGRAM DEMONSTRATES THE VALIDITY OF THIS SYSTEM. IT ULTIMATELY MAY BE DEPLOYED AS A STAND ALONE SYSTEM (MEDICAL APPLICATION) OR AS AN INTEGRATED SYSTEM (CONTROL APPLICATIONS).

ACTION ITEMS REGISTER (AIR) ITEMS 4 (# ACTIVE AIR ITEMS) 1

4 TECHNICAL TRADE-OFF STUDIES, MINIMUM SIZE AND COST TRADE-OFF NOW IN PROCESS

PROGRAM START-UP REVIEW (PSR) PSR PROCESS WAS COMPLETED OCT 15, 89

TECHNICAL REQUIREMENTS ESTABLISHED YES. DS HAS GONE THROUGH THREE REVISIONS.

TECHNICAL SPECIFICATION DS-CE-JWH-1

UNIT PRODUCT COST (UPC) MODEL ESTABLISHED PRESENTLY AT 427/500 $

PRODUCTION READINESS REVIEW (PRR) NOT APPLICABLE, BUT UPC AND PRODUCTION PLAN ESTABLISHED

DESIGN REVIEW (DR) LEVEL PEER LEVEL AND EOPS LEVEL 3 — SRR COMPLETE 10-15-89

DR ACTIVITY PEER REVIEWS AND SIZE/COST ANALYSIS NOW IN PROCESS

PRINTED WIRING ASSEMBLY PROCESS: # 3 ON SCHEDULE 3 BEHIND 0

SOFTWARE PROCESS: # 16 ON SCHEDULE 13 BEHIND 3

POWER SUPPLY NI-CAD BATTERY WITH CHARGER

SYSTEM ENGINEERING 3 APPLICATION MISSIONS NOW BEING ANALYZED

TECHNICAL PERFORMANCE MONITOR PROCESS: # 47 REQUIREMENT MET 43 BEHIND 4

1 - SIZE PLUS, 3 - APPLICATION MISSIONS PERFORMANCE REQUIREMENTS.

REPORTER JACK MONTH NOV YEAR 1989

PROGRAM REPORT # 4

Fig. **4-27b.** Program brief example

EACH ELEMENT WILL HAVE: INPUT, FUNCTION, MEASUREABLE OUTPUT, WHO IS RESPONSIBLE AND FUNDS ALLOCATION

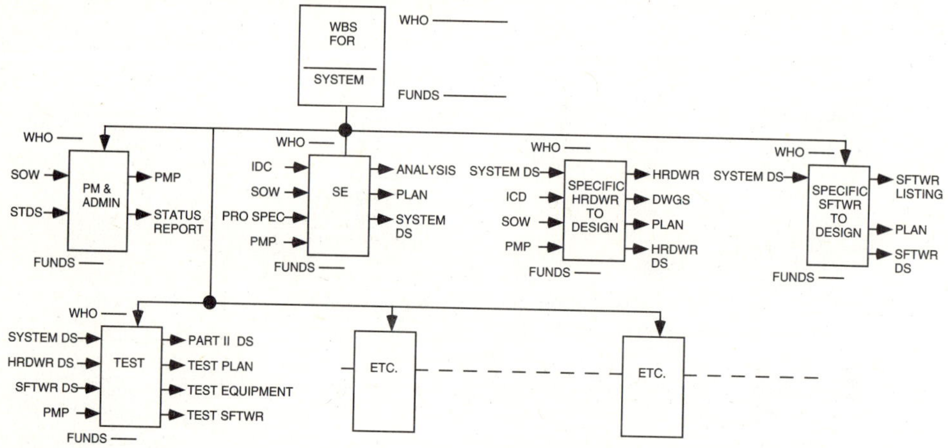

Fig. **4-28.** PMP-WBS functional block diagram. Note: the system DS is a required *input* for hardware DS and software DS process. Understanding the process sequence is critical.

PMP 3.2 specific program requirements. Specific requirements will be planned as appropriate to the program life phase, program size, and type of product or system to be developed.

PMP 3.2.1 program management process—monitor and control.

PMP 3.2.1.1 team building. What are our internal team building plans and our external (customer, peer contractors) team building plans?

PMP 3.2.1.2 communications. Plans for concise/precise communications will be provided. Define both internal and external communications, meetings, reports, development specification process, program management plan process, customer engineering letters, program directives, and coordination memos. This section can become the "alive" part of your PMP if it is used to provide a chronological file of monthly program briefs, a total action item register, and an active action item register.

PMP 3.2.1.3 risk analysis. Provide plans to alleviate schedule, costs and technical risks. This area can also be used to plan for taking advantage of possible gains in schedule, cost and technical program aspects. We are going to patent a unique proprietary design, for example.

PMP 3.2.1.4 monitoring. We will plan how we intend to monitor our program schedule, costs, and technical aspects. The program business administrator will establish the monitoring of program schedules and costs. The technical performance monitoring, however, is a key systems engineering job.

Costs—Plans to establish nonreoccurring program cost goals, cost tracking, and cost reduction must be established on all programs.

PMP 3.2.1.5 special requirements. (Provide plans when applicable.)

PMP 3.2.2 system engineering management. A *complete* technical baseline definition and the establishment of *all* design-to requirements are always required on every product/system development program. Therefore, systems engineering plans will be provided to ensure these are accomplished through the application of system engineering process and system engineering management.

PMP 3.2.2.1 semp. When contractually required, a system engineering management plan will be part of the PMP.

PMP 3.2.2.2 specifications. Plans will be provided for the timely preparation of product/system development specifications. These specifications must lead the development/build/test process. They are required as part of the product/system definition documentation regardless of program contract requirements. At a minimal level of effort, they can be released as preliminary (under releaser's control only). A tutorial with examples of how to prepare development specifications can be found in chapter SEE-5 of this book. The system development specification will be used as an opportunity to consolidate such that it becomes the single focal point for all technical issues for your system.

PMP 3.2.2.3 standards (industry or government). Part of the program start-up review is a review of all the standards contractually required on our program. The critical requirements for standards in your contract must be identified. When this review finds *critical* requirements (such as 2167-Software, 31000-Drawings, AS4613—Parts derating, etc.), plans must be made on how to tailor down or adhere to these difficult, costly requirements.

PMP 3.2.2.4 reliability/maintainability. A system engineering reliability analysis plan will be provided by the reliability analyst when reliability and maintainability requirements are necessary for a product/system.

Safety. A systems engineering safety analysis plan will be provided by the reliability analyst when safety requirements are required for a product/system.

PMP 3.2.2.5 support. (Specialties Integration) When specialists, such as thermal, structural, electromagnetic interference, reliability, logistics, and software analysts are required, specific plans for the timely utilization thereof will be provided.

PMP 3.2.2.6 configuration management. Provide a plan on how you will document and control the changes to the documentation for your product/system.

PMP 3.2.2.7 design reviews. Chapter SEE-7 establishes four levels of effort for internal design reviews. In your plan, identify which level has been selected. In addition, contractually required design reviews will be planned. The plan will establish what reviews, who is involved, a specified design review chairperson, schedule, method, location, and planned funding.

PMP 3.2.2.8 engineering parts management. When the program requires specific control of parts used in the to-be-developed product/system, a plan will be provided. It is recommended that plans include early involvement of procurement and critical suppliers.

PMP 3.2.2.9 analyses required—technical performance monitor. Systems engineers will be required to do many analyses such as trade-off studies and attributes balancing. In addition, many specialty analysts may be required. A checklist as shown in chapter SEE-8 "The Analysis Checklist," can be used to assign responsibility, determine what analyses are required, and act as a checklist to ensure an important analysis will not be overlooked.

PMP 3.2.2.10 logistics. When applicable, you should have a logistic plan for your product/system. How do we intend to support our product/system in deployment? When contractually required, a design to life cycle cost (DTLCC) subprogram will be planned.

PMP 3.2.3 software. When software is to be developed as part of any program, a software plan will be provided. A single software systems engineer will be identified as responsible for *all* categories of software developed and used on the program. The software plan will ensure that software development processes are planned to establish peer design reviews. Separate people or organizations will be assigned responsibility for software test generation and software test execution.

PMP 3.2.4 test plans. When testing is to take place on a program, this activity will be appropriately planned. A single-page example of a test plan is provided in figure 4-29. This example plan shows when the tests will be done, what tests will be done, who will perform the tests, and how much money will be spent to do the testing.

PMP 3.2.5 producibility. Plans for production are required for all product/systems that some day may be produced in production. The plan specifics will depend on what phase of the life cycle you are in as shown in figure 4-30. The two most important parts of your producibility plan are a production process flow diagram and a unit product cost (UPC) model for your product/system. Figure 4-31 is an example of a production process flow diagram. Chapter SEE-6, "Analysis Topics," provides an example of a production bottoms up UPC model.

UPC. When the product/system being developed is ultimately going to be produced in production, a bottoms up UPC model will be developed as part of the PMP. The production UPC goal will be established, the present estimate of UPC will be calculated with the support of the production department, a method to monitor the evolving design UPC, and plans to meet the goal will be established.

Value engineering (VE). When contractually required or when a potential VE subprogram is to be performed, plans for a VE program will be developed. Value engineering is a redesign of your product or system to cut its UPC, with a specific number of units to be produced in production, an estimate of the cost to make design changes, and the expected savings in cost prorated over the number of units to be produced.

PMP 3.2.6 make-or-buy (subcontracts). Programs that require make-or-buy decisions or sub-contract management will provide a plan for this activity. The early involvement of major suppliers in house for extended periods will be planned when applicable.

PMP 3.2.7 deliverables. What is to be delivered and how is it to be delivered? This includes hardware, software, test equipment, reports, specifications, drawings, and support. Reference the statement of work (SOW) and contract.

Program closeout review (PCR). How do we know when we are done? What are the criteria for completion of each WBS work element? What's the process to close them out? We make our plans to close out a program to ensure that we get all our billings in and receive all the money we have coming. We also plan on how we establish a program file with an action item register (AIR) as shown on the program brief, figure 4-27b. This file acts as a history file for future reference.

PMP 4.0 quality control plan. How will your program be audited to ensure you will have a quality program with all risks and technical performance monitors (TPM's) under control?

PMP 5.0 preparation for delivery. How do you plan on ensuring your PMP will be used?

PMP 6.0 notes. You could provide an acronym dictionary, the system perspective cube, life mission analyses, team attributes or top-level hierarchial functional analysis here.

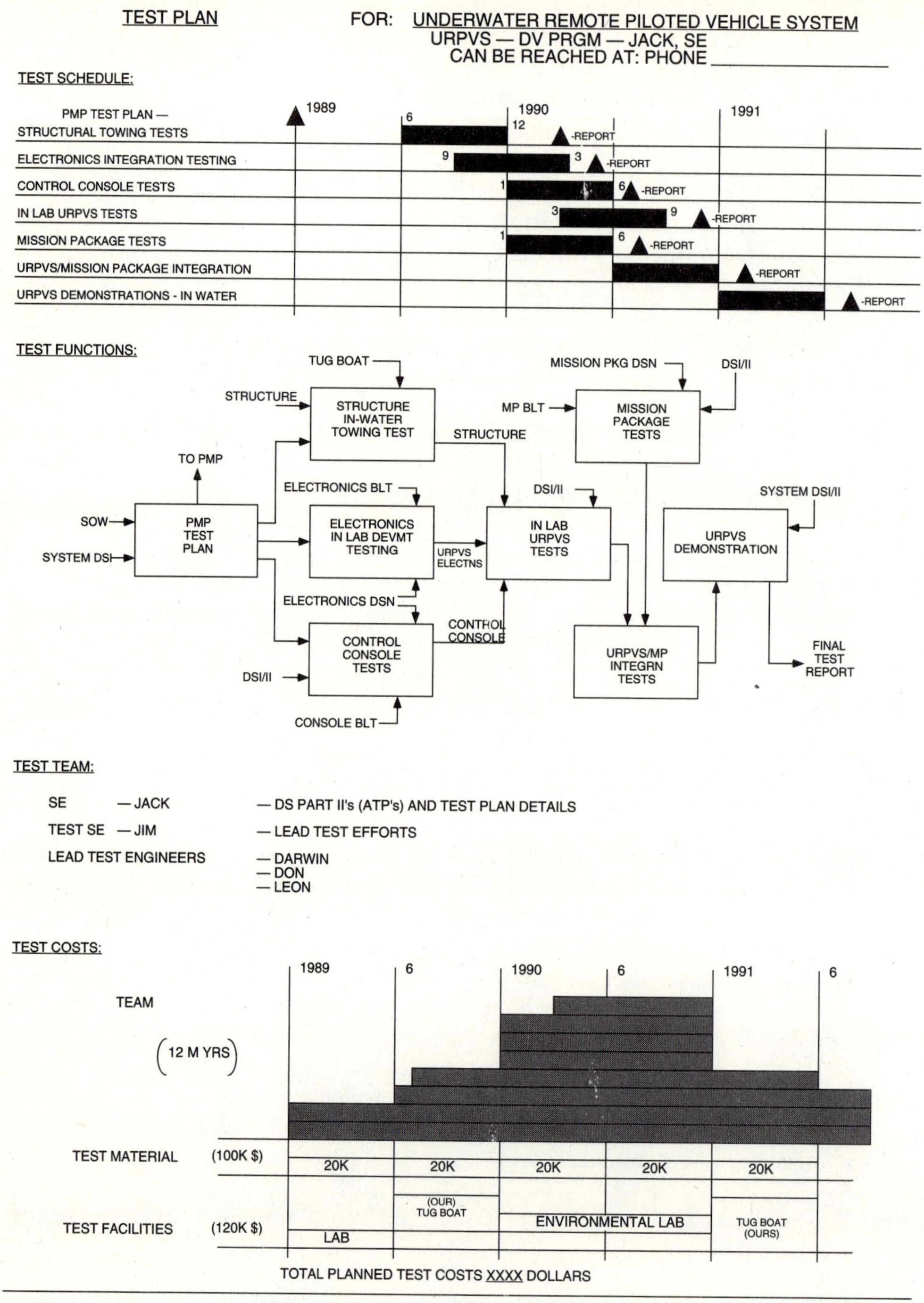

Fig. 4-29. PMP example one-page plan for a demonstration test plan

WHAT DOES PRODUCIBILITY MEAN

DURING EACH LIFE CYCLE PHASE?

CONCEPT PHASE?	UPC TOPS DOWN, PRODUCTION PROCESS FLOW DIAGRAM
DEMONSTRATION PHASE?	UPC TOPS DOWN AND BOTTOM-UP, PREPARE FOR SPC, PRODUCTION PROCESS FLOW DIAGRAM
DEVELOPMENT PHASE?	UPC BOTTOMS-UP, PRODUCTION PROCESS FLOW DIAGRAM, VE ANALYSES, SPC
PRODUCTION PHASE?	UPC TRACKING, SPC, PRODUCTION PROCESS FLOW TRACKING
DEPLOYMENT PHASE?	P^3I, VECP, SPC, FOLLOW ON PRODUCTION PROPOSALS, RE-FURBISH, REPAIR, SUPPORT

ACRONYMS:

- UPC = UNIT PRODUCT COST
- SPC = STATISTICAL PROCESS CONTROL
- VE = VALUE ENGINEERING
- P^3I = PRE-PLANNED-PRODUCT IMPROVEMENT
- VECP = VALUE ENGINEERING CHANGE PROPOSAL

Fig. 4-30. PMP producibility planning

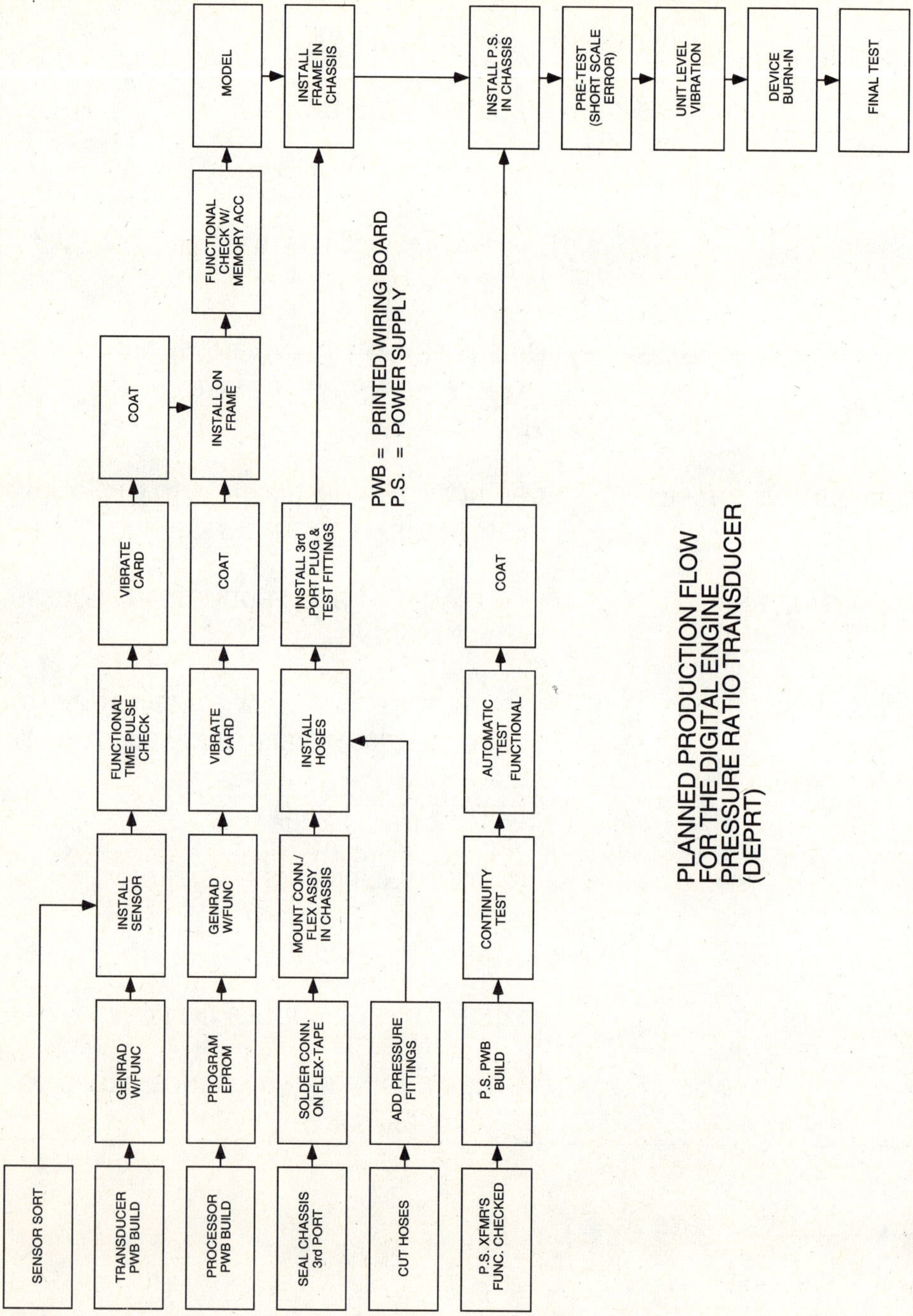

Fig. 4-31. PMP example production process flow diagram

4.3.2 System Engineering Management Plan (SEMP)

Section 3.2.2.1 of the example PMP outline calls out the SEMP. A SEMP is a requirement that is imposed on some programs by establishment of Mil-Std-499A "Engineering Management" as a requirement. Subsequent requirements may require Mil-Std-499B (and so on); if so, the systems engineer will respond appropriately. Excerpts from a SEMP example based on Mil-Std-499A follows for the "DEPRT" subsystem:

Example system engineering management plan (SEMP)

Introduction. The system engineering manager has been given a key role in the Digital Engine Pressure Ratio Transducer (DEPRT) engineering development program. He and his staff will define requirements, monitor technical progress, integrate the system, and validate its performance. This SEMP describes the process of making decisions, the techniques to be used for planning and control, and how specialty engineering disciplines will be integrated into the total engineering effort. It has been prepared to respond to the prior established customer requirements of Mil-Std-499A, the customer's Specification Control Document SCD 8427262, and Data Item DI-S-3618 per the contract data requirements list item A024.

SEMP part 1: the system engineering process

SEMP 1.0 scope. Part 1 of the DEPRT SEMP describes how our design team will apply the system engineering process to the DEPRT program. DEPRT system engineering comprises seven elements leading to a successful DEPRT engineering development program:

1. *System Engineering Analyses*—the analyses and trade-offs required to complete the system design criteria and performance allocations.
2. *System Requirements Definition*—generating specifications and interface requirements.
3. *System Design*—monitoring progress through in-process engineering reviews and assuring that specialty engineering disciplines and software engineering are a totally participative effort included in every applicable aspect of the design process.
4. *System Integration*—the process of combining the DEPRT elements design into a hardware-software integrated system.
5. *Documentation*—assuring that all system design-related activities are satisfactorily documented and that documents are controlled.
6. *System Verification*—the process by which DEPRT performance will be verified.
7. *Production Transition*—the process of evolution from the DEPRT first article to the production version.

Note: In the interest of brevity I have left out the details on how the seven specific elements will be performed.

SEMP part 2: technical program planning and control

SEMP 1.0 scope. Part 2 of the DEPRT SEMP defines the five-step approach to DEPRT technical planning and control. They are: 1) organizational responsibilities and authority; 2) control of performance and design requirements; 3) control of technical program assurance methods; 4) plans and schedules for design and technical program reviews; and 5) control of documentation.

Note: A list of specific people assigned responsibility for all of the specific system engineering tasks has been left out here in the interest of brevity.

SEMP 2.0 responsibility and authority—system engineering manager. Mr. P. is the DEPRT system engineering manager and reports to the Air Transport engineering manager, Mr. E. Mr. P. has the responsibility and authority for implementing DEPRT, the system engineering tasks defined therein, and coordinating and integrating the specialty engineering effort. In this role, he is the principal technical interface with Mr. D. at the customer's facility and is the primary driving force for accomplishing the DEPRT technical objectives.

SEMP 3.0 risk management. The objective of DEPRT risk management is to identify areas of uncertainty so that potential problems can be identified early and contingency plans established to reduce the risks. Risk analysis is performed in three areas: technical, schedule, and cost. These areas are related because technical risks can cause cost and schedule problems.

SEMP 3.1 risk analysis/management. The master program schedule identifies the major activities that must be completed in the engineering develop-

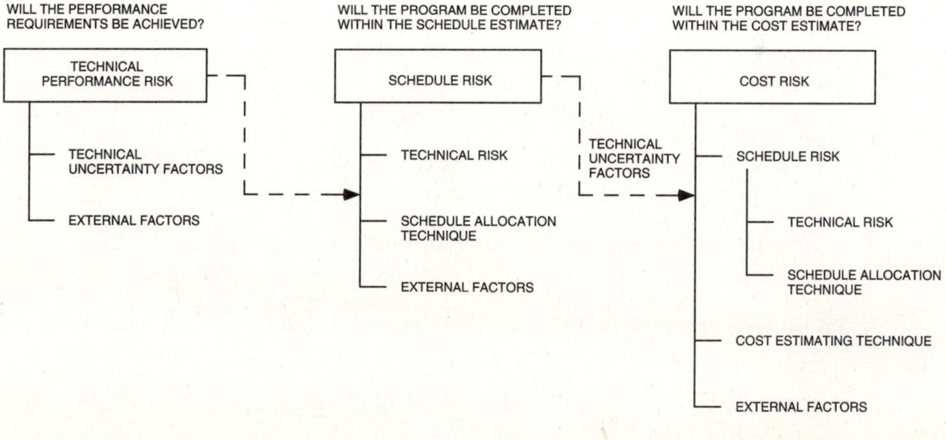

Fig. **4-32.** Risk management process

ment program. System engineering integrates risk management into the DEPRT program to ensure compliance with the master schedule requirements. Risk management follows a systematic process in resolving identified risk problems (fig. 4-32). Risk identification and mitigation will be incorporated in the DEPRT design review agendas by addressing:

- **Compliance with Schedule Performance Criteria**—Does it meet schedule and performance requirements?
- **Compliance with DEPRT Reliability/Maintainability, and Safety Consideration**—Does it meet the reliability allocation, conform to the maintainability concept, and meet the safety requirements?
- **Impact on Logistics Support Costs**—Does it impact provisioning requirements, documentation, training, and other support functions?
- **Impact on Manufacturing**—Is the design producible? Is it economical to manufacture?

The following three areas have been identified as technical risks and will comprise the initial core of the system engineering management risk assessment tasks.

1. Temperature environment
2. Vibration environment
3. Electromagnetic compatibility

These areas have not been prioritized. Their inclusion does not infer that requirements cannot be met. It identifies them as critical areas and shows that our design team intends to keep them under control through technical performance monitor (TPM).

SEMP 4.0 system integration. System engineering will be responsible for integrating the physical and electrical elements of the subsystems, including software, into a total system.

SEMP 4.1 integration plan. Our design team is currently finishing the design and starting production on other models of DEPRTs. These designs are the basis for your DEPRT.

The DEPRT system engineering manager will be the responsible authority for coordinating all functional design and engineering specialties (reliability, quality, maintainability, safety, configuration management, logistics) activities for DEPRT. With the support of our company engineering support groups plus procurement and administration, he will control this integration through the implementation of the SEMP. Design reviews, TPMs, action registers, and progress reports will be used to assess status. Specialty engineering integration will be in accordance with specialty engineering plans as discussed in part 3 of the SEMP.

SEMP 4.1.1 software integration. Software integration for a totally integrated hardware/software DEPRT system will be a system engineering responsibility. Software development, due to its importance to the program, will be accomplished by the lead software systems engineering manager. System engineering will be responsible for coordinating with software engineering and ensuring hardware/software compatibility. Software design will be on the design review agendas. The lead software systems engineer will prepare the software development specification.

SEMP 4.1.2 development and evaluation (D&E) integration. The DEPRT D&E manager is responsible for verifying that DEPRT first-article units meet the requirements of SCD 8427262 and are evaluated in accordance with the program schedule. Due to the relationship between design performance and performance evaluation, system engineering will be required to coordinate with, and support, the D&E manager in all phases of engineering development. This maximizes the chances for an effective and efficient test and evaluation of the DEPRT.

SEMP 5.0 design reviews. System Engineering has the technical responsibility for design reviews. Three design reviews will be held: an internal concept review, a preliminary design review (PDR), and a critical design review (CDR). Design reviews will be held in accordance with Mil-Std-1521.

SEMP 5.1 concept review. An internal concept review is a company requirement for all projects. This review will be scheduled by the project engineer within sixty days after contract award. The review will include senior staff from multiple disciplines who can adequately evaluate the design concept.

Note: The specific plans and requirements for the PDR and CDR design reviews have been left out in the interest of brevity.

SEMP 6.0 technical performance monitor. DEPRT system engineering is responsible for implementing the technical performance monitor process. This TPM process will manage and control the key technical risk areas. Status will be reported monthly. The parameters identified for TPM tracking are: 1) temperature, 2) vibration, and 3) electromagnetic compatibility.

SEMP 6.1 TPM implementation. The three TPMs will be continuously monitored and compared with DEPRT SCD 8427262 requirements. Thus, corrective action can be taken when necessary to ensure that all design and performance criteria are met.

The design of previous DEPRTs by our company (model 100 and model GIV) has reduced the risk for your DEPRT program. The major elements that cause concern are all related to the new environmental extremes to which the device will be subjected.

The operating temperature of 110°C will be a prime factor in determining the part types used on this design. Large case sizes and thermal overlays may be required to lower thermal impedances. We will monitor all aspects of the design that affect operation at 110°C as required by SCD 8427262-paragraph 3.1.43.2.1. DEPRT performance will be measured against the test requirements in SCD 8427262-paragraph 4.4.5: Scale Error—High-Temperature.

The vibrational environment at low frequencies is more severe than anything experienced on previous DEPRT designs. Our design approach is to minimize the impact of the specification (SCD 8427262-paragraph 3.1.43.2.2) by implementing a mechanical design that increases the natural vibration frequencies to an area of the specification where the environment is typical of those at which we already operate.

DEPRT performance will be measured against the test requirements in SCD 8427262-paragraph 4.4.9: Vibration.

The electromagnetic compatibility specification requires the DEPRT to meet the requirements of Mil-Std-461B, test no. RS03 with a test level of 20 volts/meter. This level exceeds the levels at which our previous DEPRTs have been designed to operate. We will be setting more stringent mechanical and electrical design requirements in order to meet this specification. DEPRT performance will be determined by running the electromagnetic compatibility service test as required in SCD 8427262, paragraph 4.4.15: Electromagnetic Compatibility.

Note: 7.0 Configuration Management Plans have been left out in the interest of brevity.

Part 3: engineering specialty integration

SEMP 1.0 scope.—Part 3 of the DEPRT SEMP describes the specialty engineering program that will be integrated into the DEPRT program, as required by Mil-Std-499A.

SEMP 2.0 requirements. The system engineering manager will have responsibility for integrating the efforts of the engineering specialty groups into the DEPRT engineering development program. He will provide the engineering specialty groups with the requirements, criteria, and authorized work packages for the design support effort to ensure the timely and appropriate integration of specialty engineering efforts and disciplines, in compliance with DEPRT development plans.

Each specialty has a direct contribution to the system design, and the system engineering manager will ensure their participation on a day-to-day or as-required basis. Each specialty will participate in the system engineering process as described in part 1 of this plan. The system engineering manager will ensure that groups will be represented at design reviews and will have approval responsibilities as needed for engineering documentation placed under configuration management control. Engineering specialty group representatives will assist the system engineering manager with all planning, control, and quality assurance efforts, as described in part 2 of this plan.

SEMP 2.1 specialty engineering. The specialty engineering areas listed in part 2, paragraph 1 will support the DEPRT system engineering manager in carrying out his responsibilities.

SEMP 2.2 responsibilities and authorities. This subsection defines the responsibilities and authorities of the specialty engineering groups. These definitions are summaries only.

SEMP 2.2.1 reliability engineering. The reliability engineering group will be responsible for overall reliability and standardization of the parts and material in accordance with contractual requirements, and for ensuring that the released design and changes thereto will meet maintainability and safety requirements.

SEMP 2.2.2 development and evaluation. The development and evaluation team will be responsible for conducting and reporting the DEPRT first-article built-in production testing, as described in the DEPRT master test plan.

SEMP 2.2.3 integrated logistics support. The integrated logistics support group will be responsible for establishing logistics support requirements in accordance with program requirements and applicable military standards and specification.

SEMP 2.2.4 quality assurance. This group will be responsible for the overall quality of the system within contract requirements and for verifying (during inspection and test) that the hardware and software configuration matches the technical documentation and changes thereto.

SEMP 2.2.5 parts control. Reliability engineering will be responsible for developing and implementing the parts control program plan.

SEMP 2.2.6 electromagnetic interference/compatibility. EMI/EMC engineering will be responsible for establishing EMC and emission control requirements and implementing a plan that will ensure the final design configuration meets established performance requirements.

SEMP 2.2.7 software development. The software development group will be responsible for preparing and implementing the software development plan and developing the operational software and its supporting data package.

SEMP 2.2.8 production. The production engineering and planning group will evaluate the system design to minimize design complexity with respect to manufacturing and unit product cost.

SEMP 2.2.9 configuration management. Configuration management personnel will be responsible for the control of the hardware/software design configuration throughout the program life cycle of the DEPRT program.

SEMP 2.2.10 technical publications. Technical publications personnel will be responsible for the preparation of all technical manuals and the publication of technical orders and notices.

4.3.3 Program Startup Review Process

You have done a lot of work in a hurry—defined a problem, postulated a system for solving the problem, developed the system perspective for that system, performed mission analyses, functional analysis, implementations, trade-off studies, generated a baseline system definition, birthed a development specification, program brief, program management plan, and, possibly, a system engineering plan, developed a proposal, and ultimately received a contract to perform the proposed program. Now it is time to review everything associated with the program so you can get off to a good start and successfully develop your system. The program startup review process does that. The program startup review process is usually conducted by a designee of upper management. The objectives are: free consulting, help to get off to a good start, tailoring the PMP to program size and provide synergism, that is, working together to get the program off to a good start. The program startup review transitions to design review process and program management action item register.

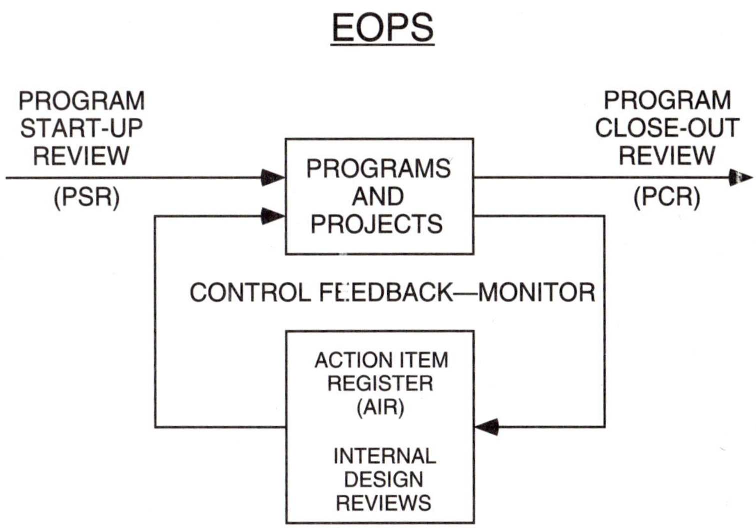

Fig. **4-33.** The Engineering Operating Procedure System (EOPS) establishes engineering management as a control system

In chapter SEE-3 a functional control diagram was constructed that established a systematic engineering management process; this is shown in figure 4-33. The engineering operating procedures system functional analysis example established engineering management as a control system. An example of functional analysis was provided that resulted in this functional diagram for the control part of the system. I said that you needed to have a process to ensure that your programs started off on the right foot, so I established the program start-up review. I also wanted a monitor process, so design and program reviews were required with an active ongoing action item register and internal design reviews. Then, to complete the control diagram, a program close-out review was established.

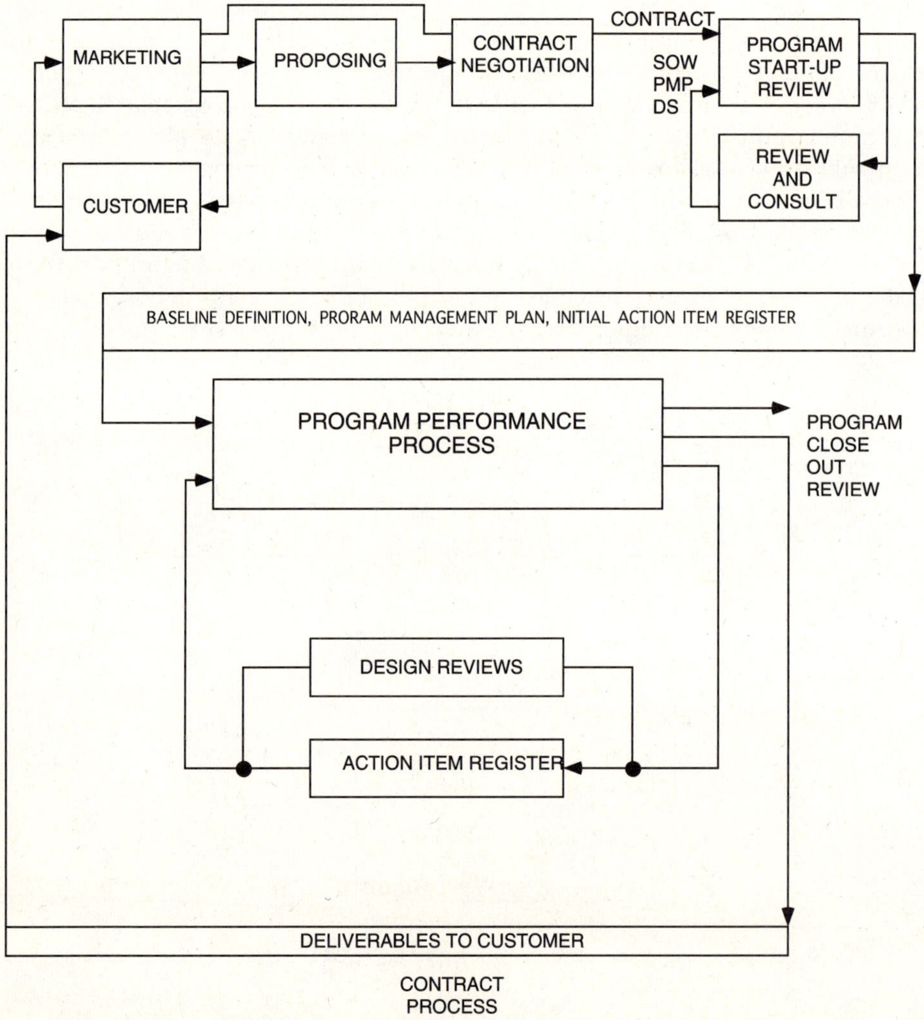

Fig. 4-34. PSR gets the program off to a good start

In figure 4-34, I show you a product producing process diagram with the PSR process in the upper right-hand corner. The PSR is also sometimes used for large proposal efforts. In those cases it becomes part of the proposal effort. The inputs to the PSR include: a contract or a request for proposal, a statement of work, a PMP, and a DS; as you can see it becomes a logical part of your SEP/SEM. You have by SEP/SEM developed the PMP and DS. Now the time is right to review your work with a PSR. You will review your program from both the management and technical perspectives. One part of the PSR is a technical design review, which requires a design to production review. Yes, the PSR is truly an integral part of the SEP/SEM process; it is the first iteration of your system design and program plan.

The term "process" for the program startup review also indicates that this review is not a singular event but an iterative interaction between participants and the reviewer over a period of time. This process could include the following features:

- Administration informs the reviewer of programs to be reviewed. The reviewer then tailors the activity to program size and risks. In low-risk, low-dollar, support type programs the reviewer may complete the review merely through a discussion with the program administrator. If program risk or size is large, the reviewer may then request a program brief. The reviewer will then meet with the program manager of the program to see how consulting can be utilized. A tailored data package will be agreed upon, and the program brief will be backed up with PMP data as appropriate.
- The PMP will act as a checklist and provide coverage of all program functional areas: work breakdown, system engineering, configuration management, design reviews, test plans, production plans, standards, schedules, costs, risks, interfaces—internal/external, skills, staffing, facilities, communications, technical performance monitors, cost performance monitors, schedule performance monitors, and all deliverables.

The program start-up process may follow an eight-step sequence such as:

1. Decision on need for program startup review
2. Program reviewer established
3. Data package (program brief, tailored PMP, customers contract, etc., and DS) submitted to the reviewer
4. Data package reviewed by reviewer
5. Reviewer provides analyses and consulting with program manager regarding questions, concerns, recommendations, and mentoring.
6. If the program has high risks, the program manager will set up a meeting with upper management, reviewer, and program team members to respond to the reviewers concerns.

7. Action items are agreed upon and are placed on the program manager's action item register.

8. Action items are worked on with action item register closeout reporting to upper management.

4.3.4 The Program Startup Review Data Package Review

The requirements sections of the development specification, program management plan, and work breakdown structure are entirely different; therefore, the reviewer must review both technical and program plan areas. The development specification requirements are derived from a flow down from the system mission technical requirements. The program management plan requirements are derived from a flow down from the specific program contract to be managed. The work breakdown structure requirements are derived from the specific customer's statement of work.

The reviewer must also determine the *Quality of* and *how the Quality will be ensured* for both the technical system to be provided and the program to evolve the system.

Quality Assurance of the development specification (Section 4.0, shown in Table 4-1) tells you how you intend to meet the requirements of sections 3.0 and 5.0 of the development specification. This is usually accomplished by providing a matrix of section 3.x.x and 5.x.x requirements against requirement met confirmed by either analysis, test, and commonalty to prior system. The following shows the form this matrix may take:

Table **4-1.** Section 4.0 QUALITY ASSURANCE—Requirements confirmation method

REQUIREMENT		**ANALYSIS**	**TEST**	**COMMONALITY**
PARA	FUNCTION			
3.2.1	10 ± 1%	X	X	
3.2.2	Less 50 Lbs.	X	X	
3.2.3	Transfer Function			X
3.2.4	Serial 1553		X	
•	•	•	•	•
•	•	•	•	•
•	•	•	•	•

The reviewer will review section 4.0 of the development specification against sections 3.0 and 5.0 and the customer's mission requirements.

The quality assurance of the program management plan is usually a statement on how the program will be periodically monitored to ensure that the schedule, cost, and technical program aspects are on plan. Section 4.0 of the program management plan will provide a statement of plan for monitor of schedule performance monitor (SPM), cost performance monitor (CPM), and technical performance monitor (TPM). The section 4.0 of a work breakdown structure will establish a matrix that shows how the specific outputs from each work breakdown structure work element are to be tracked for progress and completion.

The reviewer will review sections 4.0 of the program management plan and the work breakdown structure. Note that the program brief can be identified to be used as a monthly report form to provide monitor of SPM/CPM/TPM.

It is highly recommended that the program management plan outline and development of the PMP process be done early in a program. In fact, it should be baselined prior to the proposal.

When the development specification and the program management plan are reviewed you must review them against the criteria as shown in figure 4-35.

	Development Specification	Program Management Plan
1 — CONSISTENCY		
2 — SPEAKS TO THE DEVELOPERS AND ASSIGNS RESPONSIBILITIES		
3 — ATTRIBUTES BASELINED		(SPM/CPM)
4 — STRUCTURE		
5 — SUFFICIENTLY ESTABLISHES THE REQUIREMENTS	(TPM)	
6 — GIVES ROOM TO DESIGN/MANAGE		

Fig. **4-35.** Six review criteria for development specifications and program management plans

4.4 Summary of SEE-4

In SEE-4, you learned how to define a baseline system to support a proposal effort. You were also exposed to system engineering management and how a system engineer can organize an engineering program in a systematic way. You learned that you are dealing with process, process, process. The system engineering process is an iterative process, and so is system engineering management an iterative process, and so is the program startup review an iterative process. Simply stated, you do it over and over until you get it right. The phrase "get it right the first time," should be; "get it mostly right the first time." The more difficult your system or your program, the more likely it is that you won't get it right the first time, *so you must iterate!* The teachings of quality experts, W. Edwards Deming, A. V. Feigenbaum, and J. M. Juran, say that we develop quality by a continuous improvement process. We are talking about system engineering process, system engineering management, and the program startup review as iteratively leading to a quality system and a quality program!

There are some who believe that if you get the right organization you can be successful. You, however, know that if you have developed knowledge of the process and iteratively use system engineering processes, you can develop quality systems, and the organization structure becomes a secondary concern.

In SEE-5 you will expand on your knowledge of the development specification process.

4.5 Problems and Questions

1. True or False?
   a) My baseline sketches must be neat, complete, quickly provided and correct.
   b) A scope statement is part of the baseline definition.
   c) A baseline is a development specification.
   d) A baseline organizes the system so the designers can start to design.
   e) A baseline can be too restrictive by designing the system down to too low a level.
   f) The specification tree reflects deliverable products.
   g) The baseline must have at least a scope and five sketches to be minimally complete.
   h) The detailed block diagram is best done by a single system engineer with support from a group of experts.
   i) Our baseline is the final design in its first iteration.
   j) A detailed block diagram with a lot of working notes and extra data is not needed for a baseline definition—keep it simple!
2. Match the following:

a) 10 items required for system definition.	A. System engineering methodology parts

b) Specialty integration	B. Types of are physical, mathematical, static, dynamic
c) Risk management	C. Tailoring
d) SDF-3	D. Mil-Std-499A
e) Models	E. System integration
f) SEMP	F. Requirements allocation and system synthesis
g) Potential TPM attributes	G. Capacity, resolution, response time, temperature, vibration, electro-magnetic.
h) Appropriate level of effort	H. Planning for the work of other analysts such as reliability thermal, structural, and electromagnetic interference.
i) SDF-6	I. SEM-3
j) SEM, SDF, SIF, SVF, SSF, SAT	J. Baseline configuration

3. True or False?

a) Top-level functional requirements usually become lower-level performance requirements.

b) Mission analysis is an iterative process.

c) Requirements allocation process examples are error budgeting and power profiling.

d) System simulation, hybrid, analog, numeric, analytical, static, dynamic, functional, mockup, scale, physical, and mathematical are *all* types of models.

e) Wire lists are adequate for most interface definitions.

f) Functional analysis can be used for software allocation.

g) Mil-Std-499A defines the task of synthesis.

h) Mil-Std-490 establishes requirements for engineering management.

i) The program startup reviewer acts as a consultant.

j) A detailed system block diagram that shows all the functional interfaces and a sketch that shows the physical configuration with physical interfaces must be provided to minimally baseline a system.

4. A Problem to Chronologically Order the System Engineering Process

Complete the placement of the following items in proper sequential order:

a) Establish a concise problem definition that indicates a need for a new system.

b) System qualification.

c) Functional analysis of life mission.

d) Do system baseline definition (including scope, interface diagram, functional block diagram, DS tree, mechanical sketch, installation drawing, detailed block diagram, and other supporting data).

e) Mission analysis of your system's life.

f) Iterate the development specification Part I as necessary to guide the design of your system.

g) Develop a system perspective.

h) The development specification becomes an aid to production.

i) Provide development specification part II.

j) Mission analysis of your system sorties.

k) System implementation trade-off studies.

l) Functional analysis—collecting brainstormed mission analysis sortie actions and putting them into a hierarchy.

m) Collect baseline definition, applicable documents, and requirements derived from mission analysis to start the system development specification Part I.

n) Functional analysis—Brainstorming mission analysis sorties for all the actions required.

o) Development specification parts I and II are an aid to subsequent deployment, marketing, and future systems.

5. Questions for System Engineering Management

True or False?

a) The program management function includes system engineering management.

b) The technical lead systems engineer is not concerned with system engineering management.

c) A program management plan is required for all system development programs.

d) The program brief includes signing up for responsibility.

e) The development specification could be identified as required for a program in section 3.2.2.6 (configuration management) of the program management plan.

f) The system engineering management, system engineering management plan, program management plan, and program startup review are only of interest to the program manager.

g) Mil-Std-499A may require different approaches to your SEMP.

h) The DEPRT SEMP example uses the inputs/activity/feedback/outputs method to organize its definition of SEMP applied to the DEPRT program.

i) The DEPRT SEMP lists a specific customer person with whom our lead systems engineer technically interfaces.

j) PSR is internal consulting.

6. We have been asked to develop a solar powered total home system for geographical locations where sunshine and clear skies can be expected to be greater than 40 percent of the total hours every year. This system is not intended to be the only source of energy, but it is to be considered as the prime energy source. An auxiliary backup source is acceptable. This prime solar system will provide home heating, cooling, hot water, and electricity. This system will be in production in five years.

   Do system engineering process: study, system perspective, mission analyses, functional analyses, baseline, and a program brief.

7. Match the following:

a) Basic functional elements	A. Program management plan
b) Work element	B. (research and development costs) + (production and construction cost) + (operation and maintenance support cost) + (retirement and disposal cost)
c) Skill element	C. Concise overall objective
d) Technical requirements	D. System, hardware, software, and test equipment specifications
e) Managerial	E. An aid to team building
f) Baseline	F. 10 items that define our system
g) Scope	G. Ability to perform a work element
h) Engineering skill bank	H. Development specification
i) Life cycle costs	I. A box on the work breakdown structure
j) Production and construction cost	J. Unit product cost

8. What is the basis for using acronyms? Make a list of all the acronyms used in this chapter.
9. Do a program brief on your present project.
10. Draw a single action function element of figure 4-1, The System Engineering Process. What are the inputs, outputs, functional actions, and feedbacks?

4.6 REFERENCES

1. Honeywell Aerospace and Defense System Engineering Committee, *System Engineering Guidelines—Volume 1.* Minneapolis, MN: Honeywell Inc., October 1983.
2. Honeywell Aerospace and Defense System Engineering Committee, *System Engineering Guidelines—Volume 2.* Minneapolis, MN: Honeywell Inc., December 1983.

The Development Specification

(The Living Document—If you are not doing Development Specification Process, you are not doing System Engineering Process!)

This chapter is intended to develop your skills in the development specification process. Figure 5-1 shows the system engineering process and how SEE-5 follows SEE-4. In SEE-4, you learned that the development specification is a living document; it grows and matures as we develop the system. Its initial purpose is to guide the design of a system. Its initial constitution is derived from the work in SEE-1, establishing a system perspective, SEE-2 establishing requirements from mission analyses, SEE-3 establishing the functional requirements required to perform the mission plus starting implementation and trade-off studies, and SEE-4, establishing a baseline system definition. In SEE-5 I will show how you can gather all this data together into the system engineer's most important communication device–the development specification.

The development specification process is an iterative process, you do it over and over until you get it right. As the lead systems engineer you may have an opportunity to establish an initial development specification where one is needed, or critique and improve an existing specification to make your present program successful. The important thing is to use the development specification as a communication tool. It must be exposed to the team as soon as possible. All team members must review the development specification and assist in its iteration to keep it alive and maturing.

The examples provided in this chapter demonstrate pragmatic real-world tailoring for a specific system and subsystems. The preparation of this chapter has been very difficult. To show the wide-ranging versatility of the development specification process and to get across the "living document" concept, I initially used ten different DSs for examples. This chapter very quickly had over 100 pages of examples, with a prediction of 1000 pages! Now, however, this chapter only uses excerpts from three DSs. You will have to take my word about the wide-ranging possibilities. You can use the DS process to lead the development of almost anything—a system, subsystem, component, product, production process, a small business, a medical process, an educational process, a political process, a management process, the design process, computer-aided design process, a market analysis, a marketing process, and much more. These all can benefit from the system engineering process with the integral development specification process.

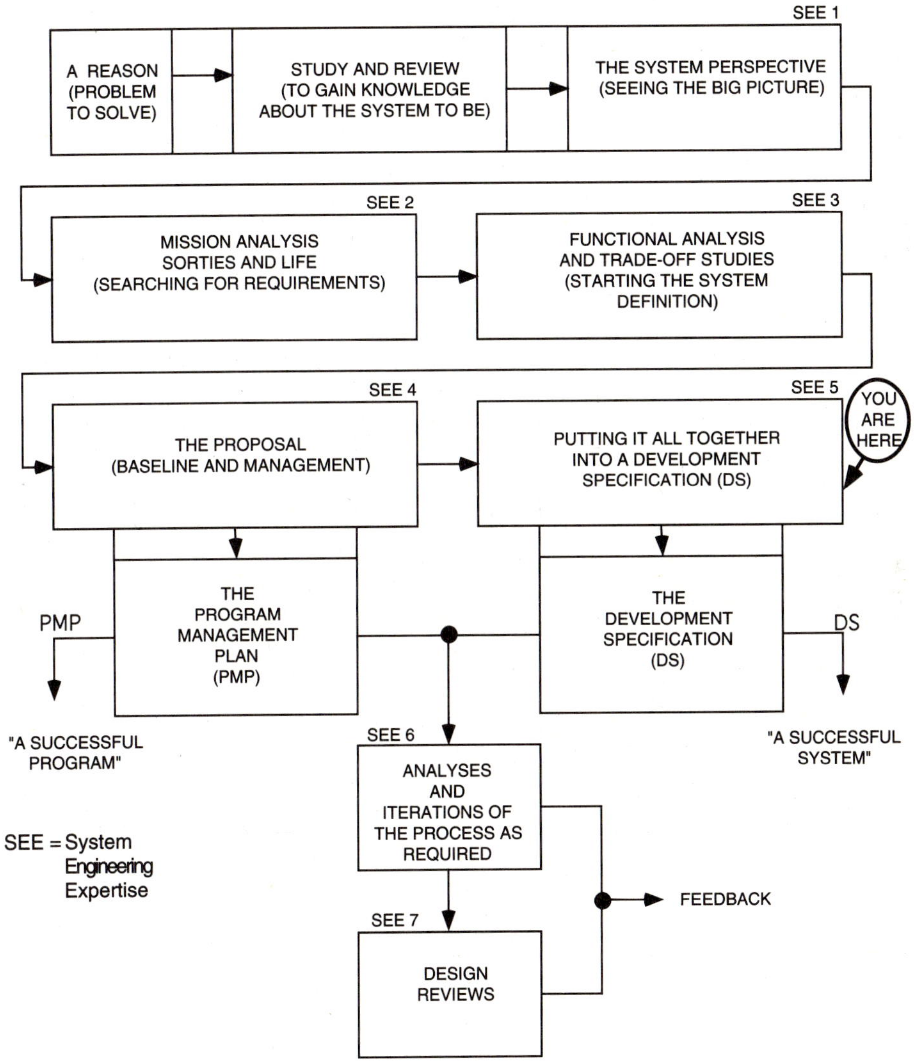

Fig. 5-1. The system engineering process

The following three DSs are being used to provide examples in this chapter:

1. The Integrated Propulsion Control System (IPCS) DS 24712-01. This is an integrated avionics system that was developed to demonstrate that an aircraft engine and inlet could be better controlled by a fly-by-wire digital computer. Fly-by-wire means the interface to the engine and inlet was by electrical signals only, no mechanical cables or linkages. IPCS was installed on an F-111 aircraft and successfully flown on many test flights.

2. The Single Board Computer (SBC) DS 28991-01 is a subsystem for a number of systems. The SBC is a computer with expandable integral memory and capability for a wide range of external input/output devices.

3. The Computer-Aided Cost-Estimating System (CA-CES) DS 32615-01 is a software tool used to do fast cost estimates of program work breakdown elements based on a data base of prior program costs.

System development specification process timing

When you spend too much time trying to establish the requirements for a system you will have too much entropy in your development process. Entropy is energy lost as a result of energy transfer from one form to another. In a chemical reaction it is lost by heat given off. In the development of a system, it is too much hot air and not enough cold hard specific requirements. An action function element like the one in figure 5-2 shows that feedback can be viewed as lost energy in the development process. When you look at these feedback loops in time on a life mission time line as in figure 5-3, you can see that a lot of small loops in the beginning of a system's life will cause much less energy loss than that late in the system's life. This is because the impact of change is limited to a small group of people and minimal tools and facilities. Software development costs also require early establishment of a good system baseline and early requirements in order to

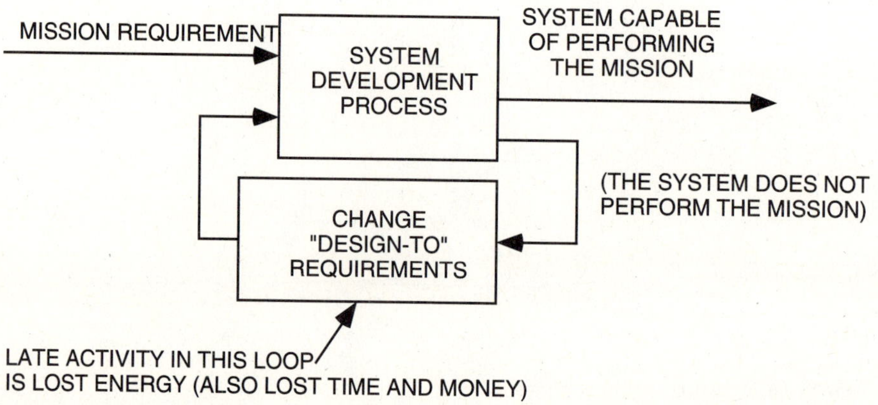

Fig. **5-2.** Too much late feedback

prevent costly changes late in the development process as shown in figure 5-4. This diagram illustrates that if you discover that you have to make a major change to the requirements at the time you are doing complete system verification, it could cost you 720 units of software development cost. The impact of late requirement changes on your system development program cost and schedule are shown in figure 5-5. You must start the system development specification process at the very beginning of your development process and force reviews to reduce lost time and effort.

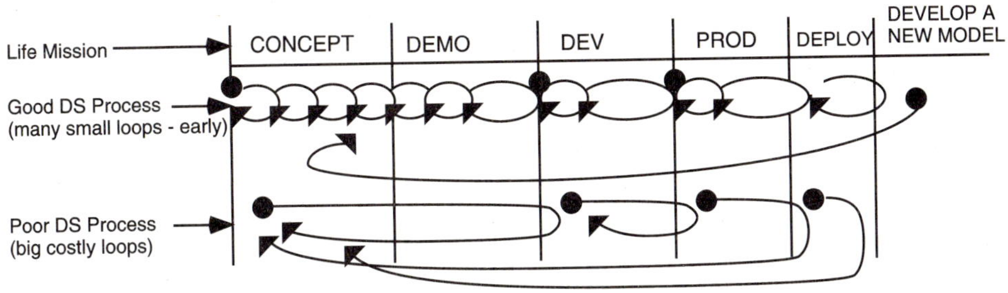

Fig. 5-3. Establish system requirements early in the life mission

SOFTWARE DEVELOPMENT—This is another example that shows the high cost of software development with poor early system requirements establishment:

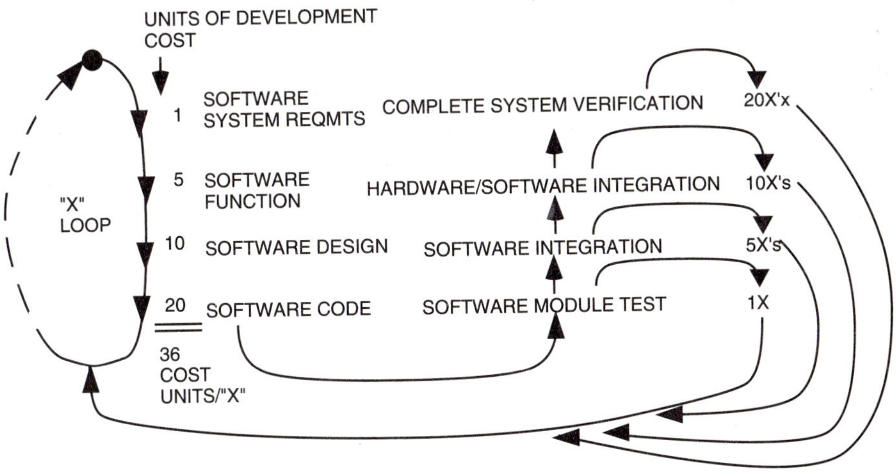

If we have to change system requirements when we are in complete system verification it may take 20 passes more through the "X" Loop. A poor job in initially establishing software system requirements can result in 20 x 36 = 720 units of cost. We must establish a good system baseline and requirements in a DS early!

Fig. 5-4. Establish software requirements early in the software development process

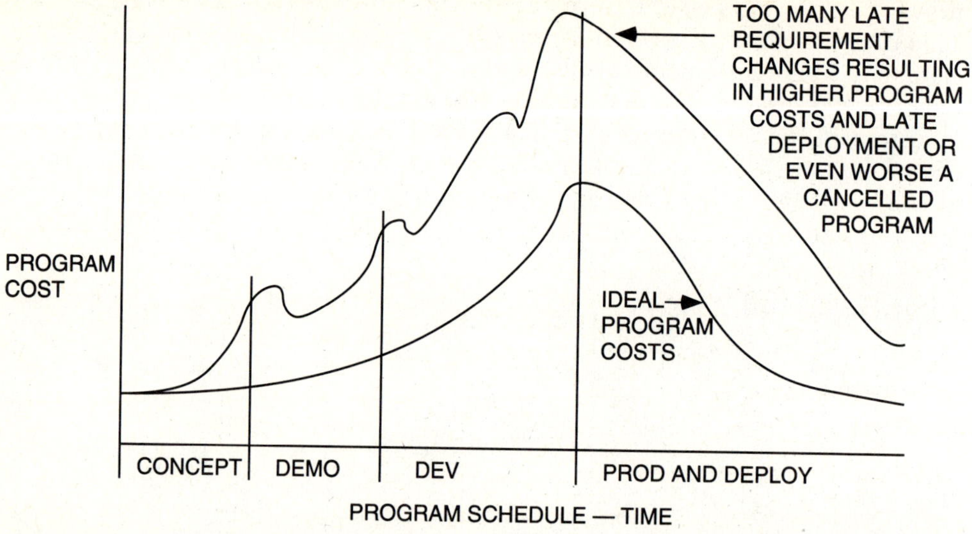

Fig. 5-5. The impact of poor DS process on your system development program costs and schedules

Writing your first DS

When you start to write your very first DS, you may ask someone for help, and he or she will give you an example of a finished mature DS. This is helpful, but they should also tell you that the first rough version of the DS was incomplete, hard to understand, probably was too big, possibly sounded and looked like a marketing proposal, did not show flow down of requirements from the mission of the bigger system above, was not concise, did not have a complete baseline definition, lacked many specific requirements, and had an inconsistent baseline. In addition, acronyms were not defined, it incorrectly allocated the specific requirements because it did not allow for test measurement device accuracy, control algorithms were undefined, interfaces were poorly defined, and it had other deficiencies. The important part of the evolution of the first DS to the mature DS is constant communication between the development team members and constant improvement of the DS during the development of the system. The first few versions required a number of informal working reviews by the initial system engineering development team. This typical team might consist of a lead systems engineer, a hardware systems engineer, a software systems engineer, a mechanical systems engineer, a control analyst systems engineer, a test systems engineer, and the program manager systems engineer. These seven can work together to evolve the DS and the PMP to lead the development of the system. This must be done prior to assigning all the other people who will be working on the system's development. Typically three months after the start of the initial system development and following many informal reviews, with in-depth communications, the first official release must be made by the team. This release must result in the first formal review of the specification by the larger team of system developers. The DS then will be revised and redistributed to the total team and to all the other people associated with the system under development, especially the ultimate user, the customer. Subsequently, it will be revised a number of times more as the system is designed. This process of review and constant improvement of the DS is what the system engineering process and the development specification process are all about. The initial small, seven member team's work in improving the DS in the first three months is critical to the success of the system and the development program. It reduces the impact of late changes in the development of the system by providing cold, hard requirements early on instead of a lot of hot air—it reduces entropy. So, as you write your first DS, make sure the rest of your team understands the DS process.

Writing the late DS

You may have to provide the DS after the system has been partially or even totally developed. This usually comes about because there is a lack of understanding of the critical timing of the system engineering process, or, in most cases, the development process is experiencing some serious difficulty. Whichever it is, it makes you realize that you do not have the requirements for the system established completely or clearly, so you better write a system development specification. In these cases the systems engineer needs to remember that the task requires a certain attitude: "I'm here to be part of the solution not the problem." "I'm to serve." "I'm to ask questions." "I'm to study." "I'm to communicate with everyone," and, ultimately, "I'm to lead." I must remember to keep a stiff upper lip. I can muddle through, and I must hasten after the team because I am their leader even if they don't believe or understand what is taking place. Experiences in situations such as this have been extremely frustrating but also have been very rewarding because of what is learned. There is a feeling of accomplishment when you can bring understanding, order, and clarity to the definition of a system and its requirements. You can also help improve the synergism of all the people associated with the system under development. The most important lesson learned from the experience of writing the late DS is that in future system developments you will do the system engineering process with the right timing so the development specification process truly leads the development.

Definitions you should know

The development specification is a document that is intended to lead the development of an item. the item may be a system, subsystem, product, process, or anything that has a defined mission.

The development specification process is a disciplined methodology of writing a DS, forcing review, improving and clarifying, rewriting, forcing review again, and continuing this process to force communication so we can lead the development of a system.

DS part I contains the requirements and implementation of what the item in question is. It is a living document that changes as the design of the item develops.

DS part II describes the fabrication of the item and the test procedure needed to prove the item meets the functional requirements of DS part I. This specification will reference the DS part I many times. The test procedure is called the "Acceptance Test Procedure (ATP)."

The ATP is an integral part of DS part II (section 3). It is a formalized, step-by-step procedure of how to test an end item. The ATP is run as the last step before a device or system is shipped to the customer. It is written using the DS part I's paragraphs 3.X (Requirements section) and 4.X (Quality section) of the system. It should contain, in addition to the test procedure, a blank test report form for the test technician to record specific item serial numbers and test data. The method of storing this data and providing it to quality assurance and the customer are indicated in section 4 and 5 of DS part II respectively. The ATP will be designed to provide data for statistical process control (SPC) so the design and production processes can be monitored and modified to optimize production yields.

Remember, the ***development*** specification is ***your tool*** to ***lead*** the ***development***. It can provide the order and discipline to lead the development of anything. In addition, as a living document, it can reach maturity and ultimately serve other purposes such as being an aid to production, and customer support.

5.1 The Development Specification in a Standard Format

The Development specification must follow some standard consistent format. This could be your established company standard, a specific industry standard, or a government standard. The MIL-STD-490 format has been shown to be excellent for many types of systems, subsystems, components, products, and process systems, both military and commercial. This chapter will use a format derived from MIL-STD-490.

I do not wish to emulate or describe all of MIL-STD-490 specification types. Your objective should be to consolidate, eliminate, and not proliferate specification types. In your simple concept you need development specifications that are living documents that are initially started by a customer's mission need and then progress and change as the system is developed, produced, and deployed to do the customer's mission.

The objective of this chapter is to develop your skills in quick baselining, establishing the development requirements for your system, and then providing them with a standard formatted DS. I recommend following the basic six-section numbering of the MIL-STD-490 format because of its completeness of coverage and the efficiency that will result from using the same basic format for all your development specifications. However, the content of the DS is yours to use as your tool to lead the development of your system, so you may wish to use a more appropriate format that relates better to the system under development and to the development team. The DS process can be used to consolidate all the communications required to technically lead your system development process. Examples will show how (part 1.0 Scope) can be used to tell a story in prose, (part 2.0 Applicable Documents) can help your team study, and (part 3.1 System

Definition) will provide a complete consistent system baseline. The specific requirements in 3.2 will be based on the baseline and will provide clear concise requirements, and 4.0 Quality Assurance will not only establish requirements to ensure quality but also include the establishment of means to provide constant improvement of your system. Part 5.0—Preparation for Delivery—can include deployment requirements to ensure the system is properly installed and is performing the customer's mission. While (part 6.0 Notes) can be used to record important development data for future reference such as trade-off study results, system change history, market analysis results, process improvement, and the Appendixes can be used in any way you may wish to help you organize your system. Each of the appendixes of the computer-aided cost-estimating specification, for example, constitute a standard formatted users manual of all the various classifications for potential user types.

Our goal in simplifying the DS is to establish a single, simple format that consolidates all pertinent technical data into a single specification, instead of a proliferation of documentation. The DS will be the one place that can answer all technical questions about a system, or at least act, as a reference source that will tell you where to go to get the answer through the section 3.1, "Specification tree." There is one system DS, and it lives and changes as a living document that leads the system through all its life phases: at the end of concept phase, the DS is mature; at the end of the development phase the DS is very mature; and during production and deployment phases, the changes to the DS reflect constant improvement of the system in producibility, cost, and mission effectiveness.

The basic recommended DS format is:

A TITLE PAGE AND A TABLE OF CONTENTS

- 1.0 SCOPE
- 2.0 APPLICABLE DOCUMENTS
- 3.0 REQUIREMENTS
    - 3.1 DEFINITION (Provide your baseline definition)
    - 3.2 SPECIFIC REQUIREMENTS
- 4.0 QUALITY ASSURANCE
- 5.0 PREPARATION FOR DELIVERY
- 6.0 NOTES

APPENDIXES

Note that this basic format can be used for any type of system! The details and specific requirements may be very different, but the basic format can be the same. I have used this basic format for hardware, software, procedural, products, and process systems.

Some comments on specific requirements

The specific requirement part of your development specification is the heart of your specification. This is the part that tells the system designers what to do. These requirements must be brief, terse, and as clear as possible. They should always refer to the baseline definition in section 3.1 to show the source flowdown of the requirement from the system missions. You will establish technical performance monitors (TPMs) so you have a basis for confirming each of the specific requirements have been met. Always provide numerical requirements (must weigh less than 12 pounds, shall not require greater than 10 watts). If you don't have a number, do some precision guessing. Identify your number as (preliminary), (under analysis), (subject to change), or (a design aim). Never give a blank or a "to-be-determined" (TBD). Any number is better than none. You can at least get started in designing and analyzing. If you calculate the worst case of "TBD," you get "TBD!" Insist on review of the specific requirements. Force feedback from all the people, especially the design team! Keep the team informed of changes to the specific requirements, make changes quickly (within the hour), and establish a revision record and a change highlighting method.

Title page and table of contents

The title page and table of contents are important parts of the DS. The title page format shown in figure 5-6 and the table of contents format in figure 5-7 are recommended. The title page is important in that it identifies the document as a specification and provides signoff and revision history. The table of contents is important in that it acts as a quick reference index, which relates to the mission of the DS as a device to expeditiously lead the development process. It is a handy index to assist and quickly lead the designers and developers of the system because they don't have to read the complete document to find what they want. The recommended table of contents also allows readers to easily identify the authors for different DS parts to expedite communication between team members.

SPECIFICATION NO. ______

PART ______ OF ______

TITLE:

SIGNATURES	TITLE	DATE
PREPARED BY		
CHECKED BY		
APPROVED BY		

☐ CUSTOMER FUNDED | CONTRACT NUMBER

☐ COMPANY FUNDED | PROJECT NUMBER

REVISIONS							
LTR	DESCRIPTION	DATE	APPROVAL	LTR	DESCRIPTION	DATE	APPROVAL

PAGE																										
REVISION																										
PAGE																				PAGE OF						
REVISION																										

Fig. 5-6. Recommended title page format

DEVELOPMENT SPECIFICATION TABLE OF CONTENTS

SYSTEM ______________________________

AUTHOR:

AUTHOR

Initials	Name
____________	____________________
____________	____________________
____________	____________________
____________	____________________
____________	____________________
____________	____________________
____________	____________________

*The items under Specific Requirements should reflect your specific system type, so if needed change 3.2.1—3.2.12 headings as they apply to your system.

Fig. **5-7.** Recommended table of contents format

DS 1.0 Scope

Your system perspective cube, when applied to a specific subsystem, resulted in a worksheet. This worksheet can be converted to a writeup to provide the scope data for the system and its development. You will use this data to establish the technical scope for your DS and also to establish the program/business scope for your program management plan (PMP).

The objective is to provide sufficient information so that anyone who reads the scope section of the DS and the PMP will establish the system perspective for the specific system. Note that the scopes of the DS and the PMP are different. The DS gives the scope of the *technical system* while the scope of the PMP gives the scope of the *program*.

The process of developing the system perspective helps you develop the scope statements. It is extremely important that you write good scope statements. *Those with the system perspective can avoid mission impossibles!*

Examples of scopes

The following scopes have been excerpted from three DSs. They were all *attempts* to inform the reader sufficiently about the system so the reader could quickly understand it. These examples are early attempts to demonstrate how the system perspective is established. You are expected to improve on these early attempts by providing concise and clear scope writeups in your DSs. We should improve our scopes by ensuring that *all data* generated on the perspective worksheet appear in either the technical scope section of the DS or the management plan scope section of the PMP, as appropriate.

Refer to the following figures:

- Figure 5-8 DS 24712-01, Scope for an Integrated Propulsion Control System (IPCS)
- Figure 5-9— DS 28991-01, Scope for a Single-Board Computer (SBC)
- Figure 5-10— DS 32615-01, Scope for a Computer-Aided Cost-Estimating System (CA-CES)

These figures demonstrate some of the flexibility of system development specification usage. Remember, the DS is intended to lead the development of a system. It is *your* tool—use it!

Figure 5-10 includes a novel chart showing how the scope for the DS is intended to change as the system evolves from the development process to ultimate production and deployment. Remember, you are communicating with the system developers and users. Tell them what the system is about and how the DS relates to the system.

SPECIFICATION NO. | DS 24712-01 |

Part I

1.0 SCOPE

The Digital Propulsion Control Unit (DPCU) is an assemblage of hardware equipment and associated software programs that constitute a Controller System. This Controller System is designed to control an entire aircraft propulsion module—engine, inlet, afterburner (A/B), and nozzle—as a unit.

This inclusion of the DPCU with the input/output hardware on the engine, inlet, A/B, and nozzle constitute a system known as an Integrated propulsion Control System (IPCS). Figure 1.0-1 shows a block diagram of the IPCS with the included DPCU. Figure 1.0-2 lists the hardware and software that constitute the DPCU—(YG#1140A01) and the specifications that govern its design. Figure 1.0-3 shows IPCS installation in the F-111 aircraft to control the left engine. The hardware can be controlled by two software programs: "BOMDIG" (Bill of Material-Digital) and "IPCS". The BOMDIG program represents a digital version of propulsion control prior to IPCS. The IPCS program provides for more control modes, which analytical studies have indicated will give significant improvements in propulsion performance.

The modularized design of the DPCU is such that it may be considered a general-purpose Controller that may be utilized for controlling various systems. It is possible to reconfigure the interface unit (IFU) merely by changing the number and types of plug-in cards. The IFU is structured such that a number of cards can be considered "computer-peripherals" such as: analog-to-digital converter (A/D), digital-to-analog converter (D/A), frequency-to-digital converter (F/D), discrete buffered input channel (DIS-IN), discrete buffered output channel (DIS-OUT), analog-input multiplexer (MUX-IN), analog-output multiplexer (MUX-OUT). In addition, other building block functions consist of: output solenoid drivers, input analog buffer amplifiers, output analog output amplifiers, and custom designed I/O.

The Digital Computer (DCU) utilized for this system is the HDC-601C which contains 16K of core memory (expandable to 32K). The computer is a 16-bit machine with direct memory access and 16 priority interrupts.

The Computer Monitor Unit (CMU) is mounted in the flight deck and allows the pilot to engage or disengage the IPCS and monitor its operation.

The Power Supply Unit (PSU) converts and switches power to the IFU and DCU. It converts 3 phase 115V 400 Hz aircraft AC power to the DC voltages required by the DPCU and switches the aircraft DC power required by the DPCU.

Ground Support Equipment (GSE) is an integral part of this system for checkout, memory loading and simulation testing. The GSE consists of: Test Set Unit (TSU), computer Control Unit (CCU), a Teletype (ASR-35) and a High-Speed Paper Tape Punch and Reader (HSPT).

Detailed design information on the various hardwares may be found in the respective design specifications as listed in Figure 1.0-2.

The intent of this specification is to describe the requirements for the DPCU controller when it functions as a "BOMDIG" Controller System or as an "IPCS" Controller System.

PAGE I-4	REV
DS 24712-01	A

Fig. **5-8a.** Scope of IPCS

SPECIFICATION NO. DS 24712-01

Part I

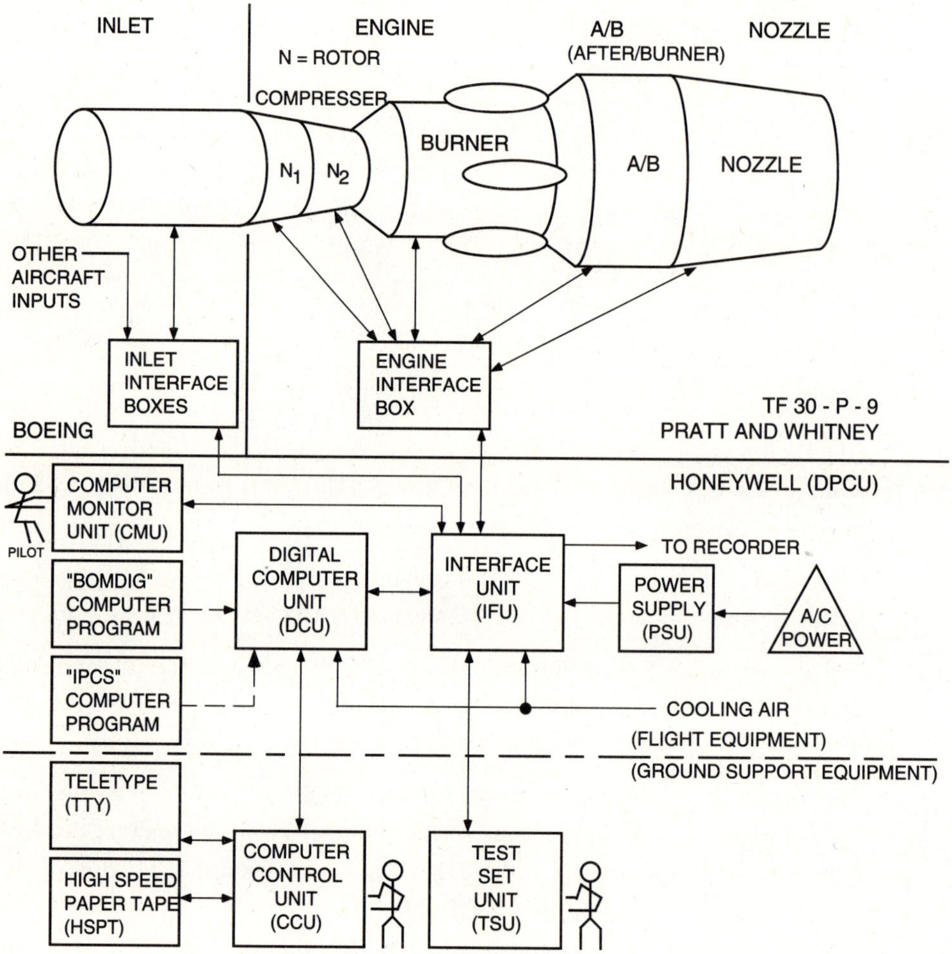

FIGURE 1.0-1 BLOCK DIAGRAM - IPCS

PAGE I-5	REV
DS 24712-01	A

Fig. 5-8b. Scope for IPCS (continued)

SPECIFICATION NO. DS 24712-01

Part I

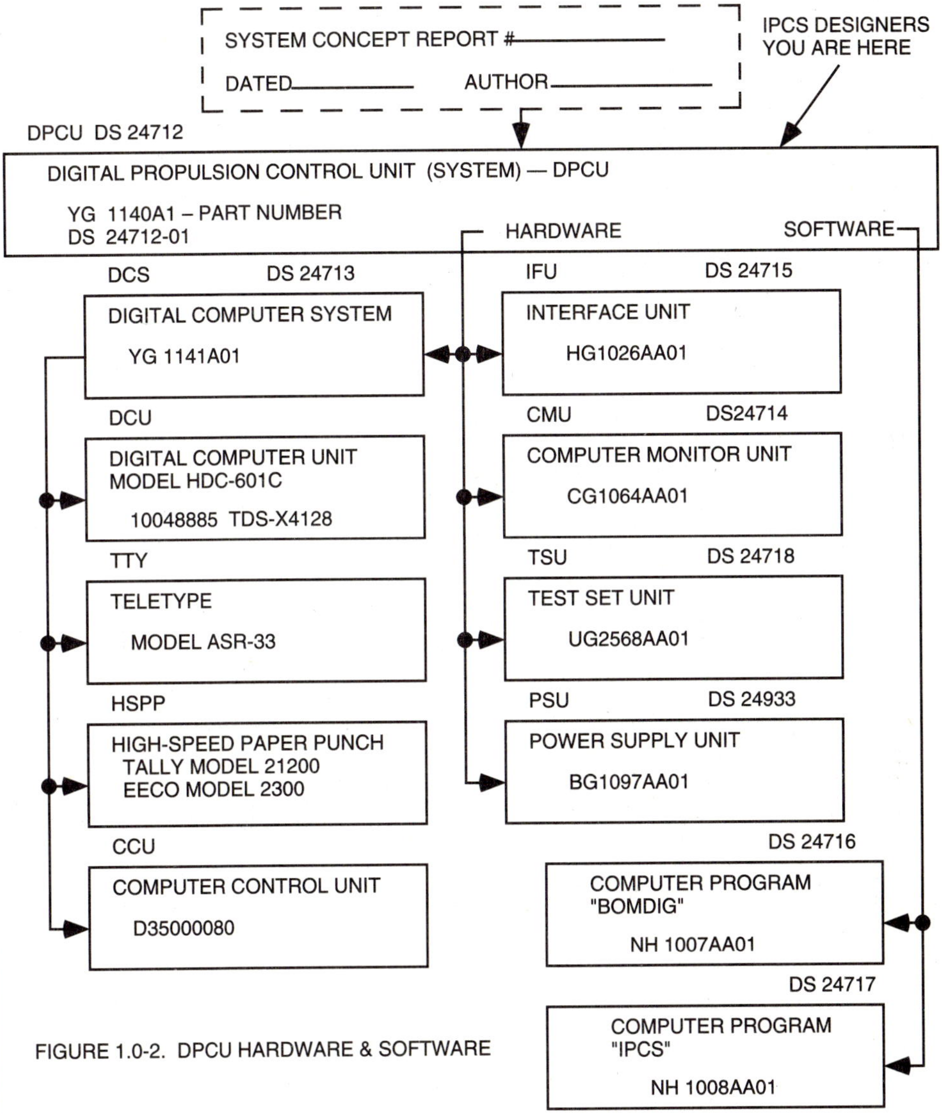

FIGURE 1.0-2. DPCU HARDWARE & SOFTWARE

(NOTE THE PARTIONING DOWN OF TOP-LEVEL REQUIREMENTS TO LOWER-LEVEL DEVELOPMENT SPECIFICATIONS)

PAGE I-6	REV
DS 24712-01	A

Fig. 5-8c. Scope of IPCS (continued)

SPECIFICATION NO. | DS 28991-01 |

Part I

1.0 SCOPE

This specification defines a Honeywell Digital Processor Board. This single board computer (SBC) shall conform to MIL-STD-1750A instruction set architecture (ISA). The SBC herein defined is initially specified for application to a flight control system. It will first be utilized in the MODEL 360 flight control demonstrator program.

The requirements provide a number of flexibility features such that the design is also usable for other system applications; however, production boards for other systems may require different packaging and minor implementation changes.

The design of this board is very dependent upon MIL-STD-1750A specifications and component 9450 specification; therefore, these should be referred to as an aid to understanding the SBC design. It was not deemed necessary to reproduce the requirements and characteristics of MIL-STD-1750A and the 9450 specification herein.

Note that a backup baselined design has been provided in the appendix of this DS. Preliminary CMOS chip set components are utilized in this backup design.

PAGE I-3	REV

Fig. **5-9.** Scope for SBC

SPECIFICATION NO. DS 32615-01

Part I

1.0 SCOPE

This specification defines and provides the requirements for a Software System "Computer Aided Cost Estimating System (CA-CES)". At the inception of the writing of this specification, cost estimating methods were uncontrolled, inaccurate, and provided poor basis for estimations. It was decided that a CA-CES should be developed to alleviate the problems associated with cost estimating. Therefore, the initial scope of this Part I specification is to provide a guide for the development of CA-CES. When the developed CA-CES is tested, the second scope (Part II) of this specification shall prove compliance to Part I with regard to accuracy, usability, and maintainability. The final scope for this specification is to establish a document as an aid for deployment of the system—(A User's Manual).

Scope of this specification:

Scope 1 — Guide for Development (Part I)

Scope 2 — Criteria for Compliance (Part II)

Scope 3 — A User's Manual (Updated Parts I and II)

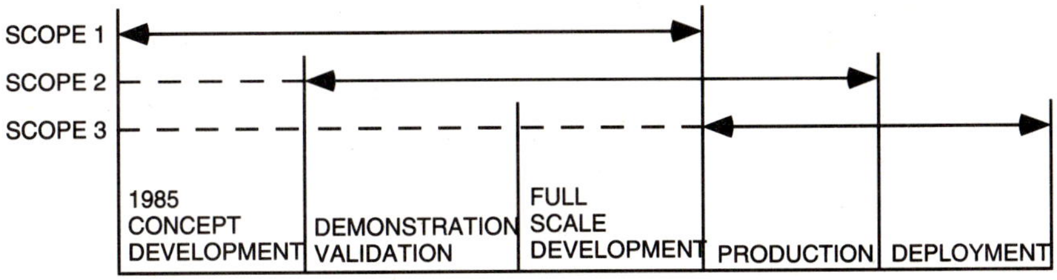

PAGE I-3	REV

Fig. 5-10. Scope for CA-CES—a computer-aided tool

DS 2.0 applicable documents

Provide a list of all pertinent documents that provide inputs and/or control the development activities to be performed. List customer, government, company, and other documents associated with the system. The objective of this listing is *to aid and expedite the developers of the system.* It is also appropriate to provide a simple sentence after a listed document to explain why this document is listed. (For example, you may say, "This report is the basis upon which this system is being developed and is therefore recommended reading by the developers of this system.") Documents will be listed under:

1. Government documents, required by the government contract to establish standards for the system to ensure safety, reliability, interface compatibility, quality, etc. (Mil-Specs, Standards, Drawings, and other).
2. Contractor documents, required by the prime contractor to ensure subsystem compatibility (Customers: procurement specification, interface control document (ICD), development specification (DS), reports and other).
3. Company documents, required by your company as part of your standards (quality standards, production standards, engineering standards, procedures, policies, etc.).
4. Reference documents, required as reference material to aid and expedite the development process (reports, papers, textbooks, prior programs, magazine articles, memos, and other).

To further help your system developers, it is a good idea to provide the actual DS paragraph number where each applicable document is applied. This is usually done at the last iteration just prior to the first formal release of your DS. Remember that the documents listed in applicable documents are a bibliography reference only. The associated specific real requirements are established in the section 3.0, "Requirements," 4.0, "Quality assurance," or 5.0, "Preparation for delivery" paragraphs.

2.X Examples of applicable documents

Refer to the following figures:

- Figure 5-11—Applicable Documents for IPCS
- Figure 5-12—Applicable Documents for SBC
- Figure 5-13—Applicable Documents for CA-CES

A review of these figures again iterates what I have said before: These figures demonstrate the flexibility of system development specification usage. Remember, the DS is intended to lead the development of a system. It is ***your*** tool—use it!

SPECIFICATION NO. DS 24712-01

Part I

2.0 APPLICABLE DOCUMENTS

The following documents, of exact issue shown, form a part of the specification to the extent specified herein.

2.1 Specifications

2.1.1 Specifications, Military

F33615-Q-2035	Statement of Work—Advanced Integrated Propulsion Control System
MIL-E-5400N	Electronic Equipment, Airborne General Specification
MIL-W-5088C	Wiring, Aircraft, Installation of
MIL-C-172C	Cases, Bases, Mounting; and Mounts, Vibration (For use with
8 December 1958 (Include Amendment 17 January 1970 and Interim Amendment No. 4 1 October 1971)	Electronic Equipment in Aircraft)

2.1.2 Specifications, NASA

NASA FRC Process Specification No. 21-1.

NASA FRC Process Specification No. 21-2.

2.1.3 Specifications, Boeing

D251-10006	Interface Control Document for the Digital

	Propulsion Control Unit (Integrated Propulsion Control System)
D180-15138-1	Integrated Propulsion Control System, November 22, 1972.
D180-10127-1	Quality Control Requirements for Boeing Suppliers, May 1972.
D251-10002	Reliability Program Plan for the Integrated Propulsion Control System
CG828081-9554	Purchase Contract, Integrated Propulsion Control System

2.2 Standards and Procedures

MIL-P-277733A	Procedures for Installation of Equipment in Aerospace Vehicles for Test Purposes

PAGE I-3	REV

Fig. 5-11. Applicable documents for IPCS

SPECIFICATION NO. DS 28991-01

Part I

2.0 APPLICABLE DOCUMENTS

The following documents, of the exact issue shown, form a part of this specification to the extent specified. In case of conflict, this specification has precedence.

2.1 Military Standards

MIL-STD-1750A	Military Standard 16-Bit Computer Instruction Set Architecture
(with Notice-1	
21 May 1982)	
MIL-STD-883B	Test Methods and Procedures for Microelectronics
MIL-STD-810C	Environmental Test Methods
(10 March 1975)	

2.2 Military Specifications

MIL-E-5400T	Electronic Equipment, Airborne, General Specification for
(16 November 1979)	
MIL-M-3;8510B	Microcircuits, General Specification for
MIL-P-55640	Printed Circuit Wiring Boards, Multilayer
MIL-C-55302B	Connector, Printed Circuit, Subassemblies and Accessories
(with Supplement	
1A, 16 August 1974)	

2.3 Honeywell Specifications

DS 28992-01	Real Time Support Unit (RTSU) for Honeywell MIL-STD-
(HG1082AA)	1750A Processors
DS 32027-01	Processor Control Unit (PCU)
(HG1083AA)	
ED-DGM-4000-910	Design Guidelines for Digital PWA Testability
(February 28, 1980)	

2.4 Other Specifications

9450 Specification	Fairchild
Preliminary CMOS 1750	(PACE)
Chip Set	

PAGE I-4	REV

Fig. 5-12. Applicable documents for SBC

SPECIFICATION NO. DS 32615-01

Part I

2.0 APPLICABLE DOCUMENTS

The following documents, of exact issue shown, form a part of this specification to the extent specified herein.

2.1 Military Standards (MIL-STDs)

MIL-STD-490	Specification Practices—30 October 1968 (this MIL-STD provides the basic format for this system specification)
MIL-STD-881	Work Breakdown Structures (WBS)

2.2 Honeywell Reference Material—(Memos)

a) 1984 Office & Technical (O&T) Productivity Improvement Projects "Cost Estimating Methods" (Military Avionics Division (MAvD) MAvD-7 Project,—Don Stevenson/Jack Hunger—February 1984.

b) Flight Management & Targeting Systems (FM&TS) Offsite Meeting Report—Productivity Improvement Projects—Don Stevenson/Jack Hunger—February 1984.

c) Parametric Pricing Data—A review of Honeywell experience with the RCA Price (Programmed Review of Information for Costing and Evaluation)—R. O. Niemann—January 1984.

d) Cost Estimating Background Discussion—Jack Hunger - November 1983.

e) CURVESAM—Lee Robertshaw - 1980.

f) Estimated Cost of Trip—Jack Hunger - November 1979.

g) DELTASAM program—P. R. Smith - 1979.

h) Market/Business Model—P. R. Smith - 1976.

i) Cost Modeling Briefing—P. R. Smith - 1974.

PAGE I-4	REV

Fig. **5-13.** Applicable documents for CA-CES

DS 3.0 Requirements

Your DS's title page, table of contents, sections 1.0 "Scope" (System perspective), and 2.0 "Applicable documents" provide an introduction to your DSs. You will now work on the most important section, "Requirements." Your baseline system definition skills developed in SEE-4 will provide subsection 3.1 "Description" of our system, and subsequent subsection 3.2 and on. Requirements will be derived from your mission analyses skills developed in SEE-2.

DS 3.1 System definition

System definition is basically the baseline developed in SEE-4. You will use the following as required in order to completely define your system. The items below are mandatory. Leave one of these out and your system is not properly baselined:

1. Mission time line
2. Functional block diagrams appropriate to your specific system
3. Interface diagram
4. Specification tree
5. Mechanical and installation sketches
6. Attributes table
7. Bill of material (or major components)

You are expected not only to provide the diagrams, figures, tables, and sketches to baseline your system, but also to provide a brief written description of each figure and table. The data provided will be consistent throughout all the figures. The block diagrams, interface diagrams, specification tree, mechanical sketches, installation drawing, attributes table, and bill of material all must show the same terminology and partitioning of the system.

Ideally the purpose and description of each of the seven (7) baseline items are as follows:

1. Mission Time Line: This figure is to provide a simple top-level time line of the basic sortie mission for the major system above this subsystem under development. This must also show that the subsystem time line is a subset of the major system time line. The designer and the user both need to know the mission chronological sequence of events and their relationship to the major system. The operational modes of your subsystem can also be shown on your subsystem time line.

2. Functional Block Diagrams: These diagrams evolve from your functional analyses. What are the hardware actions to be performed? What are the software actions to be performed? How do the hardware elements interface with each other? How do the software functional elements interface with each other? How do they relate to the mission time line? To each other? The detailed functional block diagram works to lower

level and provides additional functional partitioning. In some systems you may also provide a functional block diagram of the actions of persons to completely baseline your system.

3. Interface Diagram: This diagram will bound your system. It clarifies at a top level what is in your system, and what is external. If remote sensors and actuators are in your system clearly show it with this diagram. The cables to the sensors and actuators are then part of your system.

4. Specification Tree: The specification tree is a must. It should show from where the top-level requirements are coming and should partition out the system requirements. You should keep it simple. Usually a top-system DS, a hardware DS, and subcontracted hardware DS, a software DS, and a tester DS will cover most systems. Use a large arrow indicating "you are here" to identify the specific DS being read. Also, repeat the same specification tree figure in all the associated lower-level DSs with the "you are here" arrow in the appropriate place.

5. Mechanical Sketches: Provide a three-dimensional pictorial of the top-level hardware elements that are part of the system. These sketches will have estimated dimensions, center of gravity, and weight baselined.

   An installation drawing should be part of the mechanical sketches. Where will the hardware be installed—a building, a ship, a plane or a car? Provide a sketch showing expected location and mounting details.

6. Attributes Table: You do "precision guess work." The volume, weight, power, cost, and reliability mean time between failures (MTBF) will be estimated for each of the system functional elements as applicable. It is common to put a ± tolerance on your estimates so variance and confidence levels may be tracked as the system development process generates more and better data. Two other key attributes, which also could be analyzed are "mission effectivity" (how accurate? how consistently? how fast?) and "user acceptability" pilot rating of the control system's handling, user's consideration of the friendliness of the system, etc.

7. Bill of Material: A listing of the key components of the system. This helps the procurement and production people and unit product cost analysts do their planning and analyses.

Example system definitions

In chapter SEE-4 you provided a baseline for the Total Avionics Control System (TACS), to which you should refer to help you understand what constitutes system definition. In figure 5-14 I have included some system definition excerpts from the CA-CES DS 32615-01 to give you an example of a process system definition.

SPECIFICATION NO. DS 32615-01

Part I

3.0 SYSTEM REQUIREMENTS

3.1 SYSTEM DEFINITION

This Cost Estimating System Specification defines and provides the requirements for a Computer-Aided Cost Estimating System (CA-CES). This system utilizes generic algebraic equations for standardized work breakdown structure (WBS) elements. The parameters for these equations within each WBS element are selected to relate work element tasks to specific contract requirements. The magnitude of the parameters are accessed by the development of a series of questions for determining the magnitude of the basic parameters (parametric questions), questions are asked to select the appropriate contract type equations (classifier questions), and questions are asked to establish risk factors (modifier questions).

The CA-CES is primarily directed at yielding costs which correlate the WBS elements which are for the most part identified with items in the contract. However, there are usually some costs, such as those for marketing proposal costs, which are desired but which do not accrue towards the contract. The CA-CES WBS element methodology provides for estimating these non-contract costs also.

A function diagram of the CA-CES is provided in Figure 3-1. This diagram shows the following functions:

Function 1: Request for Proposal (RFP) is received and studied.

Function 2: Preliminary design activity is performed, this will require generation of preliminary Program Management Plan (PMP), WBS, and system Development Specification (DS) by Program Manager and Systems Engineer.

Function 3: WBS elements are selected; classifier, parametric and modifier questions are answered resulting in estimated cost/schedule/manloading/critical path data. This data is based on identified previous program actuals (calibration file record).

Function 4: This output data is provided for cost account managers review and risk analysis.

Function 5: Iterate as necessary (go back to step 1, 2, 3 as required).

PAGE I-6	REV

Fig. **5-14a.** CA-CES system definition

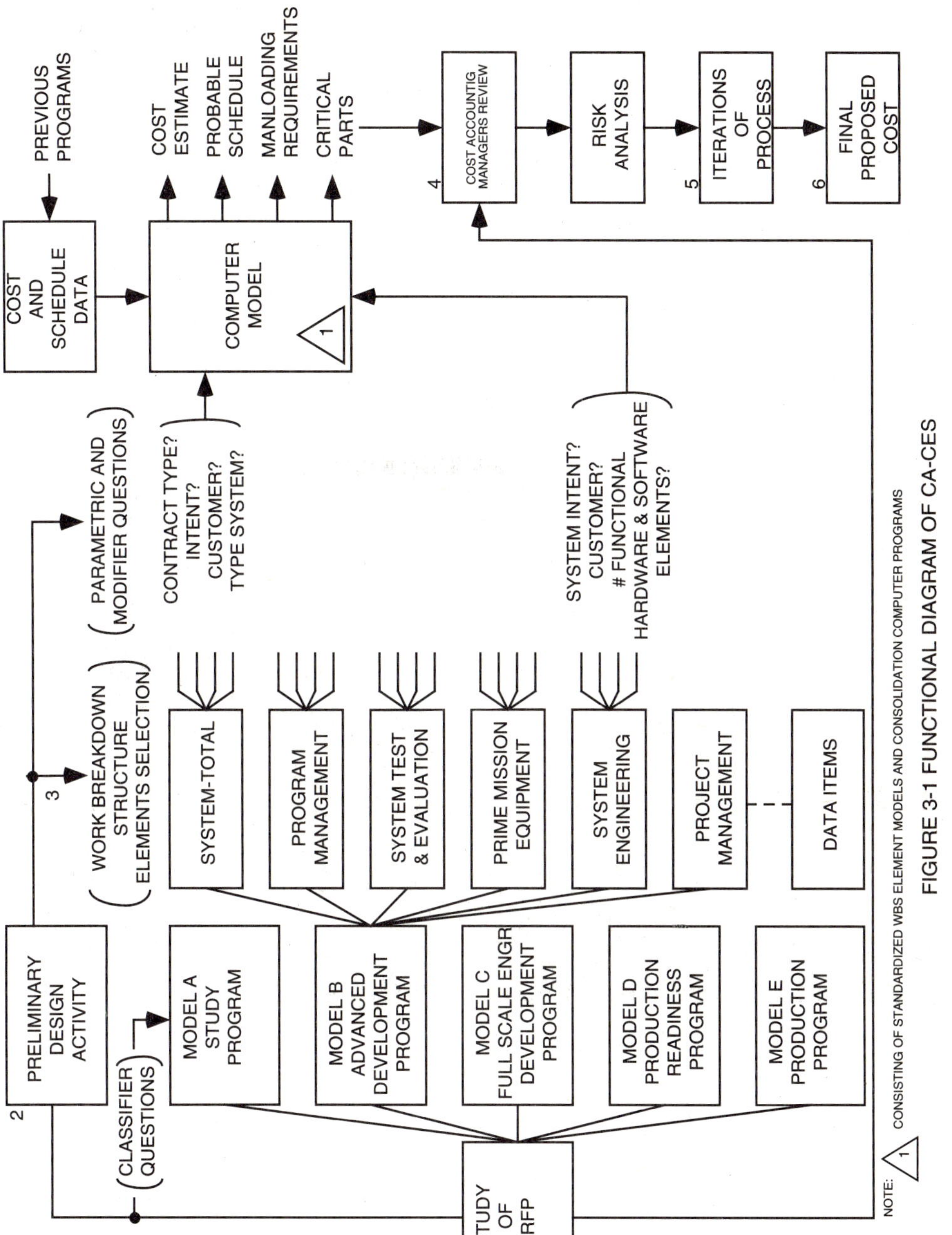

Fig. **5-14b.** CA-CES system definition

DS 3.2 Specific requirements

The specific requirements section of a development specification will vary depending on what is to be developed, customer contractual requirements, and, of course, sortie/life mission analyses.

In developing this section regarding specific requirements, I performed an analysis to determine a fair sampling of all the possible items that may be specified in the requirements section of DSs. First, I brainstormed, and then collected all the items into a hierarchial format. My brainstorming included a listing of all the requirements mentioned in Mil-Std-490, system types, the system perspective cube, development time phases, and futuristic studies. The resulting brainstormed list was collected and put into a hierarchial format that includes requirements in the following twelve areas:

- 3.2 Specific requirements
- 3.2.1 Performance requirements
- 3.2.2 Interface requirements
- 3.2.3 Mechanical requirements
- 3.2.4 Environmental requirements
- 3.2.5 Hardware design requirements
- 3.2.6 Software design requirements
- 3.2.7 Reliability, maintainability and safety requirements
- 3.2.8 Testability requirements
- 3.2.9 Documentation requirements
- 3.2.10 Producibility requirements
- 3.2.11 Installation and logistics requirements
- 3.2.12 Other requirements

You will note that the twelve specific requirement areas follow logical, developmental chronological sequence from 1-performance (mission concept) to 11-installation and logistics (deployed). These twelve areas reflect a hardware system. Your specific requirements may require entirely different areas, especially if your system is a process or procedure type of system.

When you are writing specific requirements always refer to and reference your 3.1 baseline figures. Remember your specific requirements are all based on your baseline, especially your mission time line.

All specific requirement items will be evaluated for each specific system for applicability as an external to the system requirement and also as an internal to the system requirement! "External" relates to the system's interfaces when viewed as a subsystem to a major system, while "internal" relates to the interfaces of the components that make up your subsystem. Electromagnetic interference (EMI) can be specified as an environmental requirement. The system must operate in (external). It can also be specified as well as a requirement that the system must not affect internal components of itself through internally generated EMI (power supply noise, for example).

All specific requirements will be very terse, concise, and will use the word "shall." Shall is a strong statement of a requirement. You shall meet the requirement without any deviations. Do not provide wordy long requirements. You are communicating with the individual developers of the system, so there must not be any confusion over what the requirement is.

DS 3.2.1 Performance requirements

DS 3.2.1.1 External. Requirements in the area of functional performance deal with both external and internal requirements. Externally, how must the system appear to the outside world? What must it do and how well must it do it? Figure 5-15 indicates that one of the objectives of establishing external performance requirements is to put bounds (limits) on your system.

Performance requirements must bound your system within its mission. I have provided some examples of performance requirements that will bound your system. In example 1 I discuss aircraft crashing into mountains, hills, and buildings during poor visibility. Example 2 bounds the environmental requirements to the weather conditions worldwide between ± 45° latitude. Examples 3 and 4 bound systems to limited organization mission applications.

EXAMPLE 1. A FLIGHT ENVELOPE FOR AN AIRCRAFT

This system is to fly in all weather from 5 to 5000 feet at mach 1 speed. A down looking radar has a 45° pattern-cone. This system has a bounding problem as it is defined. It will cause the aircraft to crash into a cliff at 30' with maximum pull-up after cliff detection from initial flight conditions of: mach 1, 5 feet altitude, and on a dark night. This system needs to include a forward looking radar, so its performance can be bounded to prevent such an accident. The problem was solved by requiring the radar to multiplex between down looking and forward looking antennas and solve a control algorithm that combined radar data and aircraft velocity to provide pilot warning and/or override control. A less desirable bounding solution to this problem could have been to bound the mission capability by indicating that the minimum allowable altitude the aircraft may fly is a function of aircraft speed and specific terrain. The slower the plane flies, the closer it can be to the specific minimum terrain altitude.

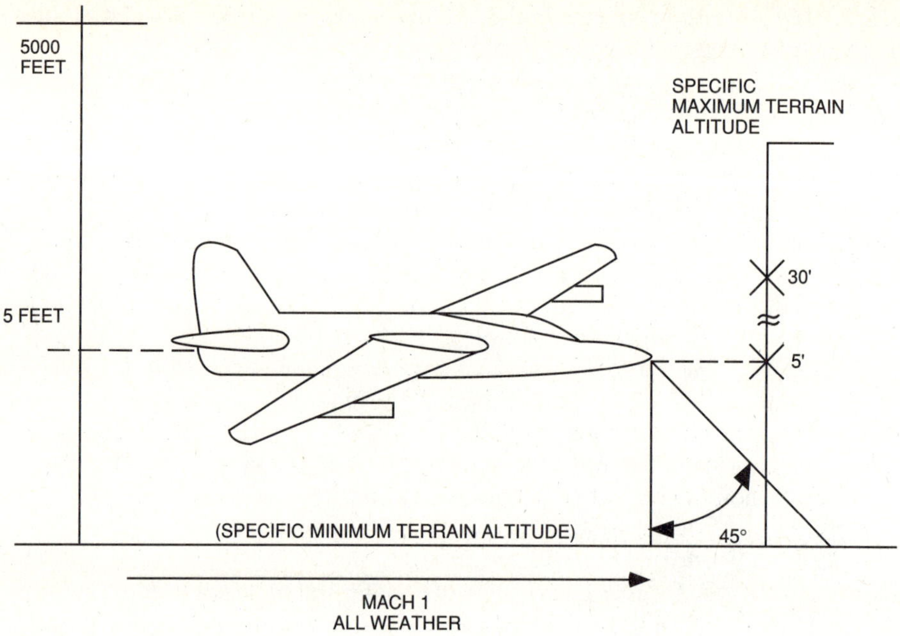

EXAMPLE 2..BOUND YOUR SYSTEM TO A RANGE OF MISSIONS: (PEOPLE MOVING IN *ALL WEATHER BETWEEN ± 45° GLOBAL LATITUDE*), THIS IS BOUNDED BY ± 45°.

EXAMPLE 3..BOUND YOUR SYSTEM TO A SPECIFIC ORGANIZATION: (ENGINEERING OPERATING PROCEDURES—EOPS—*FOR THE MILITARY AVIONICS DIVISION*), BOUNDED BY A SPECIFIC DIVISION.

EXAMPLE 4..BOUND YOUR SYSTEM TO A SPECIFIC APPLICATION TYPE: (COMPUTER-AIDED COST-ESTIMATING SYSTEM *FOR MILITARY PROGRAMS ONLY*), BOUNDED BY MILITARY ONLY.

Fig. **5-15.** Bound your system

Possible External Performance Requirement Items

The following list of possible external performance requirements shows the type of items that could be specified and some of the possible specific requirements (or typical bases for a requirement).

Basic Performance

- Range: Temperatures shall be measured between -40° F to +160°F.
- Rate: Blood pressure shall be measured once a second.
- Accuracy: Temperature shall be measured with an accuracy ±1°F.

Now in like manner you can use "shall" to establish specific requirements in the following areas:

- Velocities (miles per hour)
- Time (nanoseconds to years)
- Mission duration (time)
- Mission effectiveness (percent successes)
- Reaction time (time)
- Stability (overtime)
- Functional (perform given algorithm)
- Operational (percent mission ready)
- Latency (time delays allowed)
- Memory capacity (three little words)
- Skills (developed)
- Cost (design to meet a specific life cycle cost)

Specific System Performance

- Procedural system (time/cost)
- Quality control system (SQR—system quality rating provide criteria)
- Monitor system (sensitivity)
- Training system (in three days)
- Medical system (percent successes)
- Processing system (time/cost)
- Production system (scrap/yield)
- Test system (percent coverage)
- Activities (system usage)

Control Performance

- Control diagrams—(provide transfer functions)
- Control mode (provide list and diagrams)
- Sensor capability (accuracy/range)
- Actuators capability (accuracy/range)
- Dynamic actions (response time)

- Engine control (thrust capability)
- Flight control (flight limits)
- Navigation control (±miles per hour)
- Missile control (target kill rate)
- Remote-piloted-vehicle control (range)
- Security control (number of false alarms)
- Robotics (dynamics, range, accuracy)
- Traffic control (rate, throughput)
- Fire control (accuracy)
- Environmental control (temperature)
- Rocket control (range)
- Target acquisition (sensitivity)

Specific Device Performance

- Radar altimeter (range/accuracy)
- Helmet sight (field of view)
- Power supply (capacity, efficiency)
- Display (readability)

These are examples of overall external mission required performance requirements (randomly selected from a number of system types):

1. Temperature Control: The temperature shall be controlled at the monitored set-point to within ±1°F of the set temperature over an operating range of -40°F to +80°F.
2. Cruise Control: The cruise control shall control vehicle speed within ±3 MPH of the set-point, over a cruise control speed setting range of 40 MPH to 80 MPH.
3. Mission Success: The probability of mission success shall be greater than 80%. Mission success is defined as . . . (etc.).
4. Attitude Accuracy: The navigation system shall provide an underwater remote piloted vehicle (RPV) attitude accuracy of ±2.5° in both the elevation and azimuth directions.
5. Propulsion Performance: The Remote Piloted Vehicle (RPV) propulsion system shall provide:
    a) Thrusters that supply 150 lbs. of thrust for each thruster within a 45° cone around their mounting center-line

b) Velocity of 45 knots in any of the three orthogonal directions: up—down, forward—reverse, port—starboard

c) Acceleration of 5 knots per second in any of the orthogonal directions

d) An efficiency of 65% or higher while operating over the environment range specified in 3.2.4.

6. Navigation Position Accuracy: The navigation system shall provide the RPV with a ± 2 feet position accuracy, in all three orthogonal directions, relative to the ocean bottom. This requirement applies to depths of up to 1000 feet.

7. Altitude Hold Accuracy: The flight control computer (FCC) shall control aircraft altitude to within ± 50 feet when in altitude hold mode, with a minimum altitude hold mode engagement of 70 feet.

DS 3.2.1.2 Internal performance requirements. The systems engineer has developed a baseline system definition and presented it in subsection 3.1. Part of that definition includes the partitioning of the system into modules and the apportionment of requirements. In subsection 3.2, these translate into internal performance requirements. For example, to provide the ± 1.0% accuracy required as an external performance requirement, you must apportion this ± 1.0% among your applicable internal modules that contribute to the ± 1.0% accuracy. You may apportion the requirements in this way: The input multiplexer shall be required to provide signal translation within ± 0.10%; the A/D converter shall be ± 0.50%; the software algorithm roundoff error shall be less than ± 0.05%; the output display shall be accurate to ± 0.20%; and the test measurement to confirm your overall external accuracy shall be ± 0.1%. These are specific internal requirements that relate to a specific external performance requirement.

The following list represents the type of internal performance items that could be specified and some of the possible specific requirements (or typical bases for a requirement).

Possible Internal Performance Requirement Items

Possible internal performance requirements also can include the prior external requirements because every so-called system is really only a subsystem.

Internal basic performance

- Accuracy (± whatever)
- Algorithms (sensor signal conversion)
- Process—(input data all converted in 1 ms)
- Concepts—(self-assessment skill surveying shall be used)
- Hierarchy—(noncritical software shall be done in background)

- Gain—(10 ±1 DB)
- Time to convert (± 1 MS max)
- Sequencing (the following order)
- Frequency response—(2000 to 4000 Hz)
- Time—(less than 5 MS)
- Rate—(a multirate synchronous algorithm shall be developed)
- Executive (software mode controller)
- Range—(100 miles minimum)
- Formula—($A = B + C - D + E^2$)
- Models (transfer functions)

Internal specific device performance

- Logic diagrams (as subsets of four system block diagrams)
- Boolean equations—($A = (B \bullet C) + D$)
- Interrupts—(64 maskable shall be provided)
- Analog to digital—(11 BIT + sign)
- Digital to analog—(8 BIT + sign)
- Frequency to digital—(10 to 500 Hz = 1 MV to 500 MV)
- Analog to Frequency—(1 MV to 500 MV = 10 to 500 Hz)
- Sample and hold—(1 micro sec sample/1 milli sec hold)
- Digital to frequency—(3 MSB = 8 ranges, 5 LSB = 32 frequency ranges)
- Light coupling—(sense 0.1 lumens/sq/ft)
- Power conversion—(convert 60 Hz 115 VAC to 5 VDC with 10-watt capability)
- Power distribution—(provide charge distribution to minimize internal transients to less than ±100 ma)
- Form, fit, function replacement (shall be a direct replacement to model X)

Internal Specific Subsystem Performance

- Digital control—(shall use clock upon demand design; crystals shall not be used)
- Block diagrams—(all subassemblies shall be defined with block diagrams)

- Subsystem performance (accuracy)
- Control (internal) (software mode)
- Communications (internal) (bus controller)
- Organization (shall include single point of responsibility and monitor recorded feedback)
- Redundancy—(three channels with crossfeeds)
- BITE, BIT (built-in test equipment, built-in test software)
- Automatic control—(shall use an autonomous controller for input data conversion)
- Hybrid control—(shall include both analog and digital computers)
- Procedural sequence (shall follow this procedure)
- Functional allocation (flow down of a function to lower levels)
- Audio coupling
- Video coupling

DS 3.2.2 Interface requirements

The system definition of subsection 3.1 includes interface drawings. These show the various interfaces in general terms. You must now provide specific detailed requirements for the interfaces.

Interfaces to be defined may be electrical, mechanical, radio frequency, human, explosive, light, hydraulic, verbal, procedural, and other. All interfaces are dynamic. They also have a transmitter, a receiver, and a medium. Even the seemingly static mechanical mounting interface is dynamic because the heat source transmits, the heat sink receives, and the medium is the mounting surfaces.

The specification of interfaces and requirements thereof is of extreme importance because a casual approach to establishing interface requirements has time after time been the source of problems during system integration testing. Remember to refer to your baseline interface drawing in section 3.1. If, for example, you established five interfaces in your system definition interface diagram in section 3.1 such as 1) Power, 2) Sensors, 3) Actuators, 4) Human, 5) Test, you must now provide specific interface information and requirements for each of the five areas. Sections 3.2.2.1 through 3.2.2.5 would provide specific detailed interface data and requirements such as:

DS 3.2.2.2 sensor interfaces.

1. Pressure—A remote pressure sensor to sense outside air pressure shall be provided as part of the system.

    a. range: 1.0 to 40.0 psi

b. accuracy: ±10% of actual psi

c. cable length: 6 ± 2 FT, (2 wire shielded with characteristic impedance of 75 ohms)

d. interface type: DC bridge or digital

e. load: 1K ohm ± 10% resistive

f. cost: less than $10.00 including cable

g. weight: less than 0.25 LB including cable

h. environment: outside all world-wide weather

i. MTBF: 800,000 hours including cable and connector

Your specific interface requirements must establish source impedance $Z = (R + jX) \pm xx\%$, load impedance $Z = (R + jX) \pm xx\%$, cable type, cable length, and input voltage range. You need to know resistance, capacitance, inductance, voltage level, and frequency. The following list of interface items reflects the type of items usually specified in interface requirements. You also show some of the possible specific requirements (or typical bases for a requirement).

Possible Interface Requirement Items
Basic

- Impedance (R, L, C, f)
- Load impedance (R, L, C, f)
- Source impedance (R, L, C, f)
- Cable Length—(shall be less than 100 ft)
- Cable Characteristics—(coaxial, Z_o = 50 ohms)
- Timing (provide timing diagrams of the interface)
- Voltage levels—(+5 VDC ± 0.5 V)
- Power levels—(100 W ± 10 W)
- Transients—(±200 VDC)
- Cable list (provide a diagram and a list of cables)
- Outputs list (list all the system outputs and characteristics)
- Inputs list (as output list)
- Wire lists (list all the wires and characteristics)
- Interface control document (may be required for subsystem subcontractors)
- Signal lists (same as input and output lists)

- Power profile (provide time line in conjunction with the mission sortie time line)
- Grounding diagram
- Waveforms (provide for critical interface circuits both analog and digital)
- Interface diagrams (provide equivalent circuits showing distributed R, L, C parasitics)

Types

- Power—(per MIL-STD-704)
- Communication—(per MIL-STD-1553)
- Hardware (driving motor loads of R, L, C, f)
- Software (must be compatible with a standard or the bigger system above software)
- Human (people-print or human factors)
- Display (provide layout sketch)
- Mechanical (show details)
- Sensor (provide signal)
- Actuator (provide load model)
- Hydraulic (response)
- Pneumatic (provide response time over environmental range)
- Electrical (define)
- Electronic (define)
- Busses (define)
- People (how do different departments interface?)

DS 3.2.3 Mechanical requirements

You provided a sketch in your baseline definition, but now you must expand on the sketch by providing more specific requirements for the mechanical characteristics of your system. Remember to reference back to your baseline in Section 3.1. Your specification may read like the following example DS 3.2.3 Mechanical requirements.

Example DS 3.2.3 mechanical requirement

The baseline mechanical configuration is shown in figure 3.1-5 and figure 3.1-6 provides the interface for installation requirements. This system shall be designed to meet the mounting requirements established by figure 3.1-6, which are derived from the intent of figure 3.2.2-8 of the customer's procurement specification #X-007 dated 5–5–75.

Example DS 3.2.3.1 specific mechanical box requirements

1. Weigh less 5 LB.
2. Design to meet the environmental requirements of 3.2.4
3. Dim. per figure procurement spec #X-007 figure 3.2.2-7

Specific mechanical requirements for internal and external items such as the following must be established. Some of the items show some possible specific requirements (or typical bases for a requirement).

Possible Mechanical Requirement Items
Basic

- Weight—(shall weigh less than 10 LBS)
- Dimensions—(shall be 5" x 5" x 10")
- Volume—(less than 250 cu. in.)
- Center of gravity—(shall be within ±3 in of the dimensional center)
- Thermal—(shall not exceed 100 BTUs under all environmental conditions)
- Structural (shock and vibration)
- Materials (shall not use toxic, etc.)
- Parts (only approved list of parts)
- Mounting devices—(shall require no tools)
- Modular construction
- Maintenance access
- Fasteners (specify types required)
- Dynamics
- Movements
- Shipping
- Production limits—(shall not exceed present production handling equipment limits)
- Installation—(shall have no sharp corners)

Types

- Robotic (shall have movement range of . . .)
- Hydraulic—(shall not exceed 125 psi pressure)
- Electronic—(thermal profile shall be less than 10°C at any interface over all operating conditions)

- Crew space—(shall be able to accommodate 160 ± 40 LBS 5'8" ± 3" human operator)
- Operator station layout (a sketch)

DS 3.2.4 Environmental requirements

Your mission analyses resulted in an understanding of the environments to which your system will be exposed in production, shipping, and in usage. You will now take that information to establish specific environmental requirements that your system will be designed to withstand. In the past, some systems have been overdeveloped by insufficient study of what the real environment will be. Even worse, some have been developed with insufficient environmental capability. We can all think of mission impossible examples: designed for -65°C operational capability for a system to be permanently installed in the Sahara Desert; gun fire shock capability for equipment mounted on an aircraft without a gun; or, in the area of insufficient environmental requirements, a solid-rocket O-ring on a reusable-rocket (which becomes bent, corroded, twisted, and distorted) that can be exposed to climatic conditions outside the environmental design limits—Space Shuttle.

The establishment of environmental requirements has a direct impact on cost and risk. The importance of revisiting again and again your *mission analyses* should now be obvious.

Example DS 3.2.4 Environmental requirements

This system shall be designed to perform all its mission requirements when exposed to the following environmental requirements.

1. Operating temperature -40°F to +120°F
2. Temperature shock -60°F to +80°F in 15 minutes
3. Nonoperating temperature -60°F to +160°F
4. Shock as stated in procurement spec paragraph X-007 3.2.6.1
5. Vibration as stated in procurement spec 3.2.6.2 PC-X-007 (Ref MIL 810D)

A long list of the type of items to be included in environmental requirements follow. Always remember to refer to your baseline mission time lines (both sortie and life) and reference your section 3.1 baseline. I have provided some examples of specific requirements (or typical bases for a requirement).

Possible Environmental Requirement Items
Some of these may be internal as well as external requirements:
Basic

- Temperature—(shall operate from ambient temp -40°F to +160°F)
- Vibration (provide frequency vs. level curve)

- Shock—(less than 10 gs in defined X, Y, Z axes)
- Electromagnetic interference (per standard requirements)
- Humidity—(operate from 10% to 98% relative humidity)
- Climate—(world-wide ± 45° lat.)
- Hazards (shall not cause)
- Storage environment (-60° to +200° F)
- Transportation environment (common carrier)
- Production environment (standard room)
- Mounting orientation (can be mounted in any orientation without functional degrade)

Mission Specific

- Shelf life—(10 years in all world environment)
- Snow (cannot be lost in snow drifts)
- Wind—(80 mph max)
- Rain (must be water proofed)
- Salt spray (per standard requirement)
- Sand (per standard)
- Dust (per standard)
- Nuclear (per standard)
- Chemical (per standard OSHA)
- Toxic products (per standard safety requirements)
- Space environment (per NASA standard)
- Explosive (shall continue to operate during . . .)
- Noise—(shall not emit audio greater than 1 db from 20 to 20,000 Hz)
- Magnetic (field level)
- Overpressure (psi)
- Underwater (depth)
- Altitude—(shall operate from sea level to 20,000 feet)
- Blast (level)
- Radiation (level)
- Underwater explosions (as explosive)

DS 3.2.5 hardware design requirements

The requirements for hardware in the top system DS must establish a functional description of each of the hardware units that constitute the hardware system. If, for example, your system is made up of a digital controller unit, an interface unit, an operators panel unit, and a tester unit, each of these will be defined and the detailed requirements established in their specific DSs. You have in section 3.1 of your top system specification a specification tree that establishes the hardware development specifications.

You must, however, establish top-level hardware requirements in your top system development specification such as hardware functional units description, cables and interfaces, partitioning of functions, filters, power supplies, analog to digital converters, digital to analog converters, other signal conversion functions, overall input-output functions to perform requirements, and specific general requirements. If one of your hardware units is to be purchased from a major subcontractor you must provide a procurement type development specification, which must be negotiated with potential suppliers. The top-level system DS organizes and defines the functional hardware system and its top requirements, while the specific lower-level hardware DSs are used to lead the development or the purchase of specific hardware units.

Example DS 3.2.5 hardware design requirements

Figure 3.1-2A provides a functional block diagram of the hardware functional elements of your system. The following provide the specific design-to requirements for the hardware. (Note: on large multiple hardware unit type systems, you would provide a reference to the 3.1 spec tree to show how the hardware requirements are flowed down, similar to the software requirements methodology in 3.2.6.)

Example DS 3.2.5.1 hardware common requirements

1. Only parts established in the approved standard parts manual shall be utilized in the hardware design unless a nonstandard-part approval has previously been established with the customer per PC-X-007 3.4.1.
2. Parts derating shall be per Honeywell standard criteria as established in Engineering Operating Procedures System–DS33107-01.
3. Software programming rules shall be provided in the specific hardware specifications.

Example DS 3.2.5.2 hardware functional element specific requirements

Example DS 3.2.5.2.1 the analog-to-digital convertor shall be designed to provide:

a) Range ± 10 VDC

b) 12 Bit (11 plus sign)

c) 64 input channel differential multiplexer

d) Cost less than $200.00

e) MTBF greater than 200,000 hours

f) Accuracy ± 1 LSB

g) Fully monotonic over the full environmental range

h) Conversion time less than 25 microseconds (including multiplexer settle time)

Example DS 3.2.5.2.2 programming data

This requirement area is established to record for software designers information on the "as designed" hardware for software programming, address, word format, response time, etc.

Possible Hardware Requirement Items for the System DS:

- Specific hardware unit baselined in greater detail than in section 3.1
- Architecture (provide a diagram)
- System cabling (provide a diagram)
- Transfer functions—hardware (provide equations)
- Standards (parts, subassemblies, design)
- Computer-aided design requirements—(all circuit designs shall be in the XYZ-CAD system)
- Testability requirements (provide requirements for built-in test equipment (BITE))
- Maintainability requirements (provide requirements for hardware partitioning)
- Gains—(Eo/Ei = 10 ± 1)
- Calibration requirements—(all calibration shall be accomplished by software and internal hardware precision references)
- Power requirements—(10 watts maximum)
- Accuracy—(±1.0% of full-scale)
- Throughput—(200 MEGA BYTE per second)
- Memory size—(1 MEGA WORD with expansion to 64 MEGA WORD and maximum access time of 50 nanoseconds)

DS 3.2.6 Software design requirements

The requirements for software are usually provided in a separate DS as shown in the specification tree examples, but the top system DS must establish the top-level software requirements such as software functional elements, partitioning, algorithms, high-order language, software DS, and specific general requirements. The separate software DS will not be covered in detail herein. It will suffice to state that the software DS is used to lead the software development process.

Example DS 3.2.6 software design requirements

(In your 3.1 System Definition Figure 3.1-4 Specification Tree) you established the requirement for a software development specification, which will provide the detailed flow down of specific software design-to requirements. Figure 3.1-2B provides the functional software modules required for this system. You also established the system mission time lines in figure 3.1-1, the baseline information established in the software functional block diagram, the mission time lines programming data from the "as designed" hardware, and the customers PCX-007 section 3.6.6 software requirements. These provide the basis upon which the 3.1-4 specification tree software specification shall be developed.

Example DS 3.2.6.1 software common requirements

1. The computer has been baselined as an Intel 8086–16 bit.
2. The software may be programmed in machine language.
3. The software shall be designed for compatibility with the hardware as per the programming data specified in 3.2.5.

Possible Software Requirement Items for the System DS Are:

- Flow diagrams (provide diagrams)
- Database (a specification or how one is to be established)
- Partitioning (provide diagram)
- High-order language—(all programming shall be in "C")
- Software test requirement (provide built-in test (BIT) requirements)
- Artificial intelligence (use real intelligence instead)
- Executive structure (provide diagram)
- Software functional elements (provide "action function diagram"—input/action/output, feedback)
- Algorithms (provide equations)
- Logic diagrams (provide diagrams)
- Software data flow diagrams (provide diagrams)
- Software standards—(DoD-STD-2167)
- Modularization (provide diagram)

- Software configuration management requirements (provide process flow diagram)
- Mode diagrams (provide diagram)
- State diagrams (provide diagram)

DS 3.2.7 Reliability, maintainability, and safety requirements

Mission analysis walk-throughs include a study of the impact of system failures and the resulting effect. These walk-throughs lead you to establish requirements to eliminate unsafe conditions, to improve reliability and maintainability. These requirements may be similar to the following example.

Example DS 3.2.7 reliability, maintainability and safety requirements

See the Customer's Procurement Specification PC-X-007-3.2.9. Also provide full compliance to DS33107-01 Engineering Operating Procedures System Company Standard R, M&SS design and analysis process requirements.

1. The flight control system (FCS) mean time between failures (MTBF) shall be greater than 12000 hours.
2. No failure of the FCS shall cause a loss of aircraft.
3. The FCS shall be designed to provide a mean time to repair (MTTR) of less than one hour.
4. The Flight Control System shall be designed so that it will not cause a hazard or unsafe condition to occur in production, shipping, installation, repair, and usage.

The following list provides some of the items we may establish as requirements to ensure adequate system reliability, maintainability and safety:

Possible Reliability/Maintainability, and Safety Requirement Items:
Basic

- Human error—(no human error shall result in an unsafe condition)
- Maintenance (ease of disassembly)
- Mean time to repair—(shall be less than 30 minutes)
- Maintenance man-hours/flight/operational hours—(shall be less than one hour per 100 operational hours)
- Toxic products—(no toxic products shall be used in the production, installation, or use of this product (toxic is defined by the Occupational Safety and Health Administration) (OSHA))
- Health and safety criteria (per OSHA)
- Failure Mode and Effects Analysis—FMEA—(shall be done on all original design and subsequent design changes)

- Maintenance complexity—(a three-year-old child shall be able to fix this product)
- Life expectancy—(100 years minimum)
- Mean time between failures (MTBF)—(greater than 5,000 hours)
- Elapsed-time meter—(shall be provided by software)

Mission Specific

- Turnaround time—(after a successful mission the system shall be ready for the next mission in less than three minutes)
- Availability—(98% of the time based on the following criteria . . .)
- Fail-safe (the system shall be fail-safe)
- Noxious gases (shall not give off)
- Accessibility (no special tools)
- Interlocks—(safety interlocks shall be provided)
- Stress monitor—(over voltage monitor shall be provided)
- Tools—(no tools required for disassembly)
- Emergency operating systems (capability for a backup system shall be provided)

DS 3.2.8 Testability requirements

The requirements for test equipment are usually provided in a separate DS (see the specification tree examples), but the top system DS must establish the top-level test requirements. Separate test system specifications and test equipment (station) DSs will not be covered in detail herein.

Example DS 3.2.8 testability requirements

The system shall be designed to provide built-in test (software), built-in test equipment (hardware), test points, test isolation capability, test connections, and test coverage that will ensure capability for isolation to a replaceable module.

a. Test Points shall be provided to allow monitor of all plug-in replaceable unit interfaces.

b. Built-In Test capability shall be provided to allow fault reduction to two plug-in replaceable units.

c. Test break out of all normal input/output cables shall be allowed with a maximum cable extension of three feet and two compatible connectors.

d. Specific requirements for the test equipment to be developed for this system are provided in DS 1234.

Possible Test Requirement Items:

- Calibration—(shall not require hardware adjustments or part selection. A special test environmental chamber and software shall be used)
- Test system—(test equipment shall be developed per the following; also see the tester DS 1234)
- Qualification—(shall be accomplished utilizing the system tester)
- BITE (provide built-in test equipment specific requirements (such as the internal A/D converter shall be used))
- Facilities and equipment for system development—(no special facilities or tools shall be required)
- Computer-aided test (provide specifics)
- Demonstration—(shall be part of ATP)
- Self-test—(shall include BITE and BIT)
- Test equipment—(no special equipment shall be required)
- Test accuracy—(shall be ± 0.1% of full scale for all measurements)
- Failure isolation—(shall be to a replaceable plug-in subassembly)
- Periodic requalification—(shall not be required except after repair)
- Test points (provide specific list of points to be used as stimulus and/or monitor)
- BIT (test software)
- Test data reviews—(shall be part of ATP process)
- ATP—(shall be provided in part 2 of this DS)
- Heuristics—(self-calibration shall occur every time the system is turned on)

DS 3.2.9 Documentation requirements

Documentation required for a system is usually specified by government, industry, or company standards. There may be, however, special requirements established. For example, schematic diagrams shall be drawn so that they are directly usable in technical field manuals without redrawing. These special cases may include documentation requirement items such as the ones listed below.

Example DS 3.2.9 documentation requirements

This system shall be documented by the following:

a. Specifications as called out in figure 3.1-4 specification tree

b. Assembly drawings for all mechanical units

c. Schematic diagrams

d. Parts lists

e. System drawings—detailed functional and control diagrams for both hardware and software as required

f. Other documentation as required per contract

All documentation shall be released to full configuration management control as defined in the company configuration management manual (EOPS-CMM-1001).

Possible Documentation Requirement Items:

- Documentation (shall be per company standard XYZ)
- Computer-aided engineering—(shall be used with tools, which shall result in digitally formatted data in the Electronic Publishing Standard (EPS))
- Computer-aided design (as above in EPS)
- Configuration management—(shall maintain a record for ten years after the life of the system)
- Documentation procedures (provide as required)
- Production for second source—(all documentation shall be provided to the second source company)
- Schematic diagrams—(shall be in digital format per EPS)
- A system development specification must be provided
- Field manual—(shall be provided with a 20-page maximum limit)
- User manual—(shall be 5 inches x 7 inches, and limited to 25 pages)
- Computer-aided manufacturing—(compatible with)
- Interface control document—(shall be provided as an appendix to the top system DS)
- Computer-aided logistics system (CALS)—(compatible with)

DS 3.2.10 Producibility Requirements

Special requirements regarding production and construction are necessary. The materials and components to be used in constructing the system are many times controlled in order to reduce risk and establish standardized designs. Some examples of construction requirement items are listed below.

Example DS 3.2.10 producibility requirements

This system shall be designed for producibility by meeting the following specific requirements:

1) Unit product cost (UPC) shall be less than $1000
2) No production selection or calibration requirements
3) No welding required
4) A 160-hour environmental stress screening (ESS) test process shall be developed as part of the production process
5) No surface mounted parts shall be used
6) 95% of all purchased part types shall be standard parts selected from the standard part manual

Possible Producibility Requirement Items:

- Accessibility of adjustments—(shall be easily accessible to the repair technician)
- Standards (1750—Computer-Instruction Set Architecture), (1553—Data Bus) specific part numbers 1234 and 5678 shall be used to improve commonality to other systems in production)
- Nonstandard parts—(shall not be used unless approved by . . .)
- Second source—(components shall be selected so a second source exists or is being developed)
- Relays—(do not use)
- Transformers—(must have groundable shield)
- Connectors—(specific type)
- Impact on existing production facilities, systems, equipments—(shall not require new)
- Design—(per the company established design process procedures)
- Parts—(none from outside the USA unless approved by . . .)
- Standard parts—(shall be used)
- Construction—(shall have no sharp edges)
- Product marking—(by integral cover design)
- Unit product cost (UPC)—(shall be less then $100.00)
- Hybrid devices—(shall not be used)
- Standards of manufacture—(per company standards . . .)
- Workmanship—(per company standard . . .)
- Soldering—(per standard . . .)
- Materials—(no toxic, etc. . . .)

- Value engineering—(a cost reduction process shall be part of the production process)
- Adjustments—(none, shall use self-calibration algorithms)
- Interchangeability—(shall be interchangeable with the following prior models)
- Production plans—(shall be reviewed by design engineering)

DS 3.2.11 Installation and logistics requirements

Deploying the system to the user requires specific requirements regarding installation and logistic support. Note that logistics is sometimes called "field service." Some examples of these types of requirement items are listed below.

Example DS 3.2.11 installation and logistics requirements

a. This system shall be mechanically designed to meet the installation requirements as specified in CP-X-007 3.14.1-3.

b. The built-in test and built-in test equipment design of this system is required to provide coverage to isolate to six field replaceable units (reference section 3.2.8):

1.	Pressure sensor	PN 0071
2.	Computer subassembly	PN 0072
3.	Input/output control subassembly	PN 0073
4.	Display subassembly	PN 0074
5.	Actuator	PN 0075
6.	Cable subassembly	PN 0076

These six shall be designed for provisioning as field replaceable spare parts.

Possible Installation and Logistic Requirement Items:

- Government-furnished equipment (GFE)—(will be supplied as follows . . .)
- Transportation—(can be shipped by common carriers . . .)
- Field-support—(shall be provided by the existing company world-wide sales organization)
- Logistics—(batteries shall be included)
- Spare parts—(from factory only)
- Pallets—(compatible with standard . . .)
- Containers (provide part number)

- Training facilities (as appropriate)
- Installation, civil (user installed)
- Installation, structural (in user manual)
- Installation, electrical (in user manual)
- Installation, hydraulic (in user manual)
- Initial operational configuration (IOC)–date—(this system shall be available for its first user installation on or before 12–12–93)
- Transportability—(shall be packaged for shipment by any common carrier means to anywhere in the world)
- Training—(none shall be required; users manual shall be adequate and available in English, German, French, Japanese, and Spanish)
- Tie downs—(none required)
- Packaging—(per specific drawing and instructions)
- Field facilities—(none shall be required)
- Training equipment —(none shall be required)
- Installation interfaces (supply installation drawing in users manual)
- Installation, architectural—(shall be aesthetic . . .)
- Life cycle cost (LCC)—(shall be established by a yearly service contract with the user)

DS 3.2.12 Other requirements

There are always some specific requirements that are special for each system. Some examples of other specific requirement items are listed below.

Possible Other Specific Requirement Items:

- Design review special requirements—(all designs for this system shall follow the established design review process of chapter SEE-7)
- Personnel subsystems—(computer-aided design equipment shall be available for this development)
- Precedence of requirements (first, function; second, size; third, stress, etc.)
- Human engineering—(this system shall be designed for use by all people including those who use self-propelled wheel chairs)
- Security criteria—(the luda-fisk process is company proprietary and shall not be divulged)
- Personnel—(only Norwegians can use this system)
- Development facilities, equipment, and personnel—(will be available during the 1993 development time in Oslo)

DS 3.2.X Examples of specific requirements

It is important to recognize that prior examples of development specification trees showed the partitioning down of large systems to a number of lower-level requirement development specifications. You may be given the responsibility to write the top system DS, the hardware DS, the software DS, or the test equipment DS for your system. The specifications for a process system may require a top system DS, a management DS, a people DS, a facilities DS, a communication DS, and a support services DS. The number and type of DSs required for various system types depends on the specific mission of your system. The most important DS is, of course, the top system DS. A system without a top system DS is ultimately a nonsystem.

Specific requirement sections excerpted from DS 28991-01 (SBC), and DS 32615-01 (CACES) now follow. These provide only a small sample of the possible specific requirement specifications, but it is hoped they will guide you in your establishment of specific requirements for your system:

- Figure 5-16, SBC, DS 28991-01, a computer subsystem
- Figure 5-17, CA-CES, DS 32615-01, a software tool system.

SPECIFICATION NO. | DS 28991-01 |

Part I

3.2 Specific Requirements

3.2.1 Instruction Set Execution (Note: A Dictionary of Terms is provided in 6.2 notes)

This computer shall execute the DAIS floating point instruction mix at a minimum of 480 KOPS at 20 MHZ CPU CLK, and 400 KOPS at 16 MHZ CPU CLK. The design shall be such that it can take advantage of faster memories, when available, to improve throughput. (The 9450 is limited to 700 KOP with the DAIS mix at 20 MHZ and memories that require no wait states. The no floating point optimum for the 9450 is 1400 KOP.) See section 6 notes, figure 15, for the DAIS mix definition.

3.2.2 Memory Management and Block Protect

Memory expansion more than 64K word addressability is not required. Therefore, facility for interface to the FAIRCHILD 9451 memory (expansion) management unit (MMU) shall not be provided. Memory protection features embodied in the FAIRCHILD 9452 block protect unit (BPU) are also not required; therefore, facility for the BPU shall not be provided.

3.2.3 CPU

The central processor unit for this computer shall be the FAIRCHILD 9450 CPU. Refer to the latest FAIRCHILD specifications regarding this device (baseline design specification is "Advanced Definition November 1982").

3.2.4 Power Interface

The power interface for the SBC shall consist of:

a. 5 VDC Power (input)

b. VINJ (input)

c. RESET (input)

d. Normal Power UP Discrete (output)

e. Power Down Interrupt (input)

These interfaces have the following requirements:

a. 5 VDC—The SBC shall have full-board power and ground planes. A total of thirty pins for 5 VDC and 5 VDC-ground-return shall be provided (see figs. 6 and 7). Six power and six ground at one end of the connector, six power and six ground at the opposite end, and three power and three ground in the . . .

PAGE I-7	REV

Fig. 5-16. SBC specific requirements

SPECIFICATION NO. DS 32615-01

Part I

3.2.1 Work Breakdown Structure (WBS) Element Equations

Each WBS element developed as part of this system shall be defined by a simple algebraic equation, which can be easily understood and solved manually or be programmed on any computer. The equation and element structure must not be dependent on a specific computer for operation.

3.2.2 WBS Element Questions (SAMS = System Access Memory System)

Each WBS element developed as part of this system shall be documented on a Honeywell cost estimate (computer input) form (also known as a "SAMS" sheet) as in Figure 3-2 (example form). Figure 3-3 shows the CA-CES applied to a standard WBS structure. This generic WBS element description shows inputs, functions, outputs, and feedback. Figure 3-4 shows a collection of WBS elements combined into an example WBS tree. The Level 1 Element (system block in Figure 3-4) represents a consolidation of all lower level estimates for total contract costs. A separate "Contract" level cost estimate equation is also provided. It will be used for a quicker, rougher estimate without involving lower level estimates. The SAMS sheet form shall provide the questions to be asked to establish the values for the equation parameters. The questions on development process shall be preceded by a functional analysis of the specific WBS element (what functions must be accomplished in order to complete the work required for this element, and what questions quantify these functions?)

Typically three question types shall be provided for each WBS element:

3.2.2.1 Classifier Questions

Classifier questions shall be asked which will classify the work element to: Contract, intent, customer and type system. These questions should only provide the depth and breadth applicable to the specific functions for the WBS element; avoid the tendency to ask too many superfluous questions. These questions are used only to select an appropriate set of algebraic equations to be solved for the specific contract and system type.

3.2.2.2 Parametric Questions

Parametric questions shall be asked which will establish the work elements parameters. These questions shall result in actual numerical impact on specific terms in the algebraic equations.

PAGE I-9	REV

Fig. **5-17a.** CA-CES specific requirements

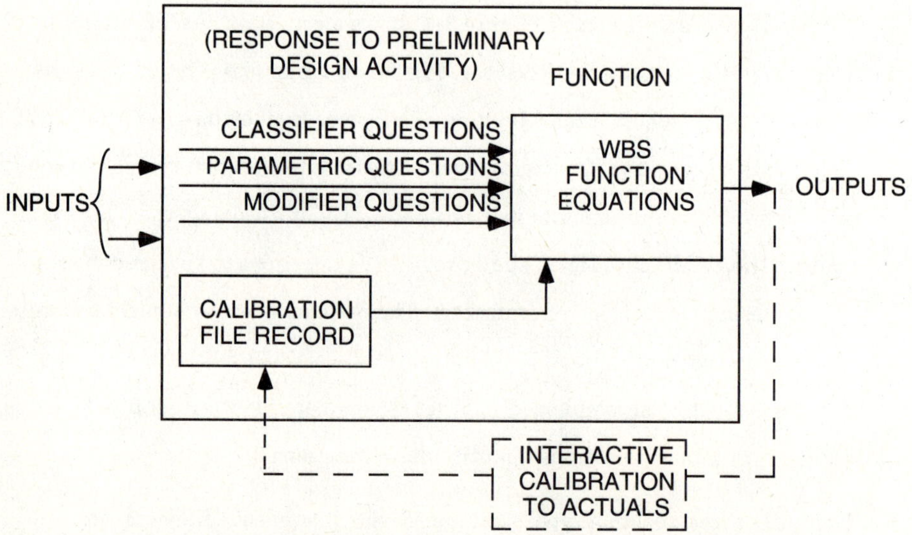

DS32615091 Figure 3-3. The basic functional element for a WBS element of the Computer-Aided Cost-Estimating System (CA-CES)

Fig. **5-17b.** CA-CES specific requirements

SPECIFICATION NO. DS 32615-01

Part I

3.2.2.3 Modifier Questions

Modifier questions shall be asked which will modify the overall algebraic equations to increase or decrease the cost estimate. Risk assessment questions may be asked which will tend to raise the estimate, while productivity improvement questions will tend to lower the estimate.

3.2.3 WBS Element Calibration-File Record

Data from prior work performed for each WBS element shall be utilized to establish the basis upon which the hours and material costs are being estimated. These prior program actual costs shall be identified as the "basis for estimate." The Program Close-Out review specified in DS 33105-01, Engineering Management Subsystem of Engineering Operating Procedures System (EOPS), shall be used in supplying cost history. A record shall be kept for each WBS element, which identifies the program by name, date, and expert person who "calibrated" the model from actual costs.

3.2.4 WBS Element Computerization

It shall be the intent to computerize the WBS elements as separate stand-alone software modules. These modules shall be programmed in "Basic" and listings of these "Basic" programs shall be provided. The decision to use Basic listings is based on the concept that it is easy to type the Basic programs into any personal computer, thus optimizing the transportability of the WBS elements. It shall be a design aim to extend the usage of the CA-CES WBS elements to all company facilities. If this "Basic listings" method can be improved upon by a better method, it shall be incorporated.

3.2.5 WBS Element Dictionary

Each WBS element shall be defined by a dictionary type definition. This definition shall provide a concise description of the work to be performed by the WBS element. Each description shall define the required inputs, work function activities, and outputs.

PAGE I-13	REV

Fig. **5-17c.** CA-CES specific requirements

SPECIFICATION NO. DS 32615-01

Part I

3.2.6 WBS Element Documentation

Each WBS element shall be defined by:

1) Dictionary which defines the element inputs, activities, outputs.
2) SAMS sheet questionnaire, with parametric, classifier and modifier questions.
3) Algebraic equations, flow diagrams, and other information which describe the mathematical features of the cost estimate process.
4) A sample questionnaire filled out with actual equation solutions.
5) The run corresponding to the sample questionnaire and a "Basic" language listing of the computer program for the element.
6) Procedural information to assist in use of the WBS element cost estimating process (if needed).

3.3 Consolidation of WBS Elements

The standardized WBS elements shall be used to construct a specific WBS tree for each project or program. See example Figure 3-4. This shall be per MIL-STD-881A format, and shall utilize the standard WBS elements as a baseline; special WBS work elements may be required for some programs, however. The cost-estimating data from the elements shall be collected and used as input data to a WBS Elements Consolidation Computer Program. This program combines the element data to provide a cost estimate report. This shall be extended to include schedule, manloading requirements, critical parts, risks listing, variance analysis (both element and total), and parameter sensitivity data in future development of CA-CES.

A methodology shall also be developed to provide for the implementation of programs to translate CA-CES WBS data to customers special non-MIL-STD-881A format.

The WBS element outputs are in terms of labor hours (by labor type) and material dollars. Material dollars are differentiated as to physical material and that for computer changes. The consolidation program applies the appropriate labor and burden rates, plus acquisition rate to the physical material. The procedure for applying the consolidation process to the WBS elements is explained in Appendix 10.

PAGE I-14	REV

Fig. 5-17d. CA-CES specific requirements

DS 4.0 Quality assurance

Section 4.0 of your specification shall include all of the examinations and tests (by reference where applicable) to be performed in order to ascertain that the system, subsystem, product, material, or process to be developed for acceptance conforms to the requirements in section 3.0 and 5.0 of the specification. In chapter SEE-4, 4.3.4, "The Program Startup Review," I showed you a matrix method of establishing how requirements were to be confirmed. You shall provide such a matrix in your DS.

The IPCS DS, 24712-01, provides an example of a different way to establish quality assurance requirements as shown in figure 5-18. Section 4.1.1 establishes tests to verify the system meets its requirements, 4.1.2 tests that the system can meet the aircraft flight environments, and 4.2 requires standard company quality assurance process monitoring.

A subsystem such as the SBC will in section 4.0, have standard workmanship visual inspection requirements specified and an ATP (acceptance test procedure) referenced for all SBC items. In addition, a single-item qualification to the environmental requirements test plan will normally be specified.

The CA-CES section 4.0, "Quality Assurance," indicates that the system has local ownership; that is, users of this software tool shall establish their own quality assurance. This is appropriate because the parametric models in the CA-CES are tuned by changing constants to reflect specific user databases.

Constant improvement requirements

The establishment of requirements to guarantee constant improvement of products, processes, and systems shall be established. You may, for example, require ongoing analysis of the data obtained during production testing to aid statistical process control methods. The objective shall be to increase production good-product yield, reduce rework in production, and provide better system mission effectiveness.

SPECIFICATION NO. DS 24712-01

Part I

4.0 QUALITY ASSURANCE

4.1 Design Verification

Verification that the DPCU meets the requirements of Section 3 of this specification shall be established by demonstration tests and inspections. This section covers the tests and inspections to be performed prior to delivery of the system for the planned evaluation programs of ground and flight testing.

4.1.1 Functional Tests

4.1.1.1 Device Tests

Each assembly and computer program shall be inspected and tested according to the criteria defined by its part II specification:

- DCU per DS 68239131
- IFU per DS 24715-01
- CMU per DS 24714-01
- PSU per DS 24933-01
- BOMDIG Program per DS 24716-01
- IPCS Program per DS 24717-01

4.1.1.2 System Tests, Hardware

The DCU, IFU, PSU and CMU shall be interconnected in a system configuration and operated by means of the TSU. The specified performance of the system, as defined by Part II of this specification, shall be verified by means of input stimula generated in the TSU using test programs that exercise each input and output of the system and each mode including the proper operation of monitoring, back-up and out-of-range signal conditions.

4.1.1.3 System Tests, Software

Verification of the proper performance of the full DPCU, including the BOMDIG and IPCS software control programs, shall be performed in a system configuration using the hybrid simulation of TF-30 engine. This test shall demonstrate the control performance of each of the system loops under specified flight conditions, including sea level static, with and without afterburner and at altitudes, under subsonic and supersonic conditions, selected to exercise all control loops in each operating mode, as defined by Part II of this specification.

PAGE I-57	REV
DS 24712-01	

Fig. **5-18a.** IPCS section 4.0—quality assurance

SPECIFICATION NO. DS 24712-01

Part I

4.1.2 Flight Assurance Tests

The DPCU flight hardware shall be subjected to and meet the following environmental test program requirements.

All units shall have successfully passed the individual and their system functional tests prior to being submitted to the environmental conditions. Abbreviated functional test and inspections shall be performed following each environmental exposure and complete functional tests shall be performed at the completion of the tests. These tests shall be performed on each system with the exception of the EMI susceptibility evaluation which will be conducted on one system only. For environmental test purposes, the DCU, IFU and PSU shall be installed in the DPCU Box (supplied by BAC) in flight configuration, and operated with cooling air simulation flight conditions.

4.1.2.1 Temperature/Altitude

The DPCU Box, containing the DCU, IFU and PSU mounted and interconnected in flight configuration and connected to the TSU, shall be subjected to temperature and altitude testing per the conditions specified for Category II equipment per FRC PS 21-2, except that the cold and hot soak temperatures shall be -64°F and +160°F and the altitude shall be that at which the DPCU box cooling air flow is minimum when supplied with 80°F cooling air at the pressure corresponding to the F-111 cooling air supply for that altitude (approximately 8,000 feet). Operation at 60,000 feet shall be demonstrated by an abbreviated functional test (approximately 3 minutes) from the cold temperature conditions only. This altitude shall be established in the test chamber prior to the test. Proper functioning shall be determined by operating the system during the test and by inspection and function test afterwards.

The CMU shall be tested in similar fashion except that it shall be performed at 60,000 feet altitude conditions (no cooling air required).

4.1.2.2 Vibration

The DPCU Box containing the DCU, IFU and PSU shall be subjected to vibration resting in accordance with FRC PS 21-2, Category II, Level B (5 g's maximum). The devices shall be monitored for normal performance during vibration using the TSU. Post vibration inspection and functional tests shall be performed for comparison with pre-environmental performance data to establish successful accomplishment of the vibration test.

The CMU shall be subjected to the same vibration test environment with operation monitoring and post vibration evaluation by inspection and functional tests.

PAGE I-58	REV
DS 24712-01	A

Fig. 5-18b. IPCS section 4.0—quality assurance

SPECIFICATION NO. DS 24712-01

Part I

4.1.2.3 Shock

The DPCU Box and operating contents shall be subjected to a shock test, as a unit, consisting of two shocks in each direction on all the principal axes. the shock shall be 15 ± 1 g's, half sine of 11 ± 1 ms duration. System performance shall be monitored during the shocks and the units inspected and tested following the series of shocks. Performance data shall be compared with pre-shock data to verify the successful survival of the system.

The CMU shall be separately subjected to the same test conditions and performance requirements.

4.1.2.4 Electromagnetic Interference (EMI)

A limited EMI test, using MIL-STD-461A, Notice 3, as a reference, shall be performed on one DPCU system. The tests shall be limited to the following.

CS01 Conducted susceptibility, power leads 30 Hz to 50 Hz

CS01 Conducted susceptibility, power leads 50 Hz to 400 Hz

CS06 Spike, power leads

RS03 Radiated susceptibility, electric field 14 KHz to 1000 MHz

A limited power transients test, using MIL-STD-704A as a reference, shall be performed.

Testing shall be limited to one set of DC power line transient tests ranging from ±600V, 10 usec, to 70V, 500 usec.

4.2 Quality Control

Honeywell Quality Assurance procedures, including testing and inspection at all levels of fabrication and assembly, shall be applicable to assure conformance to specifications, drawings, and approved processes and procedures.

PAGE I-57	REV
DS 24712-01	

Fig. **5-18c.** IPCS section 4.0—quality assurance

DS 5.0 Preparation for delivery

Per Mil-Std-490 this section is generally applicable to product specifications only and shall include applicable requirements for preservation, electrostatic discharge (ESD) protection, packaging, and packing the item, plus marking of packages and containers.

You, however, write DSs to serve your purposes and, therefore, may use section 5.0 to establish other requirements related to logistics (field manuals to be shipped with each item), restricted shipping environments (to prevent breakage in transient), and data correlation (ATP data for each specific item), etc.

My example for the IPCS is excerpted from DS 24715-01, which is the interface unit (IFU) for that system, a product or item to be delivered.

DS 24715–01—IPCS Section 5.0 Example:

5.0 Preparation for delivery

5.1 Closures

Appropriate protection caps per MS 3181 or equivalent shall be placed on all electrical connectors.

5.2 Packaging

Each interface unit (IFU) shall be cushioned and individually boxed in a substantial commercial shipping container that is constructed to ensure acceptance by common or other carrier for safe transportation to the point of delivery.

5.3 Acceptance test data

Acceptance test data shall be provided with each device shipped.

SBC example for section 5.0

There is no example for the SBC because the SBC is a subsystem that is not normally delivered to a customer by itself; it is normally an internal part of a larger system. If, however, at a later date, the SBC was sold as a separate product, it would be necessary to establish section 5.0 requirements.

CA-CES example for section 5.0

CA–CES, our software tool example, is delivered as a system by design release. The medium of the "system" is the DS and associated appendixes—User Manuals.

DS 6.0 Notes

Section 6, "Notes," of specifications shall contain information of a general or explanatory nature. *No requirements shall appear therein*. Mil-Std-490 provides the following list of possible subjects to cover in the notes:

- Intended use
- Ordering data

- Preproduction sample—pilot model or pilot lot data
- Standard sample definitions and qualification provisions
- Cross reference of classifications
- Miscellaneous notes

You can provide in your section 6.0 miscellaneous notes with some very helpful information:

- A DS chronological change record to provide the history of the development process.
- An alphabetical index to assist in finding specific topics; this may be combined with an acronym dictionary.

The "Note" section of the IPCS-DS included specific details of the interface hardware. The SBC-DS notes included a dictionary of terms and a definition table of the instruction mix. The CACES-DS provided a list of the appendixes that provide user's manual type of information for the WBS elements.

DS Appendix

Appendixes may be added to your DS to append large (multipage) data tables (NOTE: IPCS—6.1 Interface Hardware Specifications could have been added as an appendix). Market studies, classified information, trade-off studies, and bulky user manuals can all be provided as appendixes.

In my CA–CES example, I established a specific requirement for a standard outline for my appendixes. Then following this standard, I developed WBS elements similar to mini-user's manuals and added them to my DS. This methodology of constructing your system specification lends itself to modular-growth type systems.

5.2 SUMMARY OF SEE-1, -2, -3, -4, AND -5

1. You can use the DS to lead the development process of anything.
2. The DS is derived from the process sequence of problem definition, study, system perspective, mission analysis, functional analysis, trade-off studies, baseline system definition, and collecting together into a DS format.
3. You iterate the whole process as required to evolve the system definition and requirements adequately.
4. Analysis is the game. Now your analysis game will change, for you will enter into a much wider range of analyses. In SEE-6, I will provide an overview of analysis topics.

5.3 Problems and Questions

1. True or False?

   a) The development specification is your tool to lead the development.

   b) The title page of the DS identifies the document as a DS and also provides facility for revision data.

   c) The table of contents acts as an expeditor to aid the designers in finding requirements.

   d) Scope should reflect the system perspective cube.

   e) It is appropriate to list a report in applicable documents and follow it with a statement such as, "This report is the basis upon which this system is being developed and is therefore recommended reading by the developers of this system."

   f) The DS is to be used for military systems only.

   g) The interface drawing must be provided as part of the system definition.

   h) The specification tree is a must for a DS.

   i) The DS must be complete and correct before exposure to the system developers.

   j) System definition will provide system partitioning.

   k) Specific requirements can be classified as external to the system and also internal to the system. External relates to the system's interfaces when viewed as a subsystem to a major system, while internal relates to the interfaces of the components that make up your subsystem.

   l) Thermal requirements deal with external environments to the system only.

   m) Establishment of specific requirements include apportionment of external requirements such as accuracy, timing, cost, size, and weight among the components of your system.

   n) The sum of apportioned accuracy requirements may exceed the system accuracy requirement; however, root-sum-square (rss) analysis is applicable in lieu of worst-case analysis when the number of variables is greater than three.

   o) It is okay to specify a load as 50 ohms.

   p) Cable length and characteristics are important interface requirements to be specified.

   q) Environmental requirements are derived from both customer-stated requirements and from your mission analyses.

   r) System DS writing does not require much analysis.

   s) A single sortie mission analysis is sufficient to develop a system.

   t) A single sortie mission analysis is sufficient to develop a system initial baseline.

   u) Quality assurance requirements shall include requirements that establish methods for constant improvement.

v) Quality assurance applies to sections 3.0 and 5.0 requirements only.

w) Section 5.0, "Preparation for Delivery," requirements can be specified: "Use Best Commercial Practice"

x) Placing the IPCS interface hardware specifications in the notes (section 6.0) was inappropriate. This should have been added as an appendix per Mil-Std-490. This was your DS, however, so it was okay to deviate from 490.

y) Adding an index as well as a table of contents is a good idea to help large DSs lead the development process.

2. Write a DS

Your assigned problem for SEE-5 is to prepare a preliminary DS for some part of your present job. You may select to write a system DS, a subsystem DS, a product DS, or whatever seems applicable.

3. Review a DS

Assuming your present job has all the DSs complete, or your present job does not have a possible DS subject (your job is a thermal analyst, for example), then select an existing DS from any system and critique it. The critique shall consist of marking up a copy of the DS in red with questions, corrections, improvements and comments.

In SEE-4 you prepared a PMP. In this module you prepared a DS. What are the criteria for a good PMP and DS? When you review these two documents as part of the program start-up review (PSR) or rereview them through the sequence of design reviews SRR, PDR, CDR, etc., you should do the review against the six criteria listed below.

6 CRITERIA FOR DSs AND PMPs

	DS	PMP
1— CONSISTENCY		
2— SPEAKS TO THE DEVELOPERS		
3— ATTRIBUTES BASELINED		(SPM/CPM)
4— STRUCTURE		
5— SUFFICIENTLY ESTABLISHES THE REQUIREMENTS	(TPM)	
6— GIVES ROOM TO DESIGN/MANAGE		

Analysis

(The systems engineer is smarter than the collection of inanimate objects that make up a system. Unfortunately, the system engineer does not always take the time to prove it.)

In the introduction to this book, I gave some attributes of the systems engineer. I said that systems engineers have excellent analytical capabilities and love to analyze. Analysis is the number one activity that the systems engineer will be required to do. More time will be spent in analysis than any other activity, so it is good that they love to analyze! Their analyses require constant observation, utilizing all their senses (sight, touch, listening, smell, and even taste when appropriate).

While sitting still in the lab a navigation system had a spurious low frequency output. The undesirable condition seemed to be affected by the orientation of the navigation system on a rate table, but it was not always repeatable. The ring laser gyros and the accelerometers used as sensors in this navigator showed no unusual characteristics; changing them made no difference. The system had four gyros, one of which was skewed. It was suspected that the conversion algorithm for the skewed gyro was causing the "incorrect" output (notice the assumption of incorrect). Special testing, data analysis, and consultations went on for many weeks in the attempt to solve this problem. A senior staff systems engineer was called in to help solve the problem. The engineer asked many questions to try to understand the problem: How do you know that the output is incorrect? What tests have been performed? He also asked to see the accumulated data. He then asked to see the lab test setup and the actual navigator. In looking at the test set-up, he noticed that the navigation system was cooled by a fan that was sitting on the floor. A spiral wire, plastic covered tube was attached to the fan. The tube ran up to the navigation system that was mounted on the rate table. He observed the fan blade and noticed that the fan blade shaft was slowly sliding back and forth at a visually discernible low frequency. Turning the fan off eliminated the low-frequency output problem. In certain physical orientations the back pressure of the cooling system caused a low-frequency modulation of the navigation system's mechanical structure. The gyros were capable of detecting this motion and provided the appropriate "correct" output. The systems engineer used his *sight* to solve the problem. In a similar story, a systems engineer notices that the intermittent noise problem in an electronic system was in synchronization with the traffic lights on the corner outside the test laboratory.

The systems engineer touches the system. Is it hot or cold? Does my touch act as an antenna? How does the parasitic capacitance (40 pico-farads on a dry day) of my fingers affect the system, and if I insert probes or other components, how does the system respond to my touching? The system engineer listens to the system. What are the people related to the system saying? If the system is equipment, how does it sound? Systems engineers listen with instruments such as a screwdriver in the ear, a stethoscope, an oscilloscope, a logic analyzer, a spectrum analyzer, radio or television. In fact, I just tuned the AM radio next to my word processing computer so I can hear the radio frequency interference my computer puts out. I then turned my computer off to prove I was listening to its output. I was, but, unfortunately, I forgot to save my text, so now I'm typing this all over again. Now, I just turned on a television set near my computer. It really shows the interference on the video, especially when the disk drive is running. Smell: does it smell like it is overheating? Is it musty? Or, is it giving off a strange odor? Taste: as appropriate, use your own judgment, maybe all systems engineers should also be good cooks.

In addition to using all their senses, systems engineers use their minds, tools, and methods to do analysis. The range of analyses is beyond the scope of this book. My objective is to provide an overview, a checklist, awareness of the areas where the lack of analysis has caused serious problems, and some generic analysis methods. The examples provided in this book are dated. Despite changes in technology, however, these basic areas that have lacked analysis in the past will still lack it in the future and be the cause of serious problems unless we apply discipline to our analyses.

I have shown in SEE-1, -2, -3, -4, and -5 how the systems engineer initially starts the system development process by system perspective establishment (which includes a market analysis), mission analyses, functional analyses, trade-off studies, baseline system definition, and initial specific requirements establishment. These top-down analyses are used to define your system and its requirements in a preliminary development specification (DS). You will iterate this DS by doing more mission analyses, more functional analyses, more trade-off studies (by modifying the baseline), and continuously improving the specific requirements.

You will iterate as the system is developed, however. As the development progresses, the systems engineer's analyses will include a wide range of new analysis subjects while the initial analyses—mission, functional, trade-off, baseline, and specific requirements establishment will become less and less. The new analyses will be bottom-up analyses such as determining the system attributes: size, weight, power, accuracy, reliability, cost, mission effectivity, and mission readiness. Many of the analyses that the systems engineer must do are just the summation of data developed by the various subsystem designers. In addition, as the system design evolves, the systems engineer needs to do other system level analyses such as control system analyses, producibility, failure modes and effect analysis, worst case design analysis, yield in production analysis, power system analysis, interface analyses, test data analyses, product analyses, new technology analyses, and software analyses. The systems engineer may also call upon specialty analysts such as control, structural, thermal, and electromagnetic interference (EMI) analytical experts.

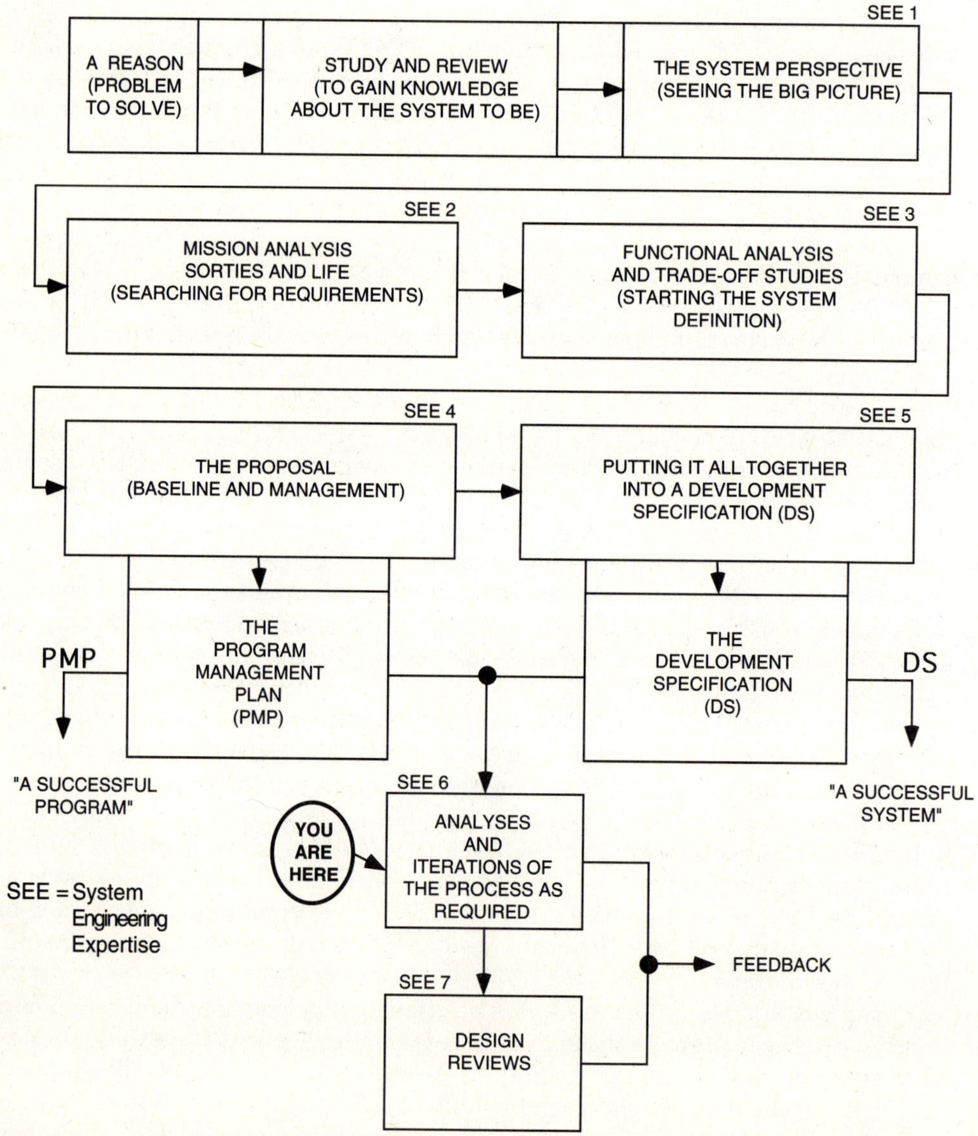

Fig. 6-1. The system engineering process

In addition to the above mentioned analyses, during the system development process the systems engineer will encounter many difficult technical problems that must be solved. These will be unforeseen, costly, and time consuming to solve. Many times they will be intermittent problems. The systems engineer needs a method to approach and solve these problems, which this chapter provides.

At the completion of this chapter, you will be sufficiently familiar with a wide range of analysis topics. You will understand the need and the timing for specific analyses. Bottom-up and top-down methods will be put into perspective. Your familiarity with these various analytical requirements will be of great enough depth so you can do some of them yourself and do a cursory job on all the others. You will know how to approach and solve difficult technical problems. It will be your goal after reading this chapter to continuously improve the scope and depth of your analytical skills for the rest of your life. Figure 6-1 shows where you are in the system engineering process.

6.1 SYSTEM—ANALYSIS BASICS

In order to develop a system capable of performing its intended mission, *you must do analysis*! First you need something to analyze. Through your top-down analyses in SEE-1, 2, 3, 4, and 5 you developed a preliminary DS with integral baseline definition, which provides the starting point for your bottom-up analysis and reanalysis. You also have to know what to analyze and what the system attributes and technical performance measures which require analysis are, such as mission effectivity, size, weight, power, cost, and accuracy.

Many of you will lament, "I don't have time to do all the analysis." In contrast, however, the systems engineer selects and tailors the analysis to fit the system needs. High-risk areas of the system receive in-depth analysis while low-risk areas receive only nominal analysis. You will recognize that tailoring your analysis means knowing when you've done enough along the sequence of nominal, worst case, root-sum-square (RSS), distribution, environmental impact, and higher mathematics! You always do a nominal analysis, but don't always do a complete environmental impact, or higher mathematical analyses.

You will also notice that the mathematical skills used to do system analysis range from arithmetic to higher mathematics, but the number of people applying these skills in the development of a system is greatest at the arithmetic end of the math difficulty spectrum. The mathematical skills required and the relative number of people using the skills to develop a typical system are shown in figure 6-2.

You sometimes do a good job of analyzing the stability, effectivity, and accuracy of a system using higher mathematics, but forget to do the *arithmetic* that tells you that your system weighs too much, costs too much, and takes too much power!

6.1.1 The System Attributes or Technical Performance Monitors (TPMs)

The program manager monitors the program by cost performance monitors (CPMs) and schedule performance monitors (SPMs). In like manner, the systems engineer uses technical performance monitors (TPMs) to monitor the technical system development process. The TPM provides the means to assess the values of essential system level performance parameters. The TPMs required for monitor are derived from the system DS definition (attributes) and requirements. TPMs are monitored through analysis and test data. Note: *Test data is never a replacement for analysis. It is used only to substantiate the analysis.* Some people think that if lab test data shows success, the system works, and they do not need to do additional analysis. They, unfortunately, find out too late that the lab success cannot be repeated in the real world application. TPMs, which must always be tracked for a system development, include size, weight, power, cost, and reliability mean time between failures (MTBF). TPMs that deal with specific system mission effectivity requirements may include: accuracy, throughput, memory usage, function time, and temperature rise. *All the specific requirements in your DS are TPMs.* You must confirm that all of the specific requirements are fulfilled. The basic system required attributes and very difficult requirements, however, need special monitoring and analyses.

6.1.2 Top-down and Bottom-up Analysis Methods

Analysis of a system starts from a top-down method (perspective, mission analysis, functional analysis, trade-off studies, baselining, establishing requirements with initial estimates and goals), but the system is confirmed by bottom-up analysis. Designers of subsystems and components initially design to ensure function can be accomplished. Then, through a process of *analysis*, redesign, and "breadboard" testing, the design evolves. Breadboarding is used to confirm analysis or to develop confidence in the design—that's all! Designers who use only breadboard or single system functional data to establish a design are asking for trouble. The next one won't work like the breadboard. Or even worse, you may get a poor yield in production.

The systems engineer insists upon, directs, establishes analysis methods, collects, and analyzes the bottom-up analysis data. This data is summed to establish TPM data and may result in modifying system requirements.

Allocating an accuracy requirement within a system

Initially a systems engineer may have established that an analog signal has to be measured with an accuracy of ±1%. In partitioning this accuracy out to subsystem designers, the A/D converter is given ±0.5%, input buffering/filtering is given ±0.3%, and ±0.2% is allotted to measurement accuracy.

Error analysis and reallocation

The A/D designer through analysis comes up with a temperature coefficient problem and would like his specification increased to ±0.7%. The test equipment engineer's analysis indicates that ±0.15% is an acceptable specification. The

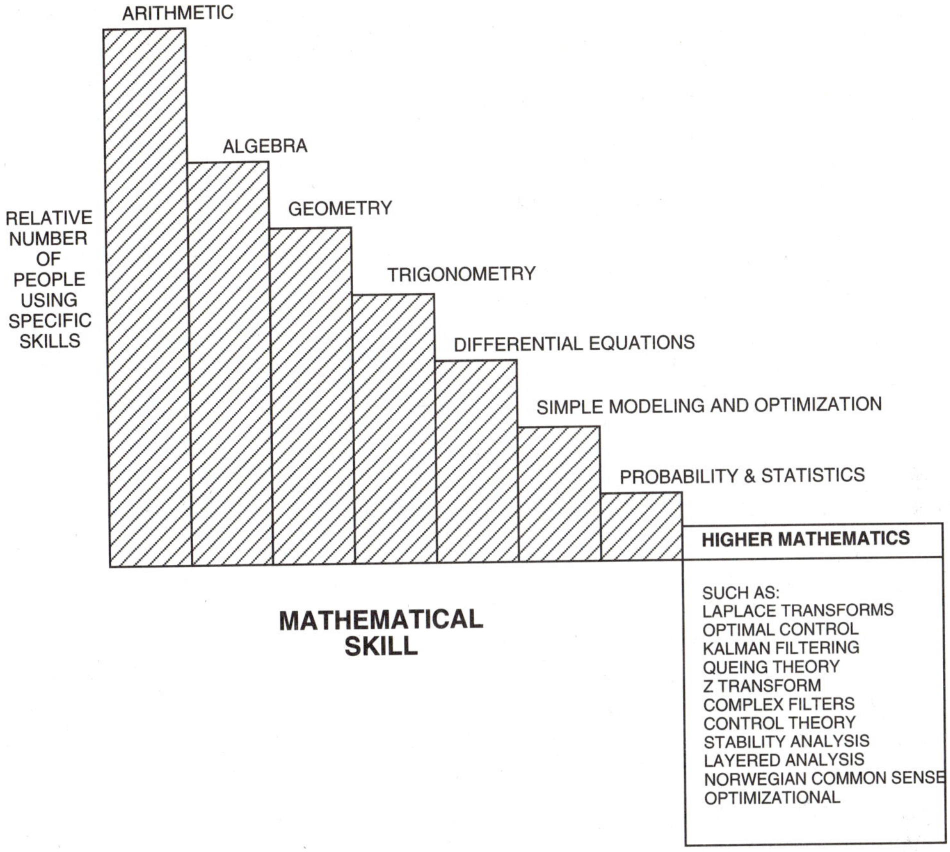

Fig. **6-2.** Mathematical skills required to develop a typical system

input buffering circuit designer is asked to provide ±0.15%. The redesigned input buffering circuit now requires more parts, etc. So this will impact other system TPMs (reliability and cost, etc.). This is an example of top-down and bottom-up analyses.

6.1.3 Nominal, Worst Case, Root Sum Square, Distribution, Characterization, and Modeling

You initially establish a nominal when you do analysis. This is always done to establish a baseline. Designs with low risk sometimes are only analyzed to the nominal. To demonstrate a nominal and then define subsequent analyses, I shall use a very simple voltage divider circuit (fig. 6-3) and a simple A/D converter measurement. These electrical circuits provide good examples of some elementary dos and don'ts when you do analysis. The same concepts and methods used in these examples are basic to all the analysis subjects overviewed in SEE-6. Through these simple examples, it is my objective to show some common basic analysis errors, which occur when you do worst case and root sum square on more complex problems.

Given:

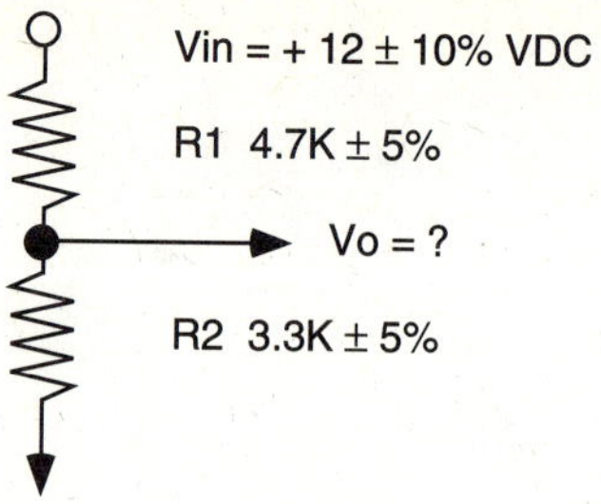

Nominal:

$$Vin\left(\frac{R2}{R1 + R2}\right) = Vo = 12\left(\frac{3.3}{8.0}\right) = 4.95 \text{ VDC}$$

Fig. **6-3.** Voltage divider

NOTE 1: The 3.3K resistor occurs in the equation twice. I have made sure it is the same in each calculation. A common mathematical error is to make it maximum on top and minimum on the bottom of the equation to numerically force an even worse result. This is not realistic, however, because a resistor cannot be two values simultaneously. Systems engineers must be sensitive to the real world. The mathematics you use must always reflect the world in real time and environment.

NOTE 2: In this example there is an unbalance of accuracy around nominal—

4.95 . . . . 5.767 (+16.5%)

4.195 (-15.25%)

If the resistors R1 and R2 had been equal, you would have equal worst case tolerances. The unbalance of tolerance must be recognized as commonplace in your analysis of various equations. Do not calculate only half and then assume that the other half will give you the same deviation in the opposite direction.

NOTE 3: A different perspective for the circuit can be obtained if you represent it as a thevenin equivalent circuit functional model as shown in figure 6-4.

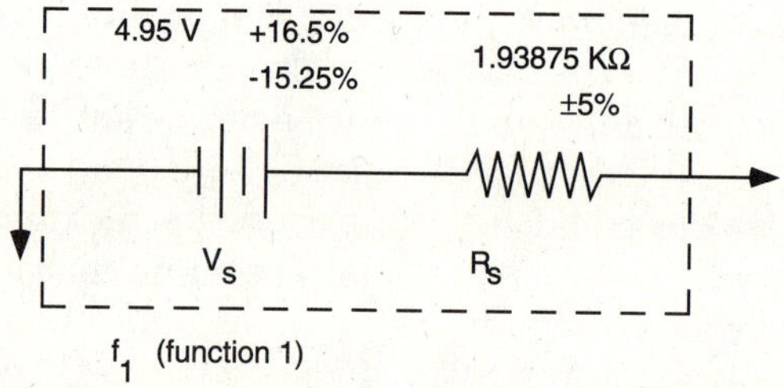

Fig. **6-4.** Functional model

You place this circuit in a box and indicate function (f1), which indicates a functional relationship exists between the elements of your f1 model. A common mathematical error here is to take 1.165 of the voltage and 0.95 of the source resistance (R_S) to calculate maximum source current. This would suggest that the real resistor R2 could simultaneously be two different values. This is incorrect (you do not get maximum voltage at minimum R_S). You need to relate to the real world and treat your equivalent models correctly.

This simple voltage divider example is used to demonstrate some very basic worst case analysis concepts and common errors made. In the analysis of complex systems (or in the use of computers to do the analysis), you must review to ensure you have not 1) mathematically established a worst case that assumes a common physical element is at one time two values, 2) assumed the absolute deviations are the same in both directions, and 3) treated a mathematical model inappropriately.

A/D converter—example

You shall provide the requirements to design the capability to measure a temperature with a sensor device and an analog-to-digital (A/D) converter (basically a meter). Your system shall include a sensor device, a buffer amplifier, a cable, an input divider, a multiplexer (switch), and an A/D converter as shown in figure 6-5. Each of these devices has undergone prior analysis that has established the accuracy of its transfer function. You shall do a worst case analysis of the system temperature measurement function.

What is the worst case measurement accuracy in terms of degrees centigrade (C)?

Temperature (measurement) tolerance

$TT = f_1$ (Sensor ±1%) + f_2 (Amplifier ±1%) + f_3 (Cable ±0.01%) + f_4 (Divider ±2%) + f_5 (multiplexer ±0.1%) + f_6 (A/D ±1%)

TT_{wc} = + (1%) + (1%) + (0.01%) + (2%) + (0.1%) + (1%) = ±5.11%

±5.11% is not good enough!

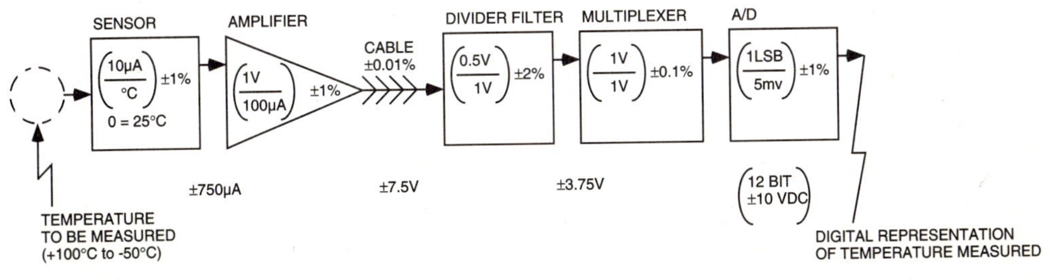

Fig. **6-5.** A/D converter

But the probability that all six functions will be at worst case is not very likely. If all six can be characterized as having a normal (Gaussian) distribution, then you can do something called "RSS'g" (root of the sum of the squares) to get a more realistic assessment of what you are likely to get.

RSS—Root of the Sum of the Squares

$$TT_{rss} = \sqrt{f1^2 + f2^2 + f3^2 + f4^2 + f5^2 + f6^2}$$

$$= \sqrt{1^2 + 1^2 + .01^2 + 2^2 + .1^2 + 1^2}$$

$$= \sqrt{7.01}$$

$$= \pm 2.648$$

If you have functions that have normal distributions, you can do the root of the sum of the squares (RSS) calculation to determine a probable result. This methodology in effect graphically overlays all the normal distributions of the functions and generates a new composite normal distribution. If you have three or more large contributing functions, the central-limit theorem states that the resultant will *tend* to a normal distribution even if the variables are not normal distributions! You must know the distribution of your major contributing variables when you use the RSS methodology. You shall also recognize that the central-limit theorem, which is the source of RSS'g, only says the resultant will *tend* to be a normal distribution. What does this mean to the systems engineer? It means that your system may require special calibration methods designed in early or you will have poor production yield because your combination of elements will not yield the desired accuracy. You may overcome this problem by specifying characterized tighter tolerance components, by designing in calibration methods utilizing both hardware and software, or in high-production large-market products you may sort and sell high- and low-grade products.

Distribution

The system engineer should develop an understanding of the realities of distributions. The normal distribution (or Gaussian) is one of the most important of all distributions. Variation is inherent in nature, and much of this variation appears to follow the normal distribution. It represents, for example, both natural phenomenon and can also describe empirically the time-to-failure distribution of a component exhibiting a constant failure rate characteristic with operating time. The pragmatic real world unfortunately never quite seems to follow the mathematical normal distribution. Human intervention, sample size, and combinational effects distort our mathematically pure distributions. The classical resis-

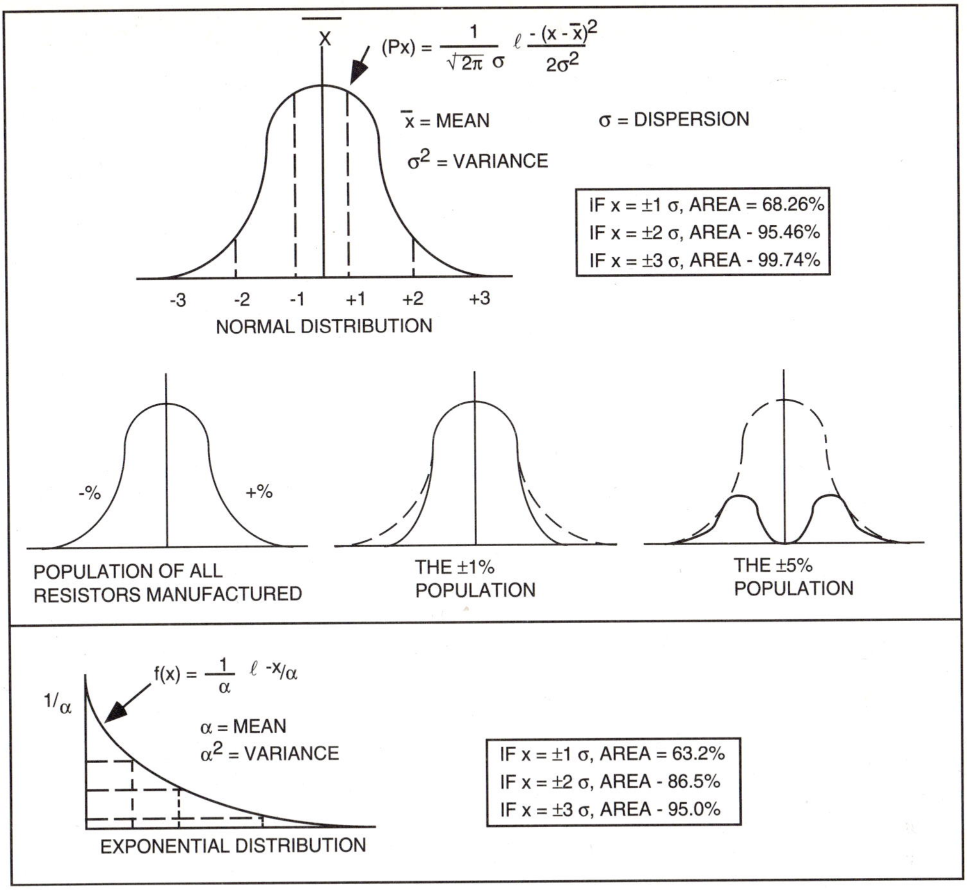

Fig. **6-6.** Distributions

tor tolerance example that selects resistors from a normal distribution by tolerance range could give different distributions for ±1% and ±5%. Other distributions also occur. Figure 6-6 shows normal and exponential distribution, which are the two most commonly encountered in system analyses.

Characterization

In order to do analysis, you need to know the distribution of the variables you are combining in a system: the tolerance distribution, the power dissipation, the weight, the speed, the accuracy, aging, drift, noise, temperature coefficient, the breakdown voltages, the failure modes, the cost trend, etc.

You shall call the process of gathering data to establish the distribution of these variables "characterization." Note, you do not attempt to characterize all the variables in your system, but you selectively characterize the most critical components. The measurement of pressure for an aircraft air-data system will require characterization of all the components associated with that measurement, for example.

How do we characterize?

Characterization—don't tell me what it is, let it tell me. By studying how it is made and from what and how it reacts to stimulus, environment, and dissection, you will know it. That is, trust and analyze your test data. Don't just believe what the supplier person tells you. You take data on as large a population as possible! On small population systems you may buy critical components in one single lot, test them to characterize their accuracy, tolerance, temperature coefficient, weight, power, speed, size, or other special attribute, and then design the rest of the system to accommodate. You also may develop some of the tests within your acceptance test procedure (ATP) to include data taking, recording, analysis and establishment of a data base for characterization. These ATP tests during the production of your system can be utilized to establish statistical process control (SPC) so you can control quality, provide for constant improvement, and increase production yield.

On high-volume production systems you may initially buy a large quantity (sometimes from different potential suppliers) of a critical part to test and characterize. Then, in addition, you may establish a production lot sample test process that ensures the critical component continues to meet the required characterization requirements throughout the production of the system.

"People systems" characterization

Just as you characterize and combine inanimate components into a system, you may, in like manner, characterize and combine animate components (people) into a system. Athletic and engineering teams can be viewed as "people systems," using the self-assessment skills survey to gather data for characterization to build teams. The "character" of the team is a reflection of the summation of the individual skills and deficiencies of the team's members.

Modeling and simulating

Modeling is a basic analysis methodology that you will use in most of your analyses. You will develop models to aid you in the analysis of risky critical aspects of your system. Physical models and mathematical models are developed. Physical model examples include mechanical mockups, circuit breadboards, and engineering model systems. These one-of-a-kind models help you to visualize your system. They also help substantiate the analysis, or even empirically develop the system. *Single physical models do not, however, ensure producibility*. To ensure that you have developed a viable producible system you must do mathematical modeling, simulations, and associated analysis. Mathematical modeling examples include: network reduction to equivalent circuits, lumped constant models, dynamic programming models, linear programming models, decision models, analog computer models, and digital computer models. Mathematical models are developed to aid analysis of a system without building a lot of costly physical systems for evaluation. Mathematical models are used to *simulate* the system, thus the term "simulation" is used. The primary use of simulation in systems engineering is to explore the effects of variations of inputs, algorithms, and outputs on total system performance without actually producing and testing each

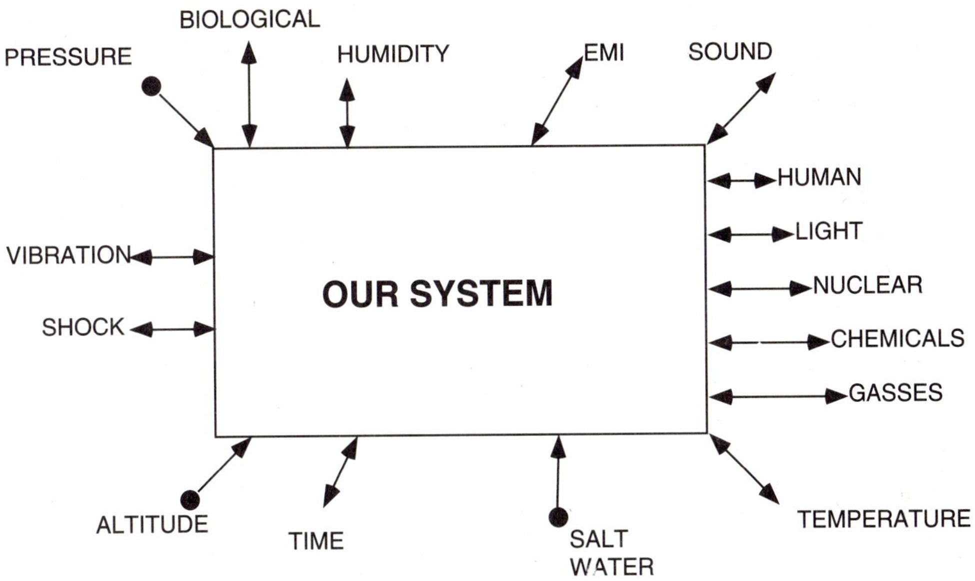

Fig. **6-7.** Environmental impact

system configuration. Your computer model simulations must be confirmed by data from actual physical models to be valid. It is expected that you will modify your computer simulations to agree with real world test data of physical subsystem models and total system physical verification testing.

6.1.4 Environmental Impact

Every system, subsystem, and component is impacted by the environment in which it must perform. You must learn as systems engineers to be sensitive to the environment to which your system will be exposed in production, shipping, storage, as well as usage. You must learn to analyze the impact of environment.

Figure 6-7 includes arrows going in and coming out of your system. This implies that you not only have to do analysis on how environment affects your system, but you must also do an analysis on how your system affects the environment.

Your worst case examples were at 25°C (normal room temperature). You shall now also include the impact of temperature at the extremes you established with your mission analyses. How will the simple voltage divider and A/D converter temperature measurement circuits be impacted by a temperature range of -40° to +90°C?

Assume that the resistors used in your voltage divider circuit are metal film resistors. They have a temperature coefficient *specification* of ±300 ppm/°C maximum (ppm = parts per million and ±300 ppm/°C = ±0.03%/°C).

The change in temperature (Δ T) from +25°C to +90°C = +65°C and from +25°C to -40°C = -65°C.

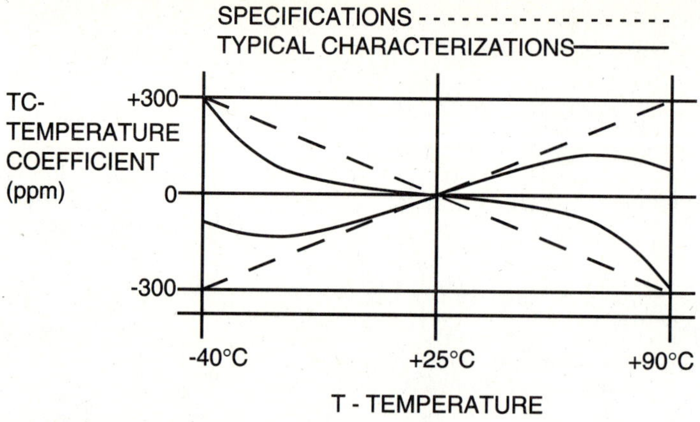

Fig. **6-8.** Specification and characterization

Therefore, the impact of temperature on your voltage divider is an additional ±1.95% at high temp +90°C and also at low temp -40°C. Note that this calculation deals with the specification of the temperature coefficient, not the actual characterization of the resistors. The actual characterization test data shows that the resistors' temperature coefficient data all fall within the specification but may have a characteristic slope or shape when plotted as shown in figure 6-8.

$$V_{OH} = (12 + 10\%)\left(\frac{(3.3 + 5\% + 1.95\%)}{(3.3 + 5\% + 1.95\%) + (4.7 - 5\% - 1.95\%)}\right)$$

$$= (13.2)\left(\frac{(3.53)}{(3.53 + 4.37)}\right) = 5.898 \text{ VDC} \qquad \text{(WAS 5.767 @ 25°C)}$$

+19.15% (WAS +16.5%)

$$V_{OL} = (12 - 10\%)\left(\frac{(3.3 - 5\% - 1.95\%)}{(3.3 - 5\% - 1.95\%) + (4.7 + 5\% + 1.95\%)}\right)$$

$$= (10.8)\left(\frac{(3.07)}{(3.07 + 5.03)}\right) = 4.095 \text{ VDC} \qquad \text{(WAS 4.195 VDC @ 25°C)}$$

-17.2% (WAS -15.25%

<u>At 25°C</u>

V_O = 4.95V NOM — WORST CASES: 5.767 (+16.5%) / 4.195 (-15.25%)

<u>At -40°C or +90°C</u>

V_O = 4.95V NOM — WORST CASES: 5.898 (+19.15%) / 4.095 (-17.2%)

This example shows the impact due to temperature change. The calculation shows the effect of a ±T.C. (temperature coefficient). Some components will have different coefficients—some positive, some negative. In critical sophisticated designs, clever designers match T.C.s to get zero T.C. circuits or predictable T.C.s for circuits or systems by utilizing characterization data. If they use characterization data, they will establish controls, monitors, and specifications that will ensure that the characterization will remain the same. This could mean that the production processes to produce the characterized component must be controlled, monitored, and specified, which could result in a higher-cost component. The systems engineer must determine if the improved performance is worth the additional cost.

In my A/D converter example I have established the T.C. specifications to be:

f_1 SENSOR ....+100 ppm/°C = (+0.01%/°C) Δ 65° = +0.65 @ +90°C, -0.65 @ -40°C

f_2 AMPLIFIER ....-50 ppm/°C = (-0.005%/°C) Δ 65° = -0.0325 @ +90°C, +0.0325 @ -40°C

f_3 CABLE ....±0 ppm/°C @ +90°C and -40°C

f_4 DIVIDER ....±300 ppm/°C = (±0.03%/°C) Δ 65% = ±1.95

f_5 MULTIPLEXER ...-5 ppm/°C = (-0.0005%/°C) Δ 65° = -0.00325 @ +90°C, +0.00325 @ -40°C

f_6 A/D ±300 ppm/°C = (±0.03%/°C) Δ 65° = ±1.95

Our worst case at 25°C was ±5.11%

at -40°C and +90°C our worst case is:

$$T_{TWC} = f_1 + f_2 + f_3 + f_4 + f_5 + f_6$$

$$@\ 25°C = 1+ 1 + .01 + 2 + .1 + 1$$

$$@\ +\Delta 65°C = (1.65) + (0.9675) + (0.01) + (3.95) + (0.09675) + (2.95)$$

Therefore: T_{TWC} @ +90°C = +9.62425

And our RSS @ +90°C:

$$= \sqrt{(1.65^2 + 0.9675^2 + 0.01^2 + 3.95^2 + 0.09675^2 + 2.95^2)} = +\ 5.29\%$$

Note: f1, 4, 6 are the three major error contributors that validate the RSS methodology. Notice that the central limit theorem establishes the RSS methodology by stating that given three major contributors to a function, they may be combined by the RSS methodology, and the result will *tend* toward a normal distribution regardless of the distribution of the three major contributors. The word tend is highlighted to reflect that the distribution *is not a normal distribution!*

Zero temperature coefficient (T.C.) design

Circuits and systems have been designed to provide no impact from temperature. Characterization of components, when matching positive T.C.s with negative T.C.s, can result in a zero T.C. circuit. For example: If a constant current

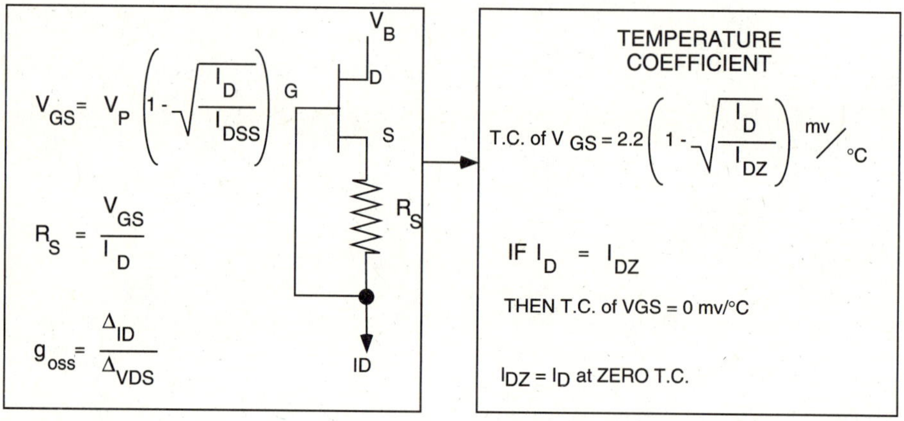

$$I_{DSS} = I_D \text{ WHEN } V_{GS} = 0$$

$$\text{and } I_D = \left(\frac{\frac{1}{I_{DSS}} + 2\frac{R_S}{V_P} \pm \sqrt{\left(\frac{1}{I_{DSS}} + 2\frac{R_S}{V_P}\right)^2 - 4\left(\frac{R_S}{V_P}\right)^2}}{2\left(\frac{R_S}{V_P}\right)^2} \right)$$

We can establish a zero T.C. I_D constant current source by characterization of a specific FET type, measuring the actual pinch-off-voltage (V_p) of each specific FET then setting the I_D to an I_{DZ} as in the following equation. K_1 is a constant determined by characterization while V_p and I_{DSS} are determined by test measurements of each specific FET.

$$I_{DZ} = K_1 \left(\frac{1}{V_P}\right)^{0.4} \text{goss } K_1 \frac{I_{DSS}}{V_p^{\,1.6}}$$

Fig. **6-9.** A zero temperature coefficient FET constant current source

source is designed from a field effect transistor (FET), it is possible to set drain current (ID) to a value that results in 0.0 T.C. of ID as shown in figure 6-9. This FET circuit provides a zero T.C. current source.

It is also possible to develop a zero temperature coefficient voltage source by matching the temperature coefficients of forward voltage drops and reverse breakdown voltage.

The temperature coefficient

$$\frac{dv}{dt}$$

of the forward voltage drop of a semiconductor diode junction with a forward current I_F is:

$$\frac{dv}{dt} = -F + G \log 10\ I_F = mv/°C$$

For most small junction diodes $\left(\frac{dv}{dt}\right) = -2.0 + 0.198 \log 10\ I_F$

This at nominal I_F currents results in a T.C. of typically -1.8 mv/°C.

The temperature coefficient of the forward voltage drop from base to emitter of a low power transistor at I_b is:

$$V_{be} = -2.3 + 0.198 \log 10\ I_b$$

where the coefficient (0.198) is:

- 0.198 for low-level diffusion (most cases)
- 0.375 for recombination
- 0.396 for high-level diffusion

The point is that by setting the forward current, you can set the temperature coefficient of the forward voltage drop of a semiconductor junction.

We also know through characterization that the reverse breakdown voltage of a transistor emitter to base junction (VEB) has a positive temperature coefficient! If you combine a diode with two transistors as shown in figure 6-10 and set the current I_F to the proper value, you can obtain matching temperature coefficients (+4.7 -2.2 -2.5 = 0.0) to get a zero T.C. voltage source.

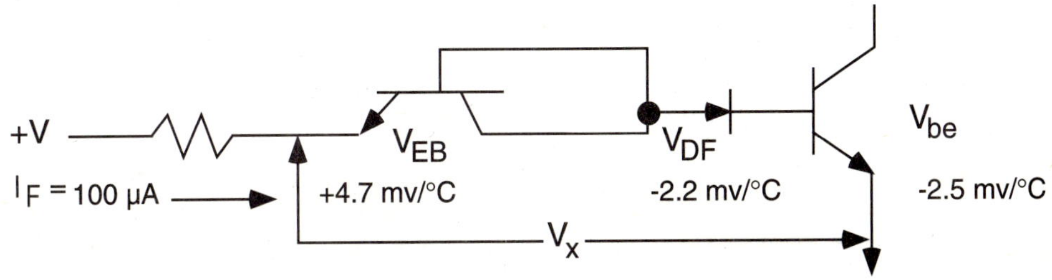

Fig. **6-10.** A zero T.C. voltage source (Vx)

The circuit in figure 6-10 has a zero T.C. of the voltage drop "V_X." The zero T.C. constant current circuit and this zero T.C volt source circuit were combined to provide a patented, highly accurate timer that worked over a very wide temperature range.

Environmental impact conclusions

The prior discussion showed an example of how your system can be impacted by temperature changes and how you can, through analysis, determine the magnitude of impact utilizing analysis and characterization of components. You also were shown how clever designers design out environmental impact. Analysis of environmental impact will, in like manner, determine the impact of other environmental factors (vibration, shock, human, EMI, etc.).

You will also determine the impact of your system on your surrounding environment. If, for example, your system generates a lot of heat through high-power dissipation, how does that impact adjacent systems by modifying their environment?

Functional analysis based layered analysis methods

One of the analysis basics that can help the systems engineer in solving difficult problems where the number of contributing functions is large is to use your functional analysis skills. A systems engineer was assigned to solve an inertial navigator problem where, during the final test phase of the development program, the navigator indicated sudden inexplicable velocity takeoffs while it was stationary in the lab. It was sometimes necessary to run the system for six hours before this phenomenon would occur. The systems engineer asked a lot of questions, reviewed all prior test data recorded in the data books, reviewed the development specification, the design, the software, and anything else to help gain knowledge about the system and the problem. Next, the systems engineer brainstormed the question: What are all the actions of persons and things that could contribute to this problem? All the brainstormed ideas were gathered into a functional equation that indicated that the problem could be the result of the following possible contributors and combinations thereof.

The velocity takeoff could be caused by: function 1 (f1), a faulty signal conversion algorithm; function 2 (f2), an error in software; (f3), electronics hardware circuitry malfunction; (f4), temperature environmental condition; (f5), specific electronic hardware; (f6), the power supply hardware; (f7), time of the day; (f8), operator error; (f9), data analysis error; (f10), ground bounce; (f11), dissimilar hardware logic circuits such as complementary metal oxide semiconductor (CMOS) and bipolar combined in the system; and (f12), the navigator really is moving but it is so fast you can't see it.

Using this equation, the system engineer then started the analysis and associated testing to evaluate the contribution of each possible contributing function to the problem. Analysis of the signal conversion algorithm showed that if a large number of upper-order bits were set from the gyro data, the algorithm would indicate a velocity takeoff, but the algorithm had no errors. The software code was walked through and had no errors. The possibility of the navigator moving so fast that you couldn't see it was considered obviously not possible. A review of the data and operator performance eliminated time of day, operator error data, analysis error, and the power supply as possible contributors.

Special software and hardware monitor data-catching tests were developed. These tests provided data that indicated that when the system was oriented a certain way, had a positive going temperature, and entered a limited range of temperatures, it would sometimes generate an erroneous signal from the sensor. The erroneous signal was in one of three upper-order bits of the digital data word and was always from the Z axis.

The next task was the study and analysis of the hardware and the software manipulation of the hardware. A system hardware transfer function diagram was developed to help study the hardware design. The analysis showed that there were dissimilar types of hardware interfacing to each other at the critical sensor data interface. It is known that bipolar computer logic gets faster as the temperature increases, while the sensor CMOS logic is slower as the temperature increases. The analysis also showed a high-resistance CMOS ground return for the CMOS hybrid circuit. This high-resistance CMOS ground return was not a problem for the CMOS current levels; however if the bipolar bus charge had to be discharged through this high-resistance ground, the result would be ground bounce and slow charge removal from the bus, especially at higher temperatures. Calculations showed that the bus ground return would generate a 9-nanosecond 1.53 volt peak ground bounce when the bipolar computer tried to discharge the bus from all 1s. It was also discovered that the Z axis CMOS signal conversion circuits were slightly slower than the X and Y axes.

The systems engineer had used the functional equation to systematically lead the analysis team in an orderly process of evaluating the problem. Based on the analyses and the combination of the functional variables to be found as contributors to the problem, the systems engineer could then provide a theory of what caused the problem. Note that the problem was intermittent, and the system engineer still had not been lucky enough to actually see the problem occur.

The theory: When the computer requests data from the CMOS sensor circuitry, it strobes for data at 1.5 microseconds. The CMOS propagation delay increases with temperature such that it is also 1.5 microseconds at the temperature at which the problem occurred. This equality can exist without a serious problem until the following data conditions occur:

- Prior data on the bus have the upper-order bits at "1"
- Data from the CMOS is attempting to set these bits to "0"

When these three conditions exist simultaneously, the calculated ground bounce will occur, thus causing the computer to strobe "1" in an upper-order bit that really should be a "0." This occurs only for 9 nanoseconds, but unfortunately this 9 nanosecond occurrence is simultaneous with the computer strobe.

The theory was first proven by inserting a software instruction that put all "0"s on the computer bus before every computer request for sensor data. The system was run for long periods of time without a reoccurrence of the velocity takeoff problem. The final fix for this problem was to lower the resistance of the CMOS ground return.

In solving this problem the systems engineer used a method that employed functional analysis skills in developing a function-of equation. Then he took one functional term at a time and independently analyzed them as layers of the problem one at a time. Then, by combining the layers, it was possible to develop a theory of what was causing the problem. It should be noted that the systems engineer was not performing this analysis alone. Support was given by test engineers, technicians, system analysts, and software engineers. The lead systems engineer was the leader, the catalyst, the driver, the trainer, and the key communicator.

This analysis scenario is also an example of a mission impossible which I talked about in the introduction to this book. If the system hardware designer had done system engineering analysis of the interfaces of the CMOS hardware to the bipolar hardware, this problem would not have occurred. It is very important that you, as systems engineers, do the analyses that follow in this chapter early in the development process so you can avoid the need to solve difficult problems such as this.

Analysis basics conclusion: In chapter SEE-6, I am faced with the same problem I had in chapter SEE-5—too much to say! The range and depth of analysis topics is so broad it is difficult to provide enough examples for all systems, all designers, and all specific analysis areas. I have decided to leave out most of the mathematical examples in the interest of brevity. There are many other basic analysis texts that will provide details for all the analyses discussed. Our objective in this chapter is to provide an overview of those analyses that have proven to be problem areas in system developments.

I have, in this brief presentation of analysis basics, tried to get some fundamental messages across:

- **Tailor your analysis**—Always do a nominal; only selectively do more complex analysis.
- **Establish TPMs**—Key system attributes shall be selected for monitor.
- **Don't forget the arithmetic—**Some of the simple (sometimes tedious) analysis must be done: weight, cost, power, etc.
- **Do both top-down and bottom-up analysis—**The bottom-up analysis must be done to confirm the top-down!
- **Apply Nominal, Worst Case, RSS, Distribution, Characterization, and Modeling—**appropriately to all your analyses. Develop a graphical perspective of real world distributions.
- **Do environmental impact analysis**—as appropriate.
- **Use your functional analysis skills—**to do the layered analysis methodology to solve difficult problems. Avoid difficult problems by doing system design analyses early in the development process.

A note about *approximations* and *guessing*: it's OK to guess; you have to get into the ballpark to play the game. When in doubt make a guess. First approximations provide a baseline for analysis. Subsequent approximations will converge on your final result.

6.2 System Design Analysis

6.2.1 System Size, Weight, and Power

Technical performance monitors (TPMs) of size, weight, and power shall be required as continuous monitors of a system during its development process. Sometimes you forget to add up the weights of your system parts as it is developed, and you find out too late that you have an overweight problem. Dimensions and volumes shall be recorded for all the subsystems in a table format. Don't forget to include the cables, the mounting racks, and support equipment. Date your data table and periodically update it as shown in figure 6-11.

Weight data for all the items on the sizing table shall be estimated, calculated, and ultimately measured as the system is developed. These weights shall be recorded on the sizing table periodically, with date of last estimate indicated. Power dissipation for the system shall be recorded on the sizing table in like manner, following the estimated, calculated, and, ultimately, actual measurement sequence. Power analysis is especially critical because it impacts power conversion, supply, distribution, and Electromagnetic Interference (EMI) design requirements for the system. Power analysis shall include the following:

1. Summation of all the nominal power dissipation of each subsystem

2. Summation of all the maximum power dissipation of each subsystem followed by an RSS calculation. Note that this is an RSS of the difference from nominal to maximum for each subsystem. This RSS of differences is then added to the system nominal to determine power supply maximum capability design requirement.

3. Development of a mission time line of system function, upon which the subsystem and total system load profiles will be plotted. This requires an analysis of peak power transients from nominals of components and subsystems, then summation on the time line.

   If peak power required exceeds the RSS maximum capability design requirement from step two, then provide an additional power supply requirement for peak power and time duration for that peak as a power supply design requirement.

4. Use of the power analysis data of step three to design the system charge distribution design. This includes determining how much power distribution capacitance is required and where it shall be located. It also includes the design of EMI filters and a ground system that reduces system power noise and ground current dynamics. Note that additional power distribution analysis methods will be discussed in 6.2.6, "System Power Distribution Analysis."

CONTROL SYSTEM SIZE, WEIGHT, AND POWER ATTRIBUTES (AS OF 10-15-87 REV D)						
				POWER - WATTS		
DEVICE	DIMENSION INCHES	VOLUME INCHES CUBED	WEIGHT LBS	NOMINAL	MAXIMUM	PEAK
Command and Cntrl Unit (CCU)	8 x 7 x 12	672	21	25	32	32
Input/ Output Unit (IOU)	8 x 7 x 6	336	12	15	24	28 (1 m S)
Control Panel Unit (CPU)	4 x 8 x 4	128	4	5	7	7
Cable CCU-CPU	92 x 0.5 diam	18.1	2 (1.5 + 0.5)	NA	NA	NA
Cable CCU-IOU	48 x 1.0 diam	37.7	4 (3.0 + 1.0)	NA	NA	NA
TOTALS for Cntrl System	NA (see sketch)	1191.8	43	45	63	67 (1 m S)
TOTALS for Tester	18 x 10 x 24	4320	60	100	120	128 (10 secs)

This table gives an example of a sizing table that provides size, weight, and power data for a system.

Fig. 6-11. Size, weight, power TPM table

Here are some *weight estimating aids* to be used for quick estimating only. Do detailed analysis and measurements when weight is critical.

Volume of cables: $V = \frac{L\pi D^2}{4}$ (L = length, D = diameter)

Weights:

(Dry air at 72°F = 0.0006912 oz/in^3)

(Water =0.58 - oz/in^3) (liq U.S.)

(8.347/gal) (liq U.S.)

(62.44#/ft^3) (liq U.S.)

(Aluminum = 1.568—oz/in^3)

(Copper = 5.168—oz/in^3)

(Typical weight of electronics

= 0.50—oz/in^3)

(depends on packaging scheme 0.50—oz/in^3 is for inline dips on a multilayer printed circuit board)

(Typical cable weight

= 1.92—oz/ft/35 lines of #22 AWG SHLD)

(depends on cable construction, also add in weight of connectors)

Note, total weight is the summation of weights of the chassis, air, internal wiring, and electronics.

6.2.2 System Cost, UPC, LCC

Cost must also be continuously monitored as a system is developed. Unit product cost (UPC) and life cycle cost (LCC) tables shall be developed and maintained in the same way as the weight, size, power table.

Definitions

UPC is the cost of a product as produced (including cost of material, production labor, and production test). The development of a specific equation for a specific product defines its cost elements. In order to do UPC analyses, you must know how many of the product are to be produced. Examples of top-down and bottom-up UPC cost models are provided in section 6.5.

LCC is the cost of a product throughout its life, which includes the cost of developing, producing, maintaining, and disposing the product. A product can be designed to be ultrareliable by using the most expensive parts and ultraconservative design; field maintenance and repair can then be negligible. The product also can be designed with the cheapest parts, resulting in a lot of field maintenance and repair costs. These statements reflect the impact of LCC analysis. The basic elements of a life cycle cost equation are: LCC = (Research and Development Cost) + (Production and Construction Cost—UPC) + (Operation and Maintenance Support Cost) + (Retirement and Disposal Cost). You must know how many products are to be produced in order to do LCC analysis.

- DTLCC—Design to LCC that could be a specific performance requirement in the DS part 3.2.1
- Value Engineering (VE)—A methodology of redesign to reduce cost. It is the cost analysis part of functional analysis. What is the lowest cost implementation to provide the required function? How much does it cost to do the redesign? How many units are to be produced? Given the number of redesigned lower-cost units, is the resulting total saving much greater than the cost to redesign? This could be a specific analysis required in the PMP part 3.2.2.9.
- VECP—A VE Change Proposal that could be a proposal to your customer to solicit support for a development program to reduce the cost of an existing product or system.

Designing for Cost

- Cost is only one of the system attributes. If you design a system based only on cost, you could have a poor-quality, low-reliability system. You must take *all* the system attributes into consideration: cost, size, weight, power, mission effectiveness, reliability, maintainability, and user acceptance.
- You as systems engineers must understand the generic top-down UPC model shown in figure 6-12 so you understand the mission requirements are the real cost determinates. It costs a lot more for a mission to take a trip to the moon than for a mission to take a trip on our world.
- You must also evolve a bottom-up UPC model as your system is developing, as shown in figure 6-13 so you can monitor the cost of your system as it is developed.
- You as systems engineers must apply value analysis methods as used in value engineering—functional analysis, number of systems to be built, cost to change.

Recommended references for further study of cost analyses are *Systems Engineering Analysis,*[1] which has a chapter on design for economic feasibility which provides life cycle cost analysis methods, and *Computer-Aided Systems Engineering,*[2] which includes a chapter on scheduling and costing involving various models, references, and the application of computer-aided tools.

UPC = f_1 (Material + Material Acqn) (Material Design Factor) (Material Requirements Factor)

+ f_2 (Build + Build Sprt) (Build Design Factor)(Build Requirements Factor)

+ f_3 (Test + Test Sprt)(Test Design Factor)(Test Requirements Factor)

UPC = f_1 (M + MA) (MDF) (MRF)+ f_2 (B + BS) (BDF) (BRF)+ f_3 (T + TS) (TDF) (TRF)

COST	= PROCESS	+ PEOPLE	+ THINGS
f_1 MATERIAL COSTS			Parts Materials
Material Acqn Costs	Parts Procurement	Procurement Incoming Test Incoming Inspn Parts Crib Technics	Cribs Incoming Insp Testers Paperwork
(The Cost Driver)	Design to Customers Mission Reqmts	Designers + Support People	
(Resulting Cost Drivers)	Customer Liason Customers Mission Driven Reqmts: Reliability, Environment, Size, Weight, Power, Accuracies	Marketing Program Mngmt Designers Customers	Specifications
+			
f_2 BUILD COSTS	Production Build	Production Technicians	Production Equipment and Facilities
Build Support Costs	Build Inspect	Production Engnrs Inspection Quality	Misc. Mtl
(A Cost Driver)	Design for Low Cost Production	Designers Value Analysts	
(Cost Drivers)	Customers Life Mission Driven Reqmts: Standards, Board Reqmts, Solder, ESD	Marketing Program Mngmt Designers Customers	
+			
f_3 TEST COSTS	Test	Test Technicians	Test Equipment
Test Support Costs	Test Monitor	Test Engnrs Inspection Quality	Misc. Mtl
(A Cost Driver)	Design for Low Cost Testability	Designers Maintainability Analysts	
(A Cost Driver)	Customers Life Mission Driven Reqmts: STD Testers, Bite, BIT	Marketing Program Mngmt Designers Customers	

Fig. **6-12.** Generic top-down UPC model

A12—FCC—UPC MODEL

$$UPC = f_1\begin{pmatrix}\text{Analog I/P}\\ \text{A11}\end{pmatrix} + f_2\begin{pmatrix}\text{Chassis}\end{pmatrix} + f_3\begin{pmatrix}\text{Digital I/O \#1}\\ \text{A14}\end{pmatrix} + f_4\begin{pmatrix}\text{Digital I/O \#2}\\ \text{A15}\end{pmatrix} +$$

$$f_5\begin{pmatrix}\text{Discrete I/O}\\ \text{A16}\end{pmatrix} + f_6\begin{pmatrix}\text{Discrete I/O}\\ \text{A17}\end{pmatrix} + f_7\begin{pmatrix}\text{Discrete O/P A}\\ \text{A18}\end{pmatrix} + f_8\begin{pmatrix}\text{Discrete OP/B}\\ \text{A19}\end{pmatrix} +$$

$$f_9\begin{pmatrix}\text{EMI Fltr Card}\\ \text{A1}\end{pmatrix} + f_{10}\begin{pmatrix}\text{Final Assy}\\ \text{and Test}\end{pmatrix} + f_{11}\begin{pmatrix}\text{IOP 1750A}\\ \text{A12}\end{pmatrix} + f_{12}\begin{pmatrix}\text{LVDT Excit}\\ \text{Ps-4}\\ \text{A5}\end{pmatrix} +$$

$$f_{13}\begin{pmatrix}\text{MPU 1750-A3}\\ \text{A13}\end{pmatrix} + f_{14}\begin{pmatrix}\text{PS I}\\ \text{A2}\end{pmatrix} + f_{15}\begin{pmatrix}\text{PS 2-3}\\ \text{A3}\end{pmatrix} + f_{16}\begin{pmatrix}\text{PS 2-3}\\ \text{A4}\end{pmatrix} +$$

$$f_{17}\begin{pmatrix}\text{Servo Amp 1-2}\\ \text{A7}\end{pmatrix} + f_{18}\begin{pmatrix}\text{Servo Amp 1-2}\\ \text{A8}\end{pmatrix} + f_{19}\begin{pmatrix}\text{Servo Amp 3}\\ \text{A9}\end{pmatrix} + f_{20}\begin{pmatrix}\text{Servo Amp 4}\\ \text{A10}\end{pmatrix}$$

$$f_{21}\begin{pmatrix}\text{Production, inspection, and}\\ \text{material support people}\end{pmatrix}$$

MATERIAL SUB-ASSEMBLIES: $f_1, f_2, f_3, f_4, f_5, f_6, f_7, f_8, f_9, f_{11}, f_{12}, f_{13}, f_{14}, f_{15}, f_{16}, f_{17}, f_{18}, f_{19}, f_{20}$.

FINAL ASSEMBLY AND TEST: f_{10}.

PRODUCTION SUPPORT AND INSPECTION: f_{21}.

Fig. **6-13.** UPC model bottom-up

6.2.3 System Accuracy and Error Analysis

The converter example in figure 6-5 A/D showed the top-down requirements for measuring temperature. A designer has now designed a circuit to meet the requirement (fig. 6-14). The systems engineer now needs to do an error analysis to confirm that the design will meet the requirements and will result in a high yield in production.

The example given in figure 6-14 provides a schematic diagram that shows components, cables, connectors, critical parasitics, and five different analog "A" ground points in the system. The transfer function equations for the basic sensor is given by:

$$T(t) = -40 + (111)\,(1 - e^{-t/l})$$

(note: l is a function of mounting)

An ERROR TERMS= (equation) is then shown. This equation lists twenty-one possible error terms that are initially established by a brainstorming session. Write down all the possible things you can think of that could possibly have an impact on the accuracy of your temperature measurement. You *save all* the possible error contributors, and then you take them one at a time and do an analysis of what they contribute to the overall error. The detailed backup analysis of each

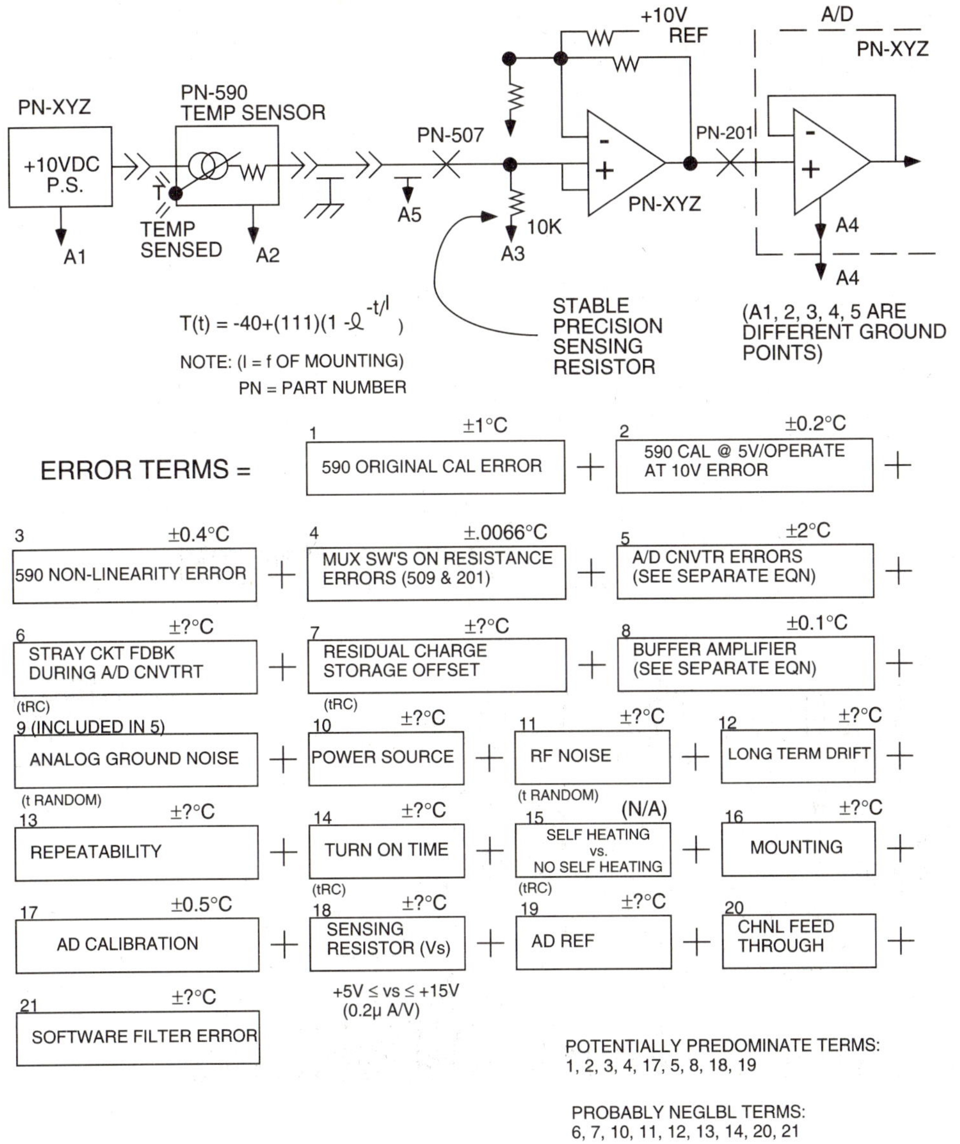

Fig. **6-14.** Temperature measurement error sources

term can be quite lengthy; therefore, for the sake of brevity, I do not include that here. Your objective is to establish the magnitude of the major error terms. In some cases this requires special in-circuit testing and data taking. You then can confidently apply a root of the sum of the squares calculation (RSS'g based on the "central-limit theorem"). This will help you establish the expected production yield of systems that will meet the specified temperature measurement accuracy requirement. In this specific case the requirement for accuracy was ±3°C. The worst case of the known predominate error terms was ±4.2066°C, while the RSS was ±2.29°C, which shows the present known error contributors do not exceed the specification for 99.74% of all the systems built. This assumes you have covered all major error contributors and you really do have 3 SIGMA error term data. NOTE! You must now also allocate ±0.5°C for production laboratory instrumentation to measure the test environment.

Initially, in section 6.1, I talked about the systems engineer allocating from the top down a required accuracy by budgeting it out within the system. Then you did an initial accuracy analysis based on designers bottom-up feedback resulting in a reallocation. Finally, in the just preceding example, I showed the systems engineer doing an analysis of the final design by generating a model and a brainstormed error source equation. Analysis of this equation with selective test data, when appropriate, then confirmed that the specification could be met with confidence of 3 SIGMA yield. I also allocated part of the error budget to production measurement instrumentation. Assume in the preceding example you are the systems engineer. The customer requests you to measure the temperature to within ±2°C instead of ±3°C! What do you do?

6.2.4 System Throughput Analysis

Throughput is a term used to define the processing capability of a computer per unit of time. The gas mileage rating of automobiles is akin to throughput because both MPG and throughput are affected by a great number of variables, thus the potential for controversy.

Throughput is affected by the type of algorithms to be processed, the computer's architecture, its control implementation, the memory speed, external system architecture, use of high order language (HOL), and environmental variables. In addition, the impact of inefficient programming during real world application of the computer can make "throughput" less then expected.

The term "throughput" of a computer is used for comparison and measurement-of-adequacy analysis during system development and trade-off analysis. I shall therefore use the term but I shall use it only as a rough measurement tool, fully understanding the variables that impact it. If you assume the use of "off-the-shelf-computers," where the internal architectural structure of the computers are out of our control we will have the following variables:

1. Type of algorithms to be processed
2. Memory speed
3. External system architecture
4. Machine language or high order language (HOL)
5. Environmental effects
6. Programmers' efficiency
7. Internal computer system architecture

1. **Type of algorithms to be processed**. How close does the instruction mix type fit the application to be performed? Is this computer designed for data processing or control algorithms, etc.?
2. **Memory speed**. Do memories exist that have the speed required? If so, is the power, size, cost, and availability of these memories within the scope of the system required attributes?
3. **External system architecture**. Is memory access affected by other devices direct memory accessing (DMA'g)? What impact does high-priority interrupts have on throughput?
4. **Machine language or HOL**. What will be the impact of HOL inefficiency on the throughput?
5. **Environmental effects**. Will system temperature extremes require running the computer at slower clock rates and/or require additional memory access time?
6. **Programmers' efficiency**. Will those who program the real-world system algorithms be code efficient?
7. **Internal computer architecture**. How does your computer operate? Multiple instructions per clock cycle may be performed.

Therefore, when you read KIPS or MIPS (thousands of instructions per second, millions of instructions per second), remember these terms are like MPG (miles per gallon) claims. They are subject to variables in the real world of computer system application.

6.2.5 System Interface Analysis

The system engineer must do analysis to ensure that *all* the interfaces of the subsystem will work together. To put *all* into perspective look again at the system perspective cube and the sortie and life mission analyses. Hardware must interface with hardware, software with software, hardware with software, people with people, documentation with documentation, and timing with timing. Also your subsystem must interface with the bigger system above. These interface analyses are done to avoid mission impossibles such as:

- Connectors that won't mate
- Poor impedance matching
- Poor frequency response
- Miswirings
- Insufficient current carrying capability
- Boxes that won't fit in mounting
- Poor thermal heat transfer

- Mounting that breaks in vibration
- External interrupts that cause software to hang up in do-nothing loops
- Lack of software function visibility
- Nonportable software
- Software that does not take finite hardware timing into consideration
- Software that assumes power up is always a clean step function to "on"
- Hardware and software people not talking to each other
- Customers not talking to suppliers
- Interface control document (ICD) or equivalent not provided or poorly done
- EMI test facilities not available when needed

I have exposed you to system engineering skills and tools to help you develop discipline in doing your interface analysis: system perspective, mission analysis, functional analysis, DS process, and PMP process skills are now to be expanded. An interface diagram is required in the DS that you will expand into specific tables and specifications as the DS develops. An interface analysis is necessary in the PMP where you are concerned with the people on the program and how they interface with each other as well as with the customer. I made you think in terms of your "subsystem" that must interface to a major system as well as the components that make up your subsystem. You started doing multiple mental walk-throughs of your sub-system as it performed its sortie missions, picking up requirements (actions to perform, constraining conditions to apply) for interfaces. You also started doing the mental walk-throughs of the life mission for your subsystem, also picking up requirements (actions to perform, constraining conditions to apply). Interface analysis for a successful system requires the right attitude, some basic engineering skills, and communication.

6.2.6 System Power Distribution Analysis

The design of systems and circuits so that they can be powered up properly is sometimes overlooked. Digital systems and circuits do not always come up in a known state, especially when they have a short duration power interruption. Analog systems and circuitry pass through instability regions and actually oscillate. The fact that one-shot circuits always fire at power up is not always taken into account. High-power drivers come up biased on, providing turn-on spikes to loads where driver source power was previously applied. Power sequencing of multiple power supplies is not designed in, resulting in random power up effects. Test power up conditions are different from actual usage of the system power conditions. The systems you interface to are powered up prior to application of power to your system or vice versa, and circuits lock up, burn out, or suffer insidious reverse breakdown failures.

The impact of power application upon your system or circuit design must be analyzed! Slow power up, fast power up, power interruptions from very short interruptions to long interruption, all sequence possibilities of multiple power supplies, and the impact of out-of-tolerance or out-of-sequence power application must all be analyzed! Draw detailed mission time line diagrams showing application of power, the appropriate slow/fast rise and fall of voltage levels, and the response of the circuit and system to these levels in time. Now on your time line turn power off. Then, relatively quickly back on, reassess how your system reacts. Continue these time line analyses until all combinations have been studied.

Power up control methods

If you discover you have potential power up problems you may design in features such as:

Master clear circuit

Provide a circuit in your system that has a turn-on threshold lower than any other control device in your system as shown in figure 6-15. This circuit shall clamp all critical logic control circuits into a known safe power up state.

Master clear timer

Provide a timer that ensures the master clear signal in figure 6-15 remains on for a sufficient amount of time to allow stabilization of the system power distribution and system circuitry.

Interlocks

To ensure proper sequencing of power, provide power sequencing interlocks such that power cannot be applied out of sequence or your system can not be powered up unless all the proper connections including safety grounds are in place. Be especially careful when testing your system. The various test configurations may require special simulated loads and interlocks to ensure your power system will not be too lightly loaded, or too heavily loaded, or that you will not experience unusual load transients.

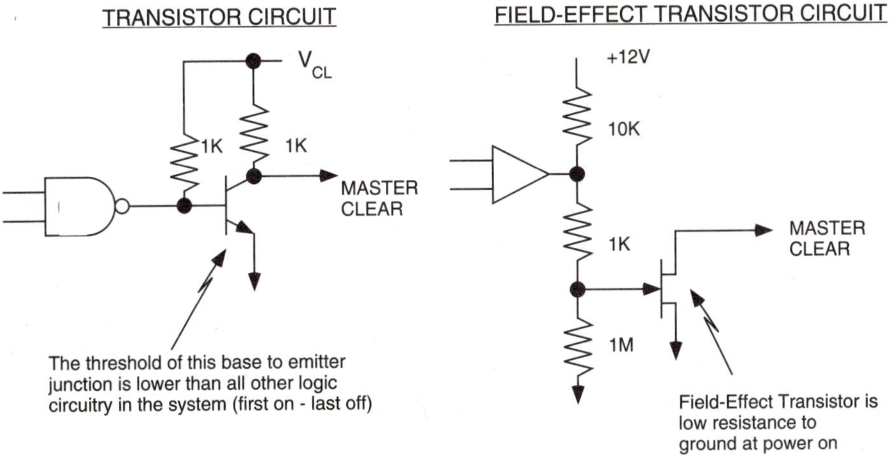

Fig. **6-15.** Example master clear circuits

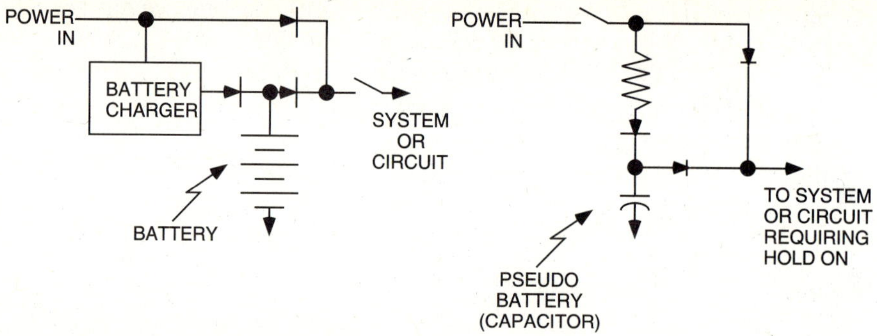

Fig. **6-16.** Example battery backup circuits

Battery backup

Provide a battery or a pseudobattery circuit as a parallel power source to overcome momentary power interruptions as shown in figure 6-16.

Power profile analysis

The interfacing of your circuit or system to a power supply requires analysis to ensure the power supply is capable of providing power as required by the load. The interface between power supply and load shall be designed to accommodate the characteristics of both load and supply. Load dynamics, power supply and load model or equivalent circuit, de-coupling rules, local circuit energy storage, surge suppression, and clamping are all areas that must be analyzed. Load dynamics or power profile is especially important with high-efficiency current mode power supplies because of their high inductance output circuit. Large sudden load changes cause current changes in the inductance generating large Ldi/dt voltage transients. Power supplies with large capacitance type outputs are also susceptible to sudden load change problems, a condition called "ground bounce" can occur in your system due to instantaneous high currents related to redistribution of charge.

To establish the power profile for your circuit or system, draw a time line and plot your load requirements as a function of time. A memory, for example, may require 1 amp at 5 volts (5 w nominal), but when addressed will jump to 2A at 5 volts (10 w) for 250 ns. The initial space shuttle engine controller power analysis showed a power profile: 170 w, 67 w, 34 w, 300 w, 180 w, 120 w, 160 w, 134 w, 177 w, 151 w, 120 w, 100 w, 0 w over a 740-second time line with some power conditions as short as 200 msecs. You must design your system or circuit to accommodate the power profile required.

Power profile design methods

Design your power distribution and energy storage capacitance so that it presents slow load transitions to the power supply. A general rule should be that in-circuit storage capacitance shall be provided for each circuit sufficient to supply energy for the load transients of that circuit.

For our memory example:

If V = Constant = 5V

And if ΔI is 1 amp: (2 amp peak – 1 amp nominal = 1 amp)

Then our Δ power is: (1A)(5V) = 5 w Δ

Applying power and charge relationships:

Q = C V = Coulombs
Joules = 1/2 CV^2 = Energy stored in a capacitor
1 Joule = 10^7 ergs
1 watt = 10^7 ergs/second
1 watt = 1 joule/second

Our 5 w= 5 joule/second

$$5 = 1/2\ CV^2,\ C = \frac{5X2}{V^2} = \frac{5X2}{25}$$

C = 400,000 UFDS for 1 second

for 0.1 msec C = 40 UFDS

In conclusion, a 40 UFD capacitor located on our memory circuit will provide 5 Joules/0.1 msec that would be more than adequate to supply energy to our memory for the 250 nsec, 5 w transient. The power supply sees a more stable load, and power/ground current transients are, in effect, filtered. In order to reduce power distribution noise in your systems you should, as part of good power distribution design, provide sufficient locally stored energy to handle local high-load variations. This, in effect, localizes the high instantaneous currents to specific small circuit areas.

System Calculation of stray parasitics in printed circuit boards, to determine impact on system performance and power profile/charge redistribution. The distributed capacitance that is present within a printed circuit board or an integrated circuit package is usually considered as parasitic because it at times has been a source of circuit malfunction, which ultimately requires costly printed circuit cuts and jumpers to fix. You shall consider these parasitic capacitors as additional circuit components that can be designed in or out of the circuit as the designer wishes. When you deliberately design them in, they are no longer undesirable parasitics but become known integral parts of your design, which can improve your circuit. The following calculations should be done where critical timing, low signal level, high-frequency, high-impedance or high-power fast switching conditions exist. The printed circuit board and integrated circuit shall then be designed in such a way that they will reduce or add capacitance as desired to improve circuit performance. The calculation of this capacitance can be done by considering the geometry of the circuit and the dielectric constant of the insulating material between the circuit layers. Figure 6-17 is an example calculation for eight-layer glass-epoxy board capacitance between parallel signal paths on adjacent planes. While Figure 6-18 shows the parasitic capacitance calculation for parallel paths on the same plane. To do quick estimating you can typically expect 7 pfd/in for adjacent planes and 0.5 pfd/in for same plane parasitic capacitance. Refer to the ground-bounce example shown in figure 6-23 to see the possible bad impact of parasitic capacitance.

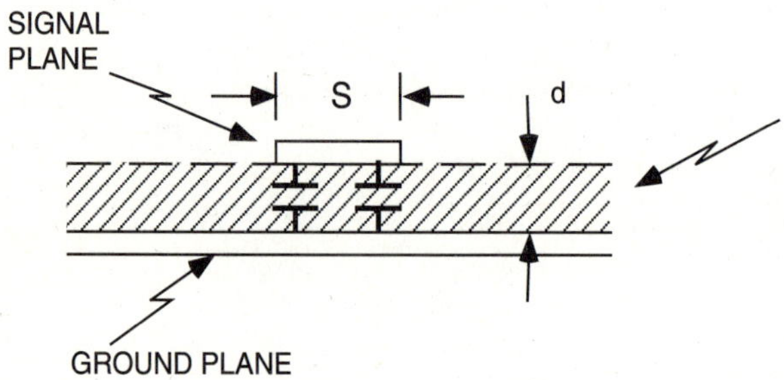

K=5 (Dielectric constant for glass epoxy boards)

N = Number of plates
N = 2
C = pfd (pico farads)
S = Area of one plate in (inches)
D = thickness of dielectric in inches
Note:

$$C = 0.224 \frac{Ks\,(N-1)}{d}$$

If "S" and "D" are in centimeters change 0.224 to 0.0885

EXAMPLE:

If d = .0093 inches (for an 8 layer multi-layer printed circuit board)

THEN:

"S" Inches	"C" pfd/inch
.013	1.6

.020	2.4
.030	3.6
.040	4.81
.050	6.02
.060	7.22

Note: As "d" becomes smaller (in 12-layer boards for example) the board parasitic capacitance will become larger.

Fig. 6-17. Distributed capacitance signal to ground plane or to adjacent signal plane

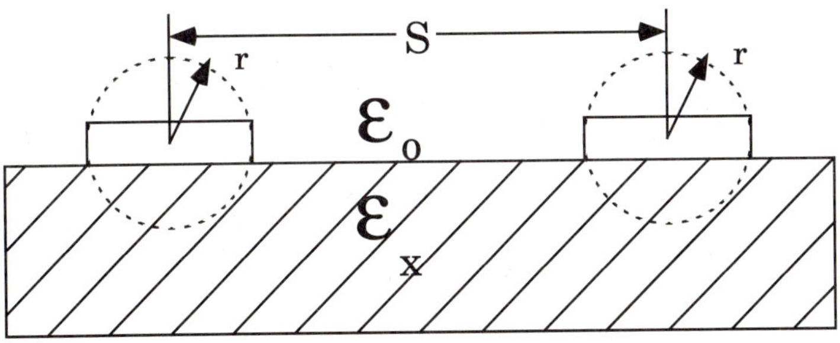

(Approximation to open two wire)

$$C = \frac{\pi \varepsilon_0}{\ln e \left[\frac{S + (S^2 - 4 r^2)^{1/2}}{2r} \right]}$$

$= \varepsilon_o = 8.85 \times 10^{-12}$ fds/meter (assume $= \varepsilon x$ also)

$\pi = 3.14$

S = Spacing center to center in meters

r = Radius of conductor in meters

If S = .714 mm	(.028 inches)
and r = .718 mm	(.007 inches)
then C = 21.5 pfd/meter	(C = 0.545 pfd/inch)

Fig. 6-18. Parallel paths same plane—parasitic capacitance

Calculation of distributed inductance

In addition to the calculation of the distributed parasitic capacitance in critical circuitry, you must also calculate the distributed inductance in critical circuitry. The self- and mutual-inductance of your printed wiring boards, connectors, integrated circuit leads, and, of course, your system hook-up wires can have a dramatic impact on how your system will work.

One foot of 30 AWG wire has 465 nanohenrys of self-inductance that can, from a 10 nanosecond 100MA current pulse, produce a 4.65-volt noise spike:

$$(e = L\frac{di}{dt})$$

If the same wire is placed over a ground plane, the noise spike can be reduced because of the impact of the parasitic capacitance. The system engineer must be sensitive to these characteristics. The goal is to calculate both capacitive and inductive features of the system packaging, then develop a model that includes these capacitances and inductances so analysis of critical circuits and interfaces can be done.

Figure 6-19, provides information for calculating the inductances for some common wiring configurations. My objective in exposing you to these low-frequency (10KHz) formulas is to develop your sensitivity and ability to ballpark these inductances. Skin effect at higher frequencies will lower the inductance and increase the resistance of wires (these formulas therefore, become safe approximations for our analyses). Note that these calculations are a must for high-speed switching circuit analysis as the impact of

$$L\frac{di}{dt}$$

can develop instantaneous voltage spikes many times larger than the circuit power supply voltage level.

High-speed switching of large inductances can cause 100V spikes from 12-volt power supplies. We sometimes forget that the self-inductance of some of the wiring in high-speed switching logic circuitry can cause noise problems in the systems.

Determining the impact of parasitics on your system

The parasitic inductances and capacitances shall be considered as an integral part of your circuit system design. The impact on circuit and system performance must be determined. Circuit and system designers are responsible for the packaging of their circuit or system. You must design your packaging; therefore, a key part of your system analysis is the analysis of your printed circuit boards, your integrated circuit packaging, your system wiring, and also your system grounding configuration.

Configuration	Formulas All dimensions in inches, inductance is in nH
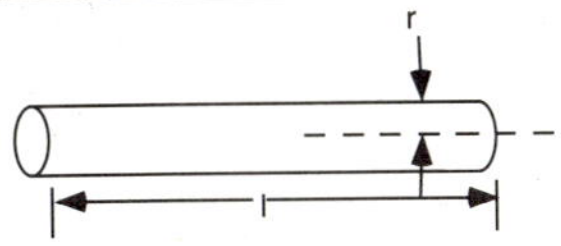 ISOLATED WIRE (l>>r)	$L_s = 5l\left[\ln\frac{2l}{r} - 3/4\right]$ ---- Low frequency, l >>r, ($\delta = 1/4$) $L_s = 5l\left[\ln\frac{2l}{r} - 1\right]$ ---- high frequency, l >>r, (limit as f → ∞, δ → 0)
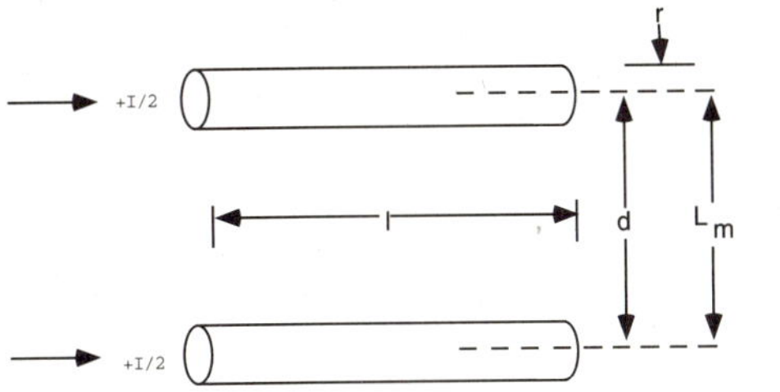 FEED WIRES IN PARALLEL	$L_m = 5l\left[\ln\frac{2l}{d} - 1 + \frac{d}{l}\right]$ Mutual inductance between thin parallel wires.
	$L_s = \frac{L_1 + L_m}{2}$ or $\frac{L_2 + L_m}{2}$ $L_s = 5l\left[\ln\frac{2l}{\sqrt{rd}} - 1 + \frac{\mu\delta}{2}\right]$ l > > d Over-all self-inductance of two parallel wires, as used in distribution cables, or bus bars, where current is in the same direction.
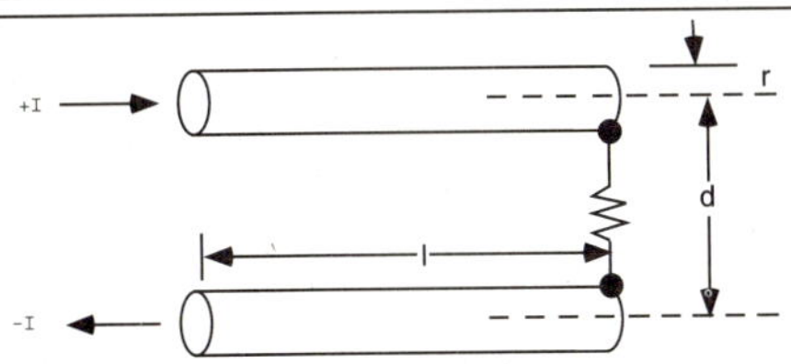 FEED AND RETURN WIRES	$L_e = L_s - L_m$ $L_e = 5l\left[\ln\frac{d}{r} + \mu\delta - \frac{d}{l}\right]$ Inductance of each of two wires in close proximity to each other, with current in each in opposite directions.
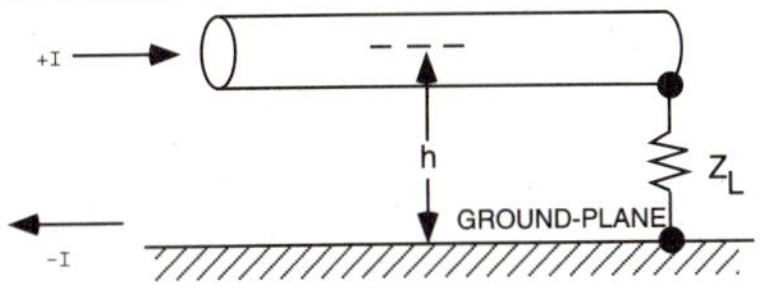 FEED WIRE OVER GROUND-RETURN PLANE	$L_e = 5l\left[\ln\frac{2h}{r} + \mu\delta\right]$ Self-inductance of a wire only, when current return is via a ground plane.

Notes: δ = wavelength,

$\delta = 1/4$ means 1/4 of wavelength

limit as f (frequency) → (approaches) ∞ (infinity) then $\delta \to 0$

μ = permeability of dry air = 1.0.

Fig. **6-19.** Inductance for common wiring configurations

System ground bounce

When your system has noise problems or your logic circuitry occasionally malfunctions, you may place your oscilloscope probes between two ground points

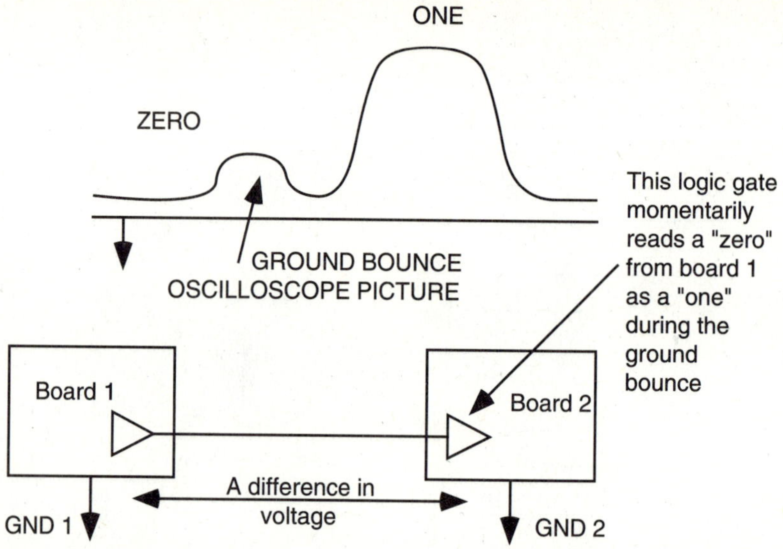

Fig. **6-20.** Ground bounce

in your system and see a varying voltage signal. This signal will have relatively high periodic voltage differences. This is referred to as "ground bounce." Our logic circuits sometimes consider a "zero" voltage level as a "one" voltage level because of this bounce.

Ground bounce can be caused by instantaneous high currents from the charging or discharging of distributed capacitance and/or inductance in your system. It can also be caused by fast switching power load changes. The instantaneous difference in potential between GND 1 and GND 2 in figure 6-20 shows ground bounce.

An example of ground bounce is provided in figure 6-21. This shows the need for the design of charge distribution and grounding methods to compensate for distributed parasitic capacitance. The example is from an actual system that had serious ground bounce. A 10 mHz logic signal sent from board 1 to board 2 would intermittently and incorrectly go from logical zero to logical one synonymous with software putting all zeros on the 16 bit bus after all ones. The ground reference of a high-frequency logic circuit on board 2 would jump up two-plus volts over the ground on board 1, so the circuit would interpret a "zero" from board 1 as a "one." This caused complete system failure. A secondary problem would also occur at a specific temperature where the all ones to all zeros transition took too much time to transfer a bus one to zero, causing the system to experience an incorrect "one" read off the bus that resulted in a velocity takeoff in a navigation system. The need for the system engineer to calculate or see that parasitic capacitance is calculated and circuit analyzed cannot be overstated.

GROUND BOUNCE EXAMPLE—SOFTWARE CAN CAUSE HARDWARE PROBLEMS:

Ground bounce is due to poor hardware design of the grounding and distributed power supply filtering. Software can establish a worst case charge redistribution condition by going from all "1's" to all "0's" on the bus, which through programming makes the ground bounce occur.

1) Take a 16 BIT Bus driver with 50 pfd distributed parasitic capacitance per line:

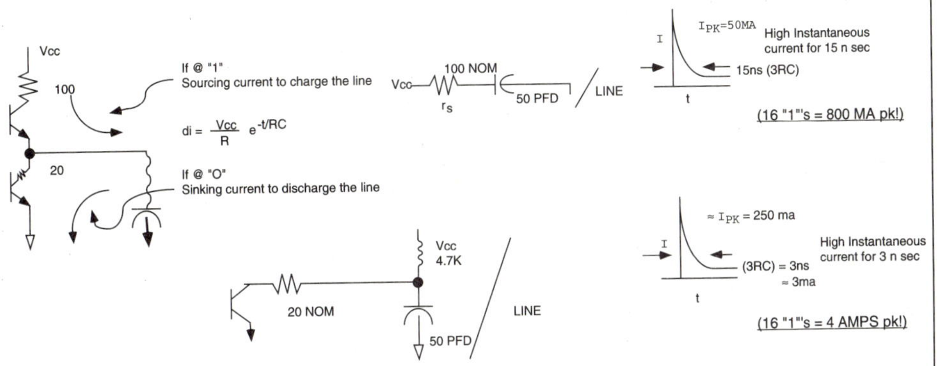

2) Charge redistribution from these high instantaneous currents can cause bounce effects where the difference in ground potential can be in volts!

3) How to fix ground bounce:

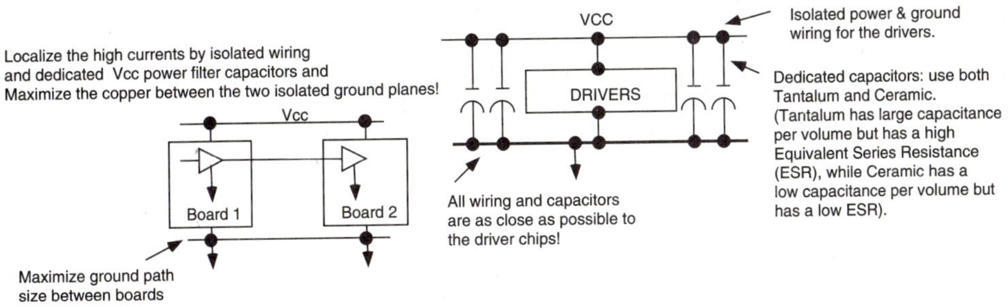

Fig. 6-21. Ground bounce example

Grounding methods and grounding design analysis

All the current that goes out must come back in the ground circuit! You must design your ground circuitry just as you design the rest of your circuit or system. The current density as a function of time should be known. You also must calculate the resistance of your ground return circuitry so that you know the voltage drops in your ground circuits. Instantaneous currents of amps can occur, resulting in instantaneous differences in potential in the grounding circuit of volts. Fortunately, most of the circuits you design are not affected by these transients; however, in radio frequency circuitry, high-bandpass amplifiers, and high-frequency digital circuitry, there have been problems related to noise in poor grounding design because these high-frequency devices can respond to the modulation of the ground.

Programming the dispenser

A very dramatic example of ground bounce occurred in the programming of a dispenser system where a 7-volt ground bounce transient caused computer upset, so it stopped because it could not read its memory. The dispenser system provided many channels of a high-energy pulse to dispense munitions by discharging 40 ufd capacitors through solid-state switches to each of the many loads. Software was provided an instruction, "charge capacitors," which was intended for a subsystem test routine without capacitors. Software, however, used this instruction because they could charge up all the capacitors with one instruction. Unfortunately, the instantaneous peak current, while charging all the capacitors, was 240 amps! Reprogramming by charging only one capacitor at a time just before dispensing a channel solved the problem. This example shows that hardware and software designers must be aware of each other's designs.

System single point ground

The ground bounce example in figure 6-21 showed that the ground can have instantaneous differential voltages due to high instantaneous currents. These voltage differences in the ground circuit are always present but are usually in the millivolt level. The ground circuit is obviously made up of finite resistances, which can be calculated. Then, by applying E=IR you can calculate the magnitude of the various ground differential potentials. In the ground bounce example 2-volt ground bounce at 4 amps relates to one-half an ohm. Sometimes these differences in ground potential are called "ground noise." You must do an analysis of this "noise" as you develop your system. Then, provide low resistance ground paths inversely proportional to the currents the ground must carry, and also provide distributed charge source capacitance to localize high instantaneous currents in small local areas to prevent modulation of the ground potential. You will carefully separate the digital circuitry ground return from the analog (sensitive) circuitry. You will provide ground decoupling circuitry between high noise grounds and grounds you wish to keep quiet. You will also choose a spot in your system that you will call the "Single Point Ground."

Where do you put this single point ground? "The single point ground for a system shall be placed as close as physically possible to that point in the system where the most sensitive signal measurement must be accomplished."

In most systems this will be your internal system voltmeter, the analog to digital (A/D) converter. Your A/D converters may have to measure millivolts or in some systems microvolts; therefore, you will try to keep the ground reference physically at the A/D as quiet as possible. Systems with microvolt measurement requirements need excellent ground system design. Also, in many cases, they need special circuit/system design techniques. These special techniques include differentially balanced input circuits, high common mode rejection, sampling methods that include settling time, ground and reference measurements, and computerized software accuracy enhancement.

Unfortunately, in trying to place the signal point ground as close as possible to your point of most sensitive measurement, you may have mechanical restraints. The chassis ground, input ground, the +15 Vdc analog ground, and 5 Vdc ground are at the power supply, while the A/D converter is on a plug-in board. In this real world condition, you will still place your single-point ground on chassis as close as possible to the A/D converter input, with as short as possible an analog ground connection to this single-point ground on chassis. Figure 6-22 is a typical system single-point ground implementation.

6.2.7 System Electrical Interference

A number of a specific type of airplane crashed in Europe while flying by the Voice-of-America transmitting tower. The radiated energy caused all three of the airplanes' "fly-by-wire" flight control channels to go hard over simultaneously, thus causing a "single-point-failure" that defeated the triple redundancy design of the airplane. This failure was caused by poor electromagnetic interference (EMI) analysis and EMI protection design. The term "EMI" is really a misnomer in this context. A better term would be "radiated energy interference," but EMI is accepted through common usage. This is only one of many examples of failures due to EMI. A key environmental analysis that must be done by the system engi-

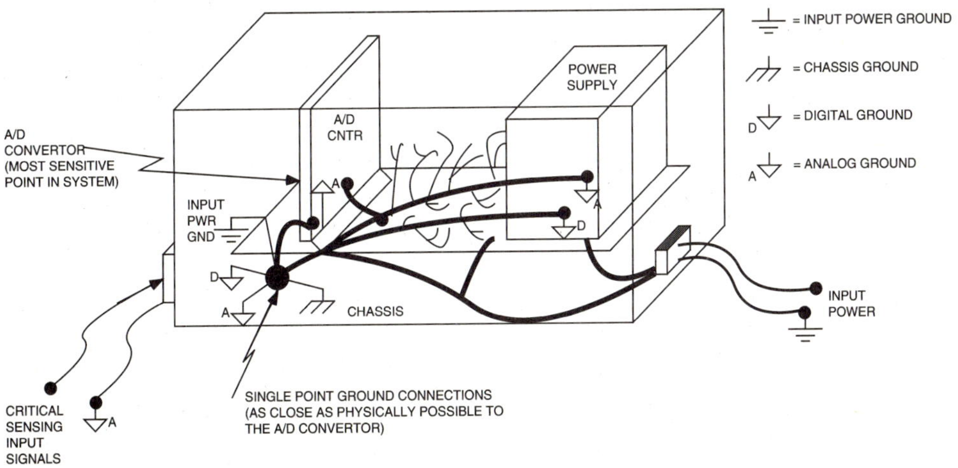

Fig. **6-22.** Single point ground example

neer is that of radiated energy from the system and also an analysis of how susceptible the system is to external radiating sources. *In addition, you must do an analysis of the radiation interference between the components internal to your system!* EMI is any electrical or electromagnetic phenomenon—man made or natural—that causes an undesirable response, performance degradation, or complete malfunction of your system. EMI varies in time and level, from a nuisance to complete destruction of mission performance. In fact, on the positive side, you sometimes develop countermeasure equipment, such as radio jammers, which are EMI sources with the objective of completely destroying someone else's mission.

The sources of EMI are seemingly unlimited. All electromechanical, electric, and electronic devices are sources of EMI. Gamma rays from a nuclear source can also be considered as a source of "radiated energy interference." The degrading effect of "total dose" on system components and the system interruption from "upset" events must be analyzed by the systems engineer. The systems engineer must be able to calculate and develop some sensitivity to magnitude and characteristics of all these radiated fields. In addition, naturally occurring phenomenon such as lightning and solar flares may affect your system, so you must analyze their impact.

EMI covers a frequency spectrum from DC to X-ray frequencies. At the DC end of this spectrum the systems engineer is aware of "interference" sources such as spurious sources: Thompson effect (temperature difference along a conductor provides a very small electrical potential difference); Peltier effect (two unlike conductors in contact at same temperature generate a voltage); Seebeck (basic thermocouple—combining Thompson and Peltier); Photo voltaic (can occur in copper to cuprous oxide boundary); and Piezoelectric effects (crystals put out voltages when mechanically stressed and vice versa). These interference sources are usually of little concern in most systems because the levels are so small, but the systems engineer does a mental scan of his system from DC to gamma ray frequencies and from pico- to megawatt power levels. The systems engineer must do an EMI analysis that gives him the perspective of a radiation frequency and power level signature (on the mission time line) of his system. He must also (on the mission time line) know the threshold of susceptibility of his system to radiated frequency/power level.

Reducing system EMI radiation and susceptibility

The systems engineer must design EMI prevention and containment into his system. Proper grounding, charge redistribution, shielding, balanced lines, twisted pairs, chassis layout, part selection, circuit design, software design, and filtering methods shall be specified in the development specification.

Grounding

Design your system with a single-point ground that is placed as close as possible to the most sensitive part of your system. Calculate the currents in the ground paths and provide sufficient stored energy around your system to isolate high-current transients. Accomplish this by doing a load profile analysis against your mission time line and then design your power distribution system with appropriate charge redistribution capability. Use ohms law and make sure that the instantaneous ground voltage drops are negligible. Ground wire current density could be specified to always be less than 400 circular mills/amp, but in some sensitive circuitry the noise immunity to ground noise (instantaneous voltage drop in segments of the ground system) could be in the microvolts. In those cases instantaneous voltage differences drops in the ground are very critical, so the current density must be limited much lower.

Shielding

Electric (E) and Magnetic (H) fields travel through free space at 3×10^8 meters/second. When these waves strike a conductor, some energy is reflected and the incident wave also sets up eddy currents in the conductor, which attenuate the signal. Thus, a shield reduces the effect of the electromagnetic wave by reflection and energy absorption. For electrically thick material:

$$R = 50 + 10 \log_{10} (PBf)—1$$

$$A = 1.7t\ (f/PB)\ .5$$

Where R = Reflected Wave in dB

A = Absorbed Wave in dB

PB = In ohms/cm

f = Frequency in MHz

t = Thickness in cm

"A" is dependent on frequency, thickness, conductivity, and permeability of the shield material. A major problem with shielding is that any discontinuity in the shield can leak EMI.

Shielding for EMI

Shielding is still the most important method of preventing or curing EMI problems. Maximum shielding effect can be realized if there are no breaks or points of entry in the shielding.

Mechanical designs

You must design your mounting boxes, chassis, and consoles as shields to prevent your systems from being affected by EMI or radiating so that they affect adjacent systems. Improperly designed openings in your mechanical designs and faulty welding can act as waveguide antennas. Above 100 MHz, leakage is serious, as the required slot size becomes smaller. The slot's length, not the width or area, determine its shielding effectiveness. For slot lengths smaller than the interference wavelength, an enclosure will shield; however, as the frequency increases, shielding becomes ineffective at the slot resonant frequency. Covers and panels must be in full contact with the body of the housing. A conductive gasket (i.e., braid or graphite impregnated rubber) must be used.

Coaxial cable

Coaxial cable provides an effective means of shielding an interconnecting line against interference and reducing radiation from the line. Proper termination is a must. The braid should be soldered so that the inner conductor is surrounded by the braid at the termination point.

Balanced lines

This method to minimize EMI susceptibility utilizes a balanced transformer at the source and load. The transformer primary and secondary windings are shielded from each other, and the transformer's shield and case are grounded. Balanced lines minimize induced voltages and currents caused by interfering fields and interference caused by a difference in ground potential.

Twisted pairs

Twisting the conductors along the length of the cable provides signal cancellation of EMI, but the length of twist is difficult to control. This method is only effective when the EMI frequency wavelength is longer than the length of the twist.

Filters

EMI can enter equipment through power and signal lines, or through the enclosure via radiation. The previously mentioned techniques are employed to minimize radiated interference, but they are ineffective against conducted power and signal line interference. EMI filters on power or signal lines are then required.

Power line filters, which are usually low pass, are designed to pass frequencies up to an appropriate frequency for your system low kHz range. Filters used on signal and power lines are classified by their type of frequency discrimination. Low-pass filters pass frequencies from DC to a specified cutoff frequency and attenuate all signals above by an increasing amount, while high-pass filters pass signals above a specified cutoff frequency and attenuate all signals below by an increasing amount. Band pass filters pass all signals in a specified band and attenuate all signals above or below cutoff, while band reject filters pass all frequencies except those in a specified band.

EMI measurement

EMI is measured in the same manner as radio wave field strength. However, peak rather than average values of noise are measured because you are interested in system momentary upset. The overall bandpass of the measuring device must be known to bound the range of your test and analysis. Higher frequencies could still be a problem, so you need to identify the limits of your measurement equipment.

6.2.8 Circuit Analysis by the Systems Engineer

All your circuitry must be analyzed by the circuit designers, but the systems engineer is responsible for analysis of all interfaces as well as the performance of the critical circuits in the system. The system engineer may not personally do all these analyses but is responsible for ensuring they are done. He also reviews them to make sure they are complete and correct. The systems engineer must be aware of critical timing, high-sensitivity, high-gain, high-frequency response, and high-power circuits in the system. These critical circuits and all interface circuits shall be modeled showing detailed input and output circuits. They shall also show inclusion of the appropriate model of interconnect wiring and spurious mechanical packaging elements.

I shall, as part of this overview of analysis, show a few circuit analysis examples. These actual circuits were designed by a designer, made to work in the laboratory, and then were analyzed by a systems engineer.

1. A high-speed clock driver
2. Differential operational amplifier—critical input
3. Operational amplifier—critical output

Note that the examples are dated, but the basic analysis methods and the circuit problems discussed remain applicable. Electronic circuitry has always and will always have elements of capacitance, resistance, inductance, gain devices, junctions, interface to the same world, feedback, frequency response, mechanical packaging, test configurations, power dissipation, currents, and voltages. These elements will exist if implemented by large-scale integrated circuits, small-scale integrated circuits, discrete components, or vacuum tubes.

I have selected the clock driver and the operational amplifier examples because the impact of parasitic elements, the criticality of the balance of the input to operational amplifiers, and the effect of loads on the operational amplifier outputs have caused very serious problems in systems. These examples should not be viewed as discussing only a digital or operational amplifier circuit, but should be viewed in the bigger context as referring to parasitics, high-gain, high-impedance, feedback, modeling frequency response, step response, and their impact on any circuit.

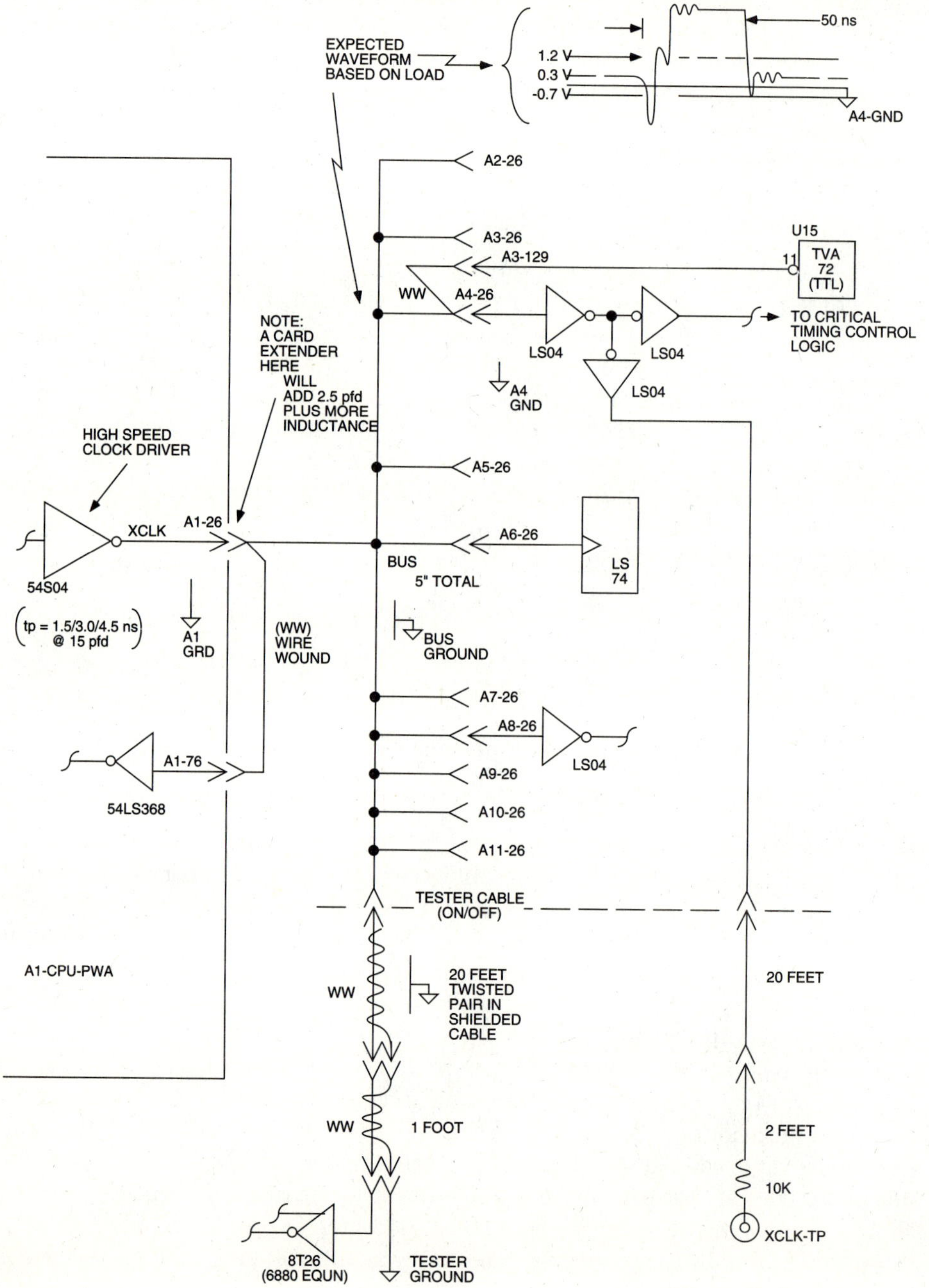

Fig. **6-23.** A high-speed clock driver circuit

Figure 6-23 demonstrates step one of the analysis of this critical timing control circuit. The first step is to draw a systems schematic diagram showing the circuit and its interface connections. You do an analysis of all the wiring and the fan out that is normally 5 but is 6 when the tester is connected.

It is obvious that this clock driver has a highly inductive load of mostly wire, and it will be dramatically affected when the tester is connected. The expected waveform will "ring" because of Ldi/dt. The amount of ringing could possibly cause the bus signal to exceed the 1.2v nominal input threshold of the transistor transistor logic (TTL) circuits. This probably will not be a problem because of the short time the inductive transient occurs and the inability of the LS04, LS74, or LS368 to respond to the transient. However, the 54S04 driver will generate internal EMI due to the fast switching, high currents in both rise and fall (the inductive load will cause ringing). The subsequent use of the clock pulse and the criticality of subsequent clocked circuitry timing must be analyzed to determine if the 54S04 (3 ns tpd) is needed for timing. I continue the analysis by drawing a model shown in figure 6-24.

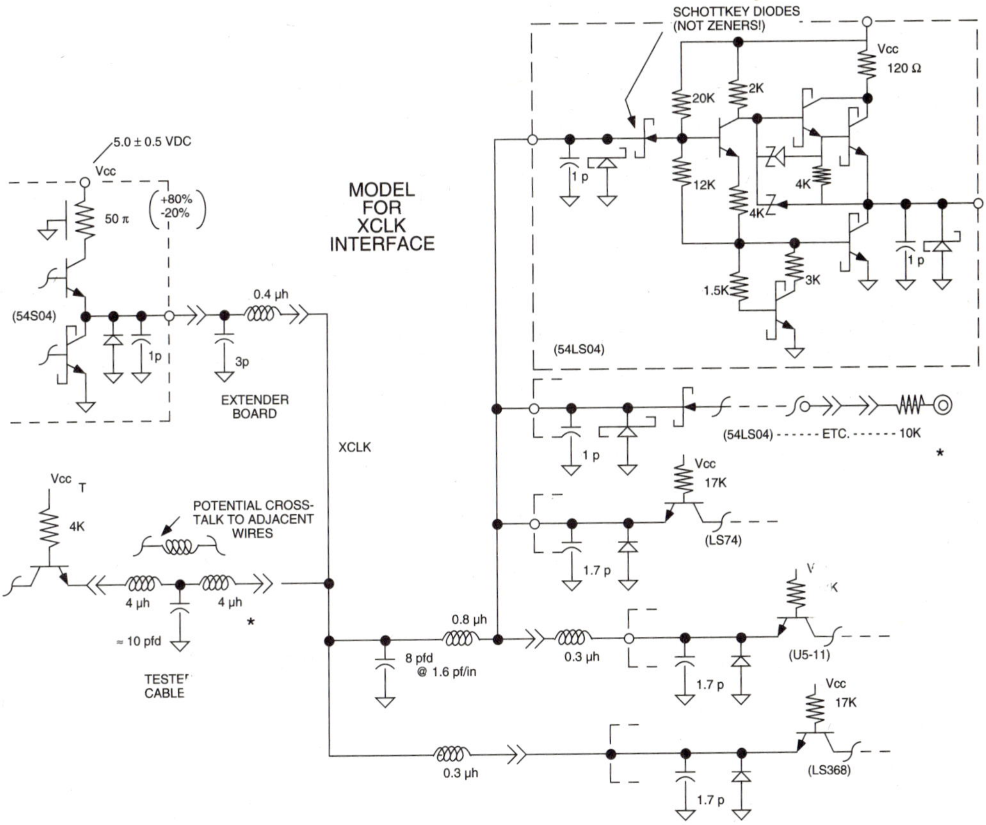

(Note: Self-inductance of 1 foot of wire = 0.465 μh.)

Fig. **6-24.** First model of high-speed clock driver circuit

Figure 6-25 shows how to reduce the model to what I am really interested in, which is the worst case lumped element model with tester and extender card connected. (Note that the systems engineer questions the need for the two tester interfaces for the clock signal.)

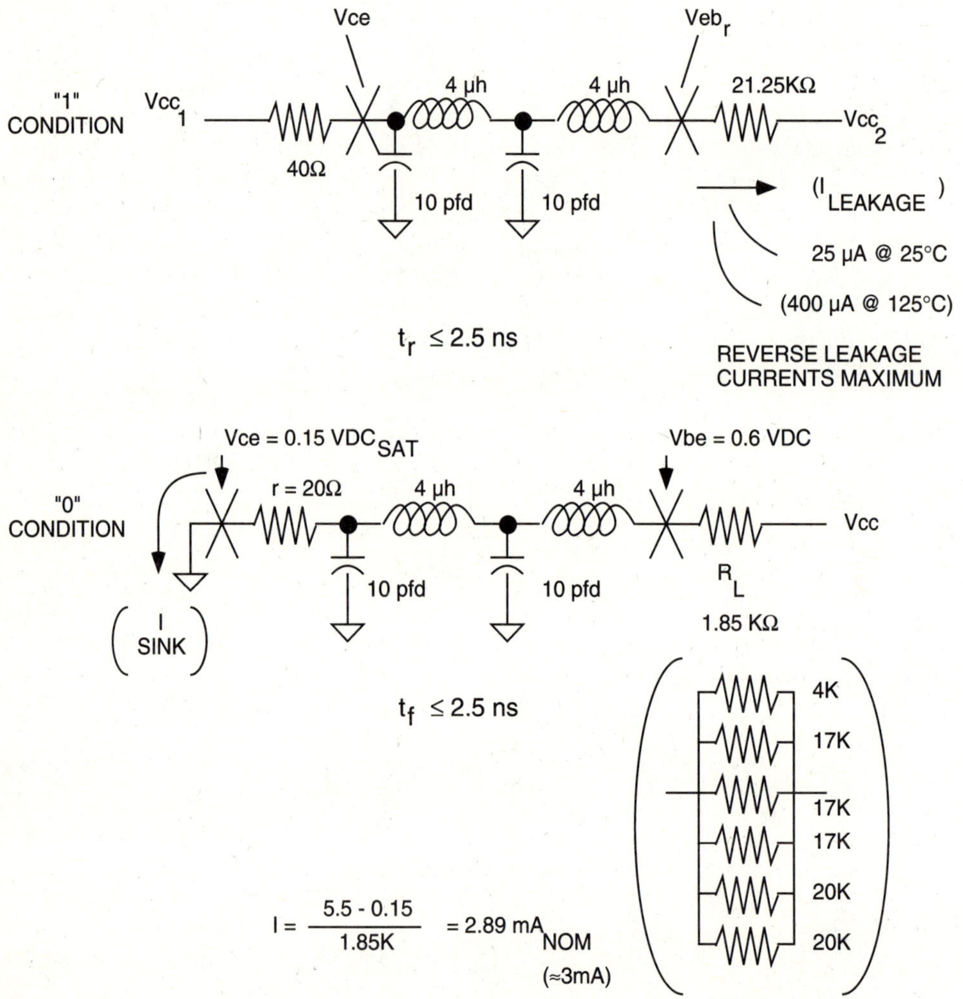

$$I = \frac{5.5 - 0.15}{1.85K} = 2.89\ mA_{NOM}\ (\approx 3mA)$$

$$\text{STORED ENERGY} = 1/2\ LI^2 = 36 \times 10^{-12}\ \text{JOULES}$$

$$\text{UNCLAMPED VOLTAGE TRANSIENT} = L\frac{di}{dt},\ dt = t_r$$

IF t_r = 1 NS, Vp = 24 V pk
= 2 NS, Vp = 12 V pk
= 3 NS, Vp = 8 V pk

Fig. **6-25.** Lumped element model for the high-speed clock driver circuit

These unclamped transients will really be clamped by the isolation diodes of the integrated circuits, but the next negative going ringing excursion will be (at 2ns)1.5V (.37x12x.37). This exceeds the 1.2V threshold of our logic, thus providing the possibility of two negative going transitions.

The post analysis recommendations by the systems engineering analyst:

1. Remove direct connection of tester cable (use LS04 buffered output).
2. Do a worst case timing analysis of tester operation. Ensure that signals are all equally delayed on the cable, and add a pulse stretcher to compensate for pulse shrinkage if necessary.
3. *Design* the tester cable.
4. Redo the driver circuit analysis for worst case timing for those circuits that use this control pulse. Note that removing the large distributed inductance will now cause the waveform to become like the one in figure 6-26. Therefore, you have to analyze pulse shrinkage and determine if minimum holdup times will be provided.

Critique of example circuit analysis #1 high-speed clock driver circuit

1. The systems engineer developed a diagram showing all the interfaces for this critical circuit.
2. Then he calculates the parasitic L and C elements.
3. Then he draws a model diagram that shows integrated circuit elements in detail.
4. Then he reduces the model to salient critical simple model.
5. Then he makes circuit design change recommendations.

Differential operational amplifier—analysis

A differential (operational) amplifier (OP-AMP), is usually the highest gain and/or most sensitive circuit in your system; therefore, the system engineer as well as the circuit designer must analyze the application of these amplifiers. I will not deal with the basics but will deal with the critical interfaces of input and

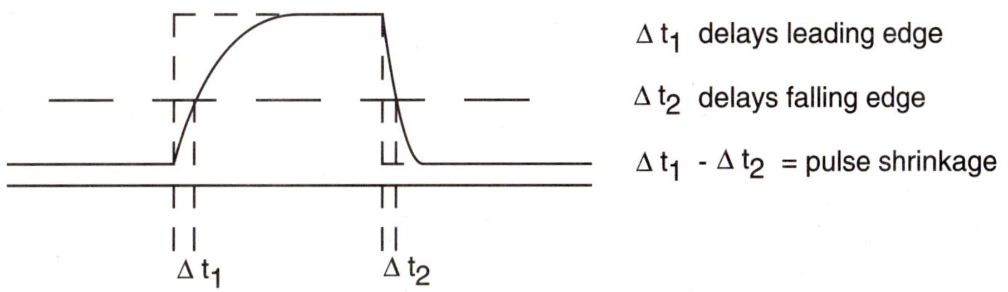

Fig. **6-26.** Pulse shrinkage

output of these amplifiers. The lack of interface design has been a serious contributing factor in OP-AMP problems. I will start with a discussion of the basic equation, analyze input interface, analyze output interface, and then cover autonull techniques.

The basic equation for the amplifier consists of the desired output term and four undesired output terms as shown in figure 6-27. The point is that this equation says that the Δ undesired terms require matching! If the V_{be} of the two inputs were perfectly matched, Δ would be zero and this term would drop out, etc. for I_{co} and ß, etc.

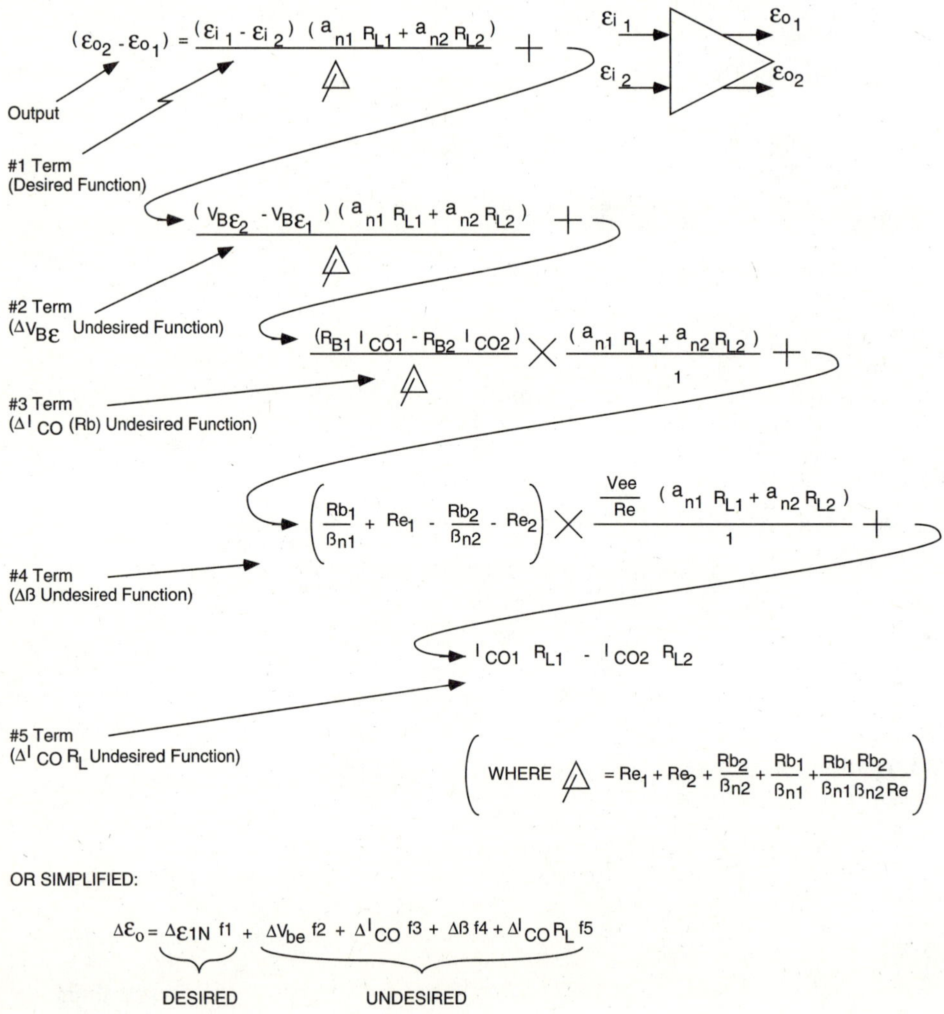

Fig. **6-27.** OP-AMP equation

Adding some real-world terms

To expand this equation into amplifier input source and output load (real world amplifier usage interfaces) will require additional terms: [(R_1 source Thev)—(R_2 Source Thev)], ΔV_{pwr}, Δ temperature, and [(R_1 Load Thev)—(R_2 Load Thev)].

Sensitivity analysis has shown that the input $\Delta R_{source\ (THEVENIN)}$ term is the largest contributor to error (ΔR_{load} has less effect assuming proper load design (see output Z); likewise Δ power supply voltage and Δ temperatures are also relatively small error contributors.

Therefore, in conclusion, to optimize a differential sensor system, it is desirable to balance out all the error terms:

1. (Amplifier) Basic amplifier—Select an amplifier with optimal match of inputs for matched and tracking ß, I_{co}, and V_{be}. This is usually easy to do by selection of an integrated circuit amplifier.

2. (Temperature) Packaging should maintain temperature balance of both sides of the input circuitry including source Thevenin circuitry. This is also easy by selection of an integrated circuit.

3. (Input Z) **The Most Important Critical Input Interfaces:** Thevenin source impedance to both sides of the amplifier must be balanced! This includes packaging and board layout considerations. This balance should include parasitic packaging capacitance and inductance. You need to mechanically design both input circuits to be exactly the same. Random and/or variable level common mode noise injection into these input source parasitic impedances can cause undesired differential input signals. Balance shall also consider that these source Thevenin impedances may be exposed to noise over a wide frequency range; therefore, the balance must be maintained over DC to a high frequency ("high" is relative, dependent upon specific systems and noise environments). In addition, the frequency response of your amplifier may be seriously reduced by a very little parasitic capacitance directly at the OP-AMP sum point. This high impedance is easily affected by very small capacitance. Packaging parasitics must be designed out (or in); therefore, include them in your model and Bode analysis.

   Serious system failures due to unbalanced input impedances picking up external EMI noise are too numerous to mention (airplane crashes in Europe, helicopter crashes, semitruck brake failures, etc.). The importance of this analysis cannot be overstated. An unbalanced input to an OP-AMP can be very sensitive to EMI. Remember, also, that the signal source (EMI source) not only may be external to your system but may be internal to your system—power supply ripple, digital circuitry, and ground noise. A young engineer came to me one day and asked how he could design a 68 pfd capacitor into the wire wrap interconnect board with only one end connected. It seemed this fixed a memory pulse strobe circuit that was malfunctioning. The capacitor acted as an antenna and injected more noise signal into one of the OP-AMPs inputs, which compensated for the unbalance at the inputs. We filtered the power supply directly at the OP-AMP power input pin, thus removing the internal noise source to fix the problem.

4. (Power) Provide a radio frequency (small ceramic) decoupling capacitance on your power inputs to your OP-AMP *close* to the OP-AMP to reduce power supply noise.

5. (Output Z) **The Critical Output Interface:** (Output impedance data from Fred R. Pfeiffer—a highly experienced control system and circuit analyst). Normally you have a relatively large load with respect to the output impedance of the amplifier. Inductive and capacitive components, however, can cause oscillations, (inductive step function positive feedback) or instability (capacitive time lagged negative feedback). A complete circuit design/analysis should include control theory analysis, developing a servo model diagram and a Bode plot as the following example from Fred Pfeiffer's excellent *Control Theory Class Notes* illustrates.[3] Circuits using wide-band gain elements should also include circuit layout capacitive and inductive parasitics in their circuit models for stability analysis.

An in-circuit board coupling capacitor of 3 pfd can provide ringing and even sustained oscillations if it couples positive feedback from a step function change to a highly inductive load. Capacitive loads can also cause problems. The example in figures 6-28 and 6-29 shows the impact of capacitive load on an amplifier. The example uses a finite 0.1, ufd (100,000 pfd) capacitor as a component, but everything being relative, the phenomenon shown could result from much smaller capacitance as you go up in system frequency. Here's the point: Do your analysis, calculate your parasitics, include them in your models, do your Bode analysis out to a frequency 10 times your highest system signal frequency and remember the capacitance could be cable capacitance. The systems engineer is responsible for these analyses that include step response as well as the phase margin due to manufacturing variances, environments, parasitic capacitance, and cables in all possible system conditions.

Effect of Capacitor Across OP-AMP Output on Stability

1) Circuit

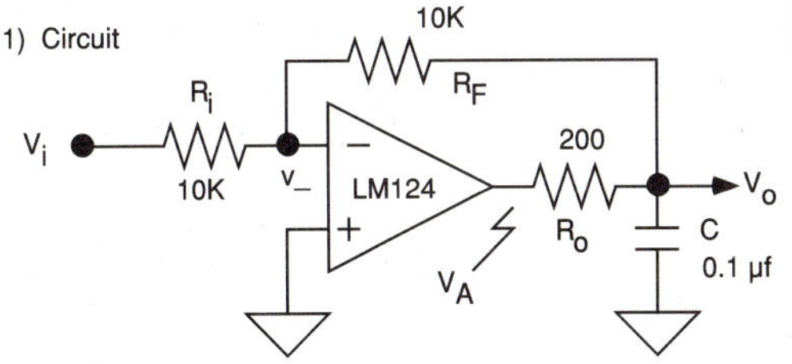

- Find servo diagram
- Find G(s), H(s) in factored form
- Find G(jw), H(jw)
- Bode plot |GH| vs. f
- Find $f_{cross\ over} = f_{odb}$
- Compute phase shift and phase margin
- Find ζ, % overshoot, frequency resp from curves

2) Servo Diagram

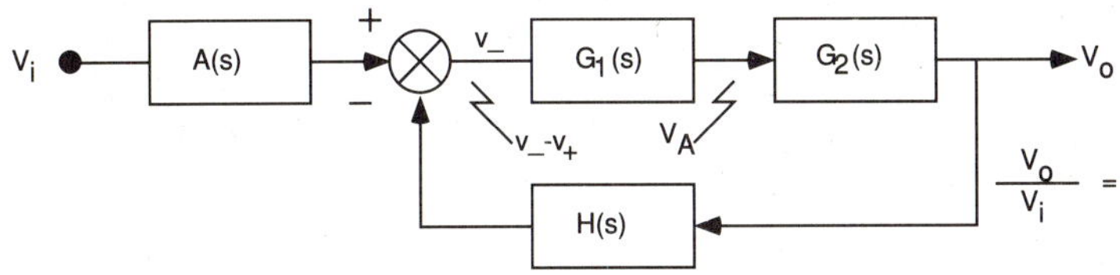

$$\frac{V_o}{V_i} = \frac{AG_1 G_2}{1 + G_1 G_2 H}$$

$$A(s) = \frac{R_F}{R_F + R_i} = \frac{1}{2} \; ; \quad G_1(s) = \frac{K_A}{1 + T_A S} \; ; \quad \frac{2 \times 10^5}{1 + j\frac{f}{5\ Hz}} = G_1(jw)$$

$$G_2(s) = \frac{1/SC}{R_o + 1/SC} = \frac{1}{1 + S\ R_o C} \; ;$$ Neglect loading by R_F since $R_F >> R_o$

$$H(s) = \frac{R_i}{R_i + R_F} = \frac{1}{2} \; ;$$ Neglect loading by $R_{amplifier}$ since 1M >> 10KΩ

3) GH(jw)

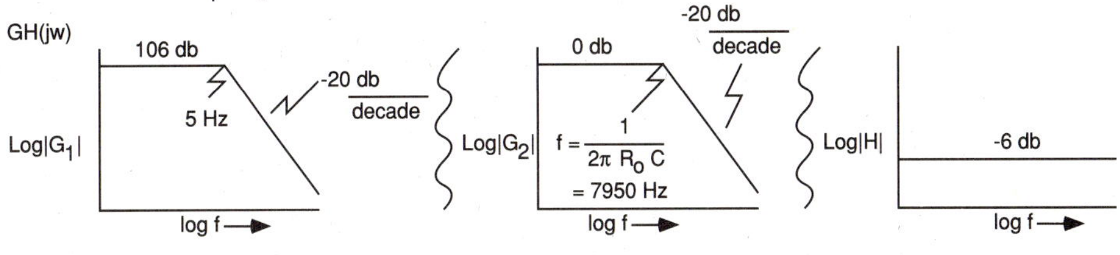

4) See Bode plot on next page for details.

5) Capacitor causes marginal stability

ζ = .07, 80% overshoot, 17 db peaking at 60 KHz

6) Must modify design to fix this!! (Reduce the value of the capacitor, also note that this capacitance may be cable capacitance).

Source: Excerpt from "Control Theory Class Notes" - Fred R. Pfeiffer, 14 February 1986 - Honeywell Microwave Systems.

Fig. 6-28. Capacitance across OP-AMP output effect

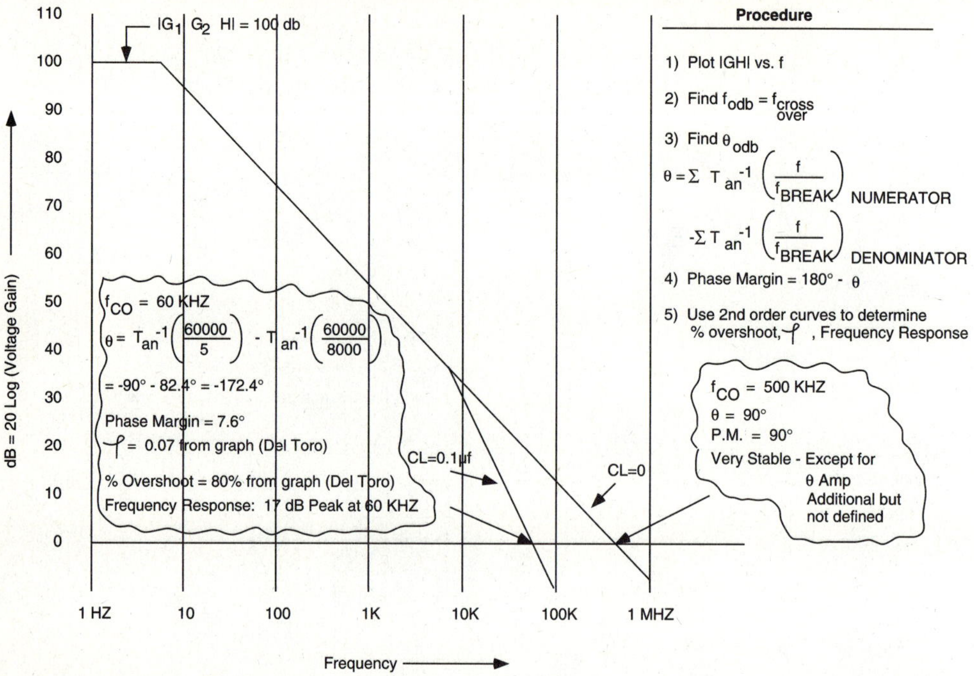

Fig. **6-29.** Stability analysis using bode plot

Some methods to improve the stability and accuracy of your operational amplifier and associated analog-to-digital converter

Autonull. Is a method for optimizing the differential sensing amplifier's zero signal input bias stability by automatically generating an error correction signal to compensate for drift due to temperature, for example.

Multiplexer switches are driven to sample the amplifier output with a switched zero voltage input to establish an offset correction voltage in an output capacitor. The charge in this capacitor then is switched into the input to provide a compensating bias correction current to maintain balance while the amplifier is used for system signals. The implementation of "autonull" and the frequency of "autonull" can be variable. Therefore, the systems engineer must analyze all uses of the autonull approach and its relationship to system signals and sampled data frequencies.

Zero and reference voltage measurement correction method. If you have a computer and an analog to digital converter (A/D) available, an additional correction mechanization to autonull could be to use a measurement of ground and measurement of the reference voltage supply. The ground measurement is used as an offset correction and reference measurement as a span correction by software. These correction measurements should be performed every iteration of the conversion algorithm.

Super A/D—analog-to-digital converters can be designed to provide "super" capability by utilization of these methods:

1. Optimize amplifier by carefully selecting an appropriate amplifier, by circuit balancing and application of autonull design.
2. Use offset and span correction measurements in software, by measuring zero and the reference voltage.

An absolute measurement A/D with a least significant bit (LSB) capability of 14.7 microvolts has been accomplished over -54°C to +71°C temperature range, by using these techniques.

Heuristic systems. Postulate a system that includes a super A/D now combined with an integral sensor, computer, feedback, and test interface. Place this system in an environmental chamber with stringent control of environment and sensor inputs. Through the test interface tell the system where it is environmentally. The system self-teaches itself at that point in time by calculating the Δ between what it thinks its environment is and what it really is and stores this heuristically developed correction term in its look-up table in memory.

Comments:

1. If the system will remain constant in its lifetime, no change.
2. If parts of the system are changed that will affect its calibration, reprogram it heuristically.
3. If aging will affect it, periodically reprogram it or provide ratio-metric conversion to a nonaging model or standard stimulus.

6.2.9 System Reliability, Maintainability, and System Safety (RM&SS) Analyses

Most of the following discussion has been provided by Lee I. Carver, a highly experienced reliability analyst. The lead systems engineer will generally find that reliability, maintainability, and system safety are contractually required to be treated equally with other system parameters such as cost, performance, weight, and size. In order to do this, the lead systems engineer may need the assistance of trained RM&SS engineers, as well as an understanding of the specified RM&SS parameters, the tasks that incorporate them into the design, and the analyses by which their estimated values are determined.

Typically required analyses specified are forms of the following:

- Mission concept including average utilization per day
- Mean time between failure (MTBF)
- Operating life, service life
- Mission reliability, flight safety reliability
- Maintenance concept, including available test equipment

- Mean time to repair (MTTR) at various levels of maintenance
- 90th (or 95th) percentile maximum time to repair
- Skill levels, facilities.
- Built in test equipment (BITE) ratios for detection, isolation, and false alarm
- Test point ratios and other requirements
- Permitted hazard categories, and required analyses

Reliability is achieved through the elimination and/or control of failure modes; maintainability, through the control of access, testability, and repairability; system safety, through the elimination and/or control of hazards.

Failure modes and effects analysis (FMEA)

The FMEA, if properly implemented, is the analysis by which the lead systems engineer determines what states of degraded operation will be experienced, what the consequences of operating in these states will be, and what self-test coverage, annunciation, and system reaction will be provided. It starts with a functional block diagram of the system and a technical description of each block, with its functional operation, its inputs and outputs, ground rules, assumptions, failure modes, and failure rates. Generally, this is easier to perform if each functional block is limited to only one output. The effect of each failure mode is then analyzed to determine its propagated effect on each system output. Hazard classifications for these output effects have to be determined, and then a determination made as to which are tolerable, and which must be eliminated or controlled. The identification of failure modes at the functional level provides the information needed in order to design an effective built in test subsystem, which must itself be subjected to an FMEA.

Failure mode and effects analysis—example

(The following is exerted from a system level FMEA)

FMEA Worksheets

The DPCU (Digital Propulsion Control Unit) FMEA worksheets are presented in the Appendix of this report. The functional elements covered in the FMEA are presented in figures 2 through 9; they make it possible to see exactly how each functional element fits into the system mechanization.

FMEA Failure Classifications

The following criticality classifications were assigned by the customer for our use in this FMEA.

I Catastrophic (loss of aircraft)

II Critical (shutdown of engine)

III Marginal (shutdown of IPCS—(Integrated Propulsion Control System))

IV Negligible

The classifications that appear on the worksheets represent the combined technical effort of the three top team companies: the aircraft, engine, and controller developers.

- The aircraft developer supplied classification for inlet spike and cone circuitry
- The engine developer supplied classification for engine main fuel, 7th and 12th stage bleed, afterburner and nozzle circuitry
- The controller developer supplied classification for DPCU power supply, power off, and control monitor unit circuitry.

Failure Mode and Effects Analysis

Methodology:

- Provide technical description of system, subsystem, or component to be analyzed
  - — Functional detailed block diagram
  - — A writeup that defines the diagram, gives ground rules and assumptions and provides failure rates and methods
- Establish hazard classifications
- Set up FMEA worksheet tables (see fig. 6-31)
- Perform analysis
- Design out hazard conditions
- Provide formal report

PREPARED BY / DATE	REVISED BY / DATE	DRAWING NO.	SYSTEM/SUBSYSTEM TITLE	ASSEMBLY/PART NO.	SHEET
PREPARED BY: R.F. RAPPEL — DATE: 1-21-74	REVISED BY: — DATE:	YG1140A01	DIGITAL PROPULSION CONTROL UNIT (MAIN FUEL CONTROL LOOP)	INTERFACE UNIT (IFU)	SHEET 1 OF M
APPROVED BY: J.W. HUNGER — DATE: 1-23-74	APPROVED BY: — DATE:			ASSEMBLY/PART: HG1026AA01	

FUNCTIONAL ELEMENT	FUNCTIONAL DESCRIPTION	FAILURE MODE AND CAUSE	EFFECT OF FAILURE ON SYSTEM	FAILURE DETECTION METHOD	DESIGN CONSIDERATIONS	CLASS	CAT I OR II FAILURE RATE
DIGITAL TO ANALOG CONVERTER	AN ELEMENT OF THE MAIN FUEL CONTROL LOOP THAT CONVERTS DIGITAL SIGNALS RECEIVED FROM THE DIGITAL COMPUTER UNIT INTO ANALOG VOLTAGE LEVELS WHICH ARE SUBSEQUENTLY SENT THROUGH THE MULTIPLEXER TO THE SAMPLE AND HOLD.	A DIGITAL SIGNAL IS CONVERTED TO AN INCORRECT ANALOG VOLTAGE LEVEL DUE TO THE FAILURE OF AN ELECTRICAL COMPONENT	THE VOLTAGE LEVEL WILL TRAVEL THROUGH THE MULTIPLEXER TO THE SAMPLE AND HOLD WHICH WILL PROVIDE A COMMAND TO THE MAIN FUEL VALVE TORQUE MOTOR TO AN UNDESIRED POSITION.	AT THE SAME TIME, THE MAIN FUEL CONTROL OUTPUT FROM THE SAMPLE AND HOLD IS SENT THROUGH ELECTRICAL CIRCUITRY TO THE DIGITAL COMPUTER UNIT WHERE IT IS COMPARED WITH THE COMMANDED VALUE. AN EXCESSIVE ERROR TURNS CONTROL OVER TO HYDRO-MECHANICAL BACK-UP CONTROL		III	

CLASS (FAILURE CLASSIFICATIONS)

I. CATASTROPHIC (LOSS OF AIRCRAFT)
II. CRITICAL (SHUTDOWN OF ENGINE)
III. MARGINAL (SHUTDOWN OF IPCS)
IV. NEGLIGIBLE

Fig. 6-30. FMEA—example worksheet

Fault tree analysis

A fault tree analysis is similar to an FMEA in that they both deal with failure modes and probabilities, but they approach the problem from different directions. The fault tree is a tree diagram, or graph, usually constructed from a single unsatisfactory event at the bottom, that traces each possible, sequential, precursor event and condition backward up the tree. The fault tree is sometimes constructed as a upside down tree, with the unsatisfactory event at the top. In either construct, each branch starts with AND/OR logic to permit calculation of the probability of the bad event occurring. Fault trees are usually used more for the analysis of hazard sources than for the analysis of fault effects and reactions.

MTBF analysis

The lead systems engineer will use MTBF data for two purposes: to determine if the design meets specific MTBF requirements, and to evaluate the relative frequency of failure events, as in the FMEA. The proper failure rates to be used will be different in each case, and this is not a simple matter. It should be left to the reliability systems engineer to determine what the appropriate failure rates are, based on part selection criteria, environmental stress, qualification, burn-in, age, and program considerations.

In order to improve MTBF, the lead systems engineer should stick with proven mechanizations, simplify the design to reduce parts count, reduce electrical, mechanical, and thermal stresses—both from extremes and from transition rates—and build safety margins into all interfaces, including the human interface with maintenance personnel. In addition, the systems engineer can use redundancy to improve the MTBF of a system. Refer to *System Engineering Methods* by Harold Chestnut for a tutorial on the analysis and design of redundant systems.[4]

With regard to evaluation of an MTBF, a few things are worth bearing in mind. There is no such thing as "The MTBF"; there are many MTBFs, each appropriate to the need for which it is intended. The value of MTBF depends more upon how it is going to be used than how it is determined. When evaluating MTBF, nothing beats historical data for accuracy. Similarity to existing equipment comes in second, and prediction from a parts list is dead last. However, as lead systems engineers you need to do some top-down MTBF calculations, once again, "just to get in the ball park." You must realistically apportion your overall desired MTBF out to subsystems.

Stress analysis

It is the circuit design engineer's responsibility to select (specify), and apply parts in accordance with the criteria specified for the program. The reliability systems engineer, with the assistance of components engineering, will determine the minimum acceptable parts quality level required to satisfy program requirements, generally in terms of a company standard parts manual. The reliability engineer will publish this in a parts requirement detail sheet signed by the program manager.

In general, the designer should limit his selection of parts to the standard parts manual to the maximum extent because of the advantages provided (commonalty, proven performance, established multiple vendors, volume buys, availability, etc.). Where a program requirement forces the selection of a nonstandard part, the designer must work closely with components engineering to make sure that the part ultimately selected performs the required task at minimum risk. Exotic parts should be avoided to the fullest extent possible until a history of satisfactory performance and multiple sources of supply have been developed. When this is not avoidable, then the lead systems engineer must assess the risk in terms of reliability, cost, schedule, producibility, and performance. This trade-off should include the system design, since an alternate design may permit the use of lower-risk parts.

The reliability systems engineer will also identify the design margin derating criteria. In the absence of other requirements, your company design margin derating criteria should be used. When more than one component source is imposed, and individual requirements differ, the one providing the greatest derating should be used. The derating is necessary to ensure that your components will perform sortie missions reliably over their life mission for your system.

The purpose of derating is to provide a safety margin between the unknown stress of the application and the equally unknown strength of the part. Part capabilities are specified in the documentation in terms of absolute maximums, beyond which performance is not guaranteed. It is common practice to design for operation at 70 to 80 percent of these absolute maximums, or even lower at elevated stresses. Parts vary from manufacturer to manufacturer and even from lot to lot for any one manufacturer. The changes in the component as it ages must also be considered. Also, the application environment is not often known with certainty. The lead systems engineer should attempt to reduce the electrical, thermal, and mechanical stress to acceptable levels by proper use of requirement relief, filtering, buffering, clamping, cooling, etc. An efficient and reliable thermal path to the best available heat sink will always pay dividends in terms of enhanced reliability due to lower component temperatures.

Although the customer's specification may state that the maximum temperature will be 71 degrees Celsius, there is absolutely no guarantee that this limit will not be exceeded. Further, if it is, there is little likelihood that you as a supplier will be aware of it unless it is catastrophically higher. The impact of additional heat from other components internal to your system or the heating from other systems external to your system must be monitored by the systems engineer as your system is being developed. Therefore, to help your systems reliability you derate the component vendor's ratings, and design to operate well within the safe area for electrical, thermal, and mechanical stresses allowable for the specific component. Confirm that you are operating in the safe area during design with the reliability analysis task called "Part application analysis" (PAA), or "Part stress analysis." It is also confirmed physically by measurements during qualification testing and, ultimately, during actual use. The cost of finding out you are wrong goes up geometrically with each advancing step of the design process. PAA pays for itself by critically examining all risky areas before the design is released, when changes are cheap.

The PAA is an independent analysis of the electrical, thermal, and mechanical stresses of the application, performed by someone other than the responsible design engineer, generally a reliability systems engineer. It may not cover all aspects of stress, and it may not cover all items of the design since it is often felt to be sufficient to limit the analysis to high-risk/high-stress areas, such as power supplies, analog driver circuitry, and the like. The PAA is usually performed in parallel with or slightly lagging the design, and the analyst is usually in direct and intimate association with the designer so that differences of opinion can be resolved before the design has solidified. The relationship is not adversarial but rather is cooperative, with the designer ensuring achievement of the performance requirements, and the PAA analyst ensuring proper control of part stress.

The design margins are not generally taken to be limits that can never be exceeded, but rather, they are taken as danger flags. You can generally operate within the limits safely, but if your design requires operation outside of the derated margins, then you should examine the application more closely and be sure that you know what the stress really is. Such an examination will generally result in an obvious answer to the question: Is this risk justified? For proven parts in a familiar application, the risk is probably small and acceptable, but for a new part in a new application, the risk is substantially higher and probably unacceptable. Other factors must be considered, also, such as ease of rectification if the gamble goes sour. What if your custom-designed, sole-source, "does-everything" special very large silicon integrated circuit doesn't meet its requirements during system qualification testing? If you are two months from production, you have a very serious problem that will impact development schedule and cost, as well as technical system mission capability.

Your reliability predictions are predicated on proper derating. If the design uses proper derating for the worst case, steady state, operating environments in order to control the stresses, you have shown that you can predict the field MTBF with reasonable accuracy. But if stresses are not controlled, then there is no rational way to predict before the fact what failure rates will result so that MTBF is not predictable. If the program has guarantees and/or warranties tied to MTBF, then meaningful risk assessment requires proper derating.

Failure analysis

Company engineering procedures shall require that a formal failure reporting, analysis, and corrective action (FRACA) process be implemented on all programs where penalties are tied to reliability performance. FRACA is a closed-loop process that provides the means for identifying deficiencies in equipment performance, determining the root cause of each deficiency, developing corrective action to remove or eliminate the cause, and documenting the significant facts of the situation. It is through FRACA that we listen to what the equipment has to say about its own performance in its environment.

FRACA is usually conducted by a failure analysis committee (FAC), which takes control of the fault isolation, repair, analysis, retest, and documentation associated with failures that occur during any formal (contractually required) testing. This usually includes acceptance tests, burn-in, qualification tests, and demonstrations. There is usually some latitude in the degree of formality required, with a balance being struck so that facts are gathered without undue disruption to schedules. If the situation requires it, every step after identification of failure occurrence can be controlled by formal committee action, although this is time consuming, and seldom necessary. For most programs, it is sufficient to have the committee chairperson advised of each failure occurrence, with subsequent actions determined by informal consultation. The necessity to get concurrence from each FAC member before sign-off of the failure analysis report (FAR) is usually sufficient to ensure coordination and objectivity. The FAC could consist of a team of specialists and systems engineers that are also doing the design reviews discussed in chapter SEE-7. The FAC chairperson could be the program start-up reviewer (PSR) we discussed in chapter SEE-4.

The key to effective FRACA is an analysis that identifies all the pertinent facts. The committee will call in experts from whatever sources or disciplines required in order to pinpoint the cause of the failure. The risk to the system, represented by the failure in terms of overall, long-term system performance will usually determine the extent of the analysis. It pays to determine the cause of a failure if the risk is substantial, and an effective corrective action can be developed that will remove the cause and thereby reduce the risk of subsequent failures.

Sometimes the politics of a failure situation dictates that the cause must be identified, or at least localized, even though the economics do not justify it. Credibility with the customer is an important part of the committee's responsibility, and this comes mainly through the presentation of timely, thorough, understandable, and believable failure analysis reports.

The reputation for honesty and thoroughness tempered by practicality must be sought, earned, and defended as the essential characteristics of a FAC that are relied upon by both management and the customer. The FAC develops factual data, which program management presents to the customer.

The FAC must objectively review all the data so that realistic decisions can be made by the responsible managers. The FAC must constantly remind itself that the facts come before the solution, and that burying or covering up uncomfortable facts or prefiltering the facts in light of "acceptable solutions," is counterproductive to the function of the FAC. It is the first responsibility of the FAC to determine the cause of failure. What corrective action then results is up to the program manager and is a business decision.

Once all of the facts are in hand so that a rational decision can be made based on knowledge of the real risks, the FAC must determine what actions would remove the cause of the failure. The FAC should develop viable candidate actions, and make recommendations, assisting in the risk analysis and decision process. The FAC is not to select and then direct the corrective actions to remove the cause of the failure. That is the responsibility of the program manager.

Worst case analysis

Worst case analysis is a design analysis task that is performed during design development phase. The worst case analysis is to verify that the design will function properly even with all tolerances taken to their limits. This analysis is also sometimes referred to as design margin analysis.

Sneak circuit analysis (SCA)

The systems engineer must be aware of potential sneak circuit problems, especially as part of interface analyses during the total life mission. This analysis requires checking all possible powered up/powered down conditions of all the subsystems and interfaces of your system to ensure you have no possible hazards or malfunctions of the system. Sneak paths have overstressed components, caused explosions, drained batteries, and killed people. These "sneak" circuits may occur when your system is unpowered but connected to a system that is powered. They may also occur by random test equipment connection sequences. A navigation system had many blown (like a fuse) bus interface circuits. These circuits were destroyed by a "sneak circuit." It was possible to inadvertently apply 115VAC power to the navigation system without a proper chassis ground while connecting the system to test equipment.

Maintenance analysis

The maintenance analysis task is performed by the maintainability systems engineer and identifies all of the maintenance skills and resources needed to repair the equipment. It involves identifying the maintenance tasks required in order to detect, isolate, repair, and verify each fault that can occur in the hardware at each level of maintenance. It will identify the tools, test equipment, skills, training, spares, manuals, special equipment, and active maintenance times associated with each task, and permit determination of the average repair time, or mean time to repair (MTTR). It is an input to deployment support and forms a basis for the deployment logistics support analysis. The systems engineer is involved to the extent that he should verify that the analysis confirms meeting the quantitative and qualitative maintainability requirements of the program.

Safety analyses

It is common for a program to require one or more of the following system safety analyses: preliminary hazard analysis, subsystem hazard analysis, and operating and support hazard analysis. The analyses will be performed by the system safety engineer and identify hazards to equipment and personnel so that they may be eliminated or controlled. The lead systems engineer must monitor the results of these analyses so that we may be aware of the potential hazards present, and how they are controlled. As with all of the other analyses, the results are influenced by many other system parameters and must be traded to achieve the maximum level of compliance at minimum cost.

Environmental effects analysis

An environmental effects analysis will identify the stresses induced, and the protective features required, for all environments while the system is in use, test, transportation, storage, etc. This is part of the system mission analysis and can be performed by a reliability systems engineer. The lead systems engineer then needs to be aware of all necessary and unusual requirements that come out of this analysis so they can be quickly established in the system DS.

Basic reliability calculation

$$R = e^{-\lambda t} = e^{-t/m}$$

R = Reliability or probability of successful operation for the time t

λ = Failure rate (failures/hour)

m = inverse of failure rate or mean time between failures (MTBF)

$$MTBF = m = \frac{1}{\lambda}$$

$$\text{if } R = 0.9 \text{ m} \quad m \approx 10^{t}$$

$$\text{if } R = 0.99 \text{ m} \quad m \approx 100^{t}$$

Prediction calculation

1. List the quantities of each different part used.
2. Multiply each quantity by its failure rate and sum the total, to obtain the overall failure rate.
3. Calculate the resulting reliability of the system from the expotential law, using the overall failure rate.

$$R_{SYSTEM} = R_1 \times R_2 \times R_3 \times R_4 \times \text{———} \times R_n$$

$$R_{SYSTEM} = (e^{-\lambda 1t}) \times (e^{-\lambda 2t}) \times (e^{-\lambda 3t}) \times (e^{-\lambda 4t}) \times \text{———} \times (e^{-\lambda nt})$$

$$\lambda_{SYSTEM} = \lambda_1 + \lambda_2 + \lambda_3 + \lambda_4 + \text{————} + \lambda_n$$

Basic reliability distributions

The following power distribution functions are utilized in doing reliability analyses:

Weibull and Exponential Distributions are the times to failure of many electronic components, motors, textile fatigue, ball bearings, leakage of batteries, corrosion resistance, marketing life, life expectancy of drugs, return of goods after shipment, and number of down times per shift. The exponential distribution represents assemblies and systems whereas the Weibull distribution covers components and parts.

Poisson distribution is used to determine the probability of occurrence of a specific number of failures during a mission, to determine standby reliability, or to calculate the number of spares required when units have constant and identical failure rates.

Gaussian is the "normal" distribution that realistically models many physical phenomena as well as the distribution generated by physical MEASUREMENTS. It can and is used to APPROXIMATE other probability functions, particularly discrete forms when the problems become cumbersome. It is also known from the Central-Limit Theorem that observations drawn from nonnormal distributions, as the number of observations become large and remain INDEPENDENT, will APPROXIMATE the normal distribution.

Recommended references for further study of reliability analysis are *System Engineering Methods* by Harold Chestnut,[4] *Probability System Analysis* by Arthur Breipohl[5], *Reliability Engineering in Systems Design and Operation* by B. Dhillion[6], and chapters 13 and 14 of *Systems Engineering and Analysis* by Blanchard and Fabrycky[1].

6.2.10 System Software Analysis

Software analysis is an application of your mission analysis skills. You must do timing studies of your software. For example, how long does it take a perpherial such as an analog-to-digital converter to respond to an instruction from your computer? Should you build in a timer while you wait for the external to respond? What happens to your software if the external is failed or not there so it will never respond? How are internal software routines impacted by external interrupts? These timing analyses basically deal with the interface of your software to the world outside. A large number of analyses other than the hardware interfaces are the human interfaces. Is your system user friendly? In your mission analysis, have you done sufficient scenarios to cover all possible human interfaces? Have you considered both normal and abnormal situations? Obviously, both FMEA and safety analyses are part of the analysis of your software.

The impact of high-order language (HOL) to critical software in your system brings about a potentially serious software problem. Just like a hardware designer who does not know the specific components that make up a purchased hardware functional element, you do not know how a specific software high-order instruction is programmed in machine language. To ensure hardware will interface and not cause critical FMEA problems, you need to dig into the specific design and analyze it by looking at a detailed schematic and a physical mechanical layout, and by doing a FMEA, with special concentration on the interface areas. In like manner, you need to do the same when you analyze critical parts of your HOL software. You need to know how the specific HOL instruction is programmed in machine language. You need to know how the computer hardware functions; does it have a background coprocessor? Does it have a queue of instructions that may follow? How does it handle random events such as interrupts and power interruptions? Yes, the analysis of your critical software does require obtaining very detailed knowledge of how the hardware functions. The example, "programming the dispenser," in section 6.2.6, which was given as an example of ground bounce, shows what can happen if hardware and software designers do not work together.

Recommended references for further study of software development and analysis are *Computer-Aided Systems Engineering* by Howard Eisner.[7] It has a chapter on software development and analysis. Also *The Mythical Man Month* by Frederick P. Brooks Jr., is must reading for software systems engineers, lead systems engineers, software development managers, and system engineering managers.[8]

6.2.11 Control System Analysis

All systems can be analyzed as if they were control systems. You can consider their inputs, transfer functions, outputs, and feedbacks as the elements of a control system. If you do this you are forced to analyze the critical input, output, feedback interfaces, and also the response characteristics of your transfer functions. It has been a lack of analysis with the control system perspective in these areas that has caused mission impossibles.

Control system input noise

Planes have crashed because of external noise (EMI) from the Voice of America transmitting tower. Trucks have crashed because they lost their brakes from external noise. Computers have experienced malfunctions such as not being able to load their memory because of internal noise from their own internal power supply. Teachers have not been able to teach because of internal classroom noise and external micromanagement noise from administration. The analysis of your control systems requires brainstorming, listing, and analyzing all the inputs, both good and bad. You then must design your system so that it can tolerate and perform its mission with all of these inputs. In some cases you can eliminate them. In other cases you can shield, filter, time multiplex, or provide the concept of autonull as feedback to prevent system failure.

Control transfer functions

Digital microcomputers have brought about faster responding systems with complex software. This has resulted in some interesting system problems because the designers of the hardware are not communicating with the designers of the software and vice-versa. Programming the dispenser, the navigator velocity takeoff, and the ground bounce examples show how software can program hardware to cause the system to fail. These failure conditions had appeared to be intermittent and the result of the combination of many factors, but they really are the result of a lack of basic design-to bounding requirements in the initial DS. The control system analysis of these systems shall result in requirements such as these: This hardware shall be designed such that all possible software programming shall not cause a system failure; the software shall be programmed within the bounding rules established by the hardware design; the system shall have designed-in interlocks to prevent use of the special production test instructions while the system is being used to perform its mission.

The impact on the control system frequency response (Bode), step response, stability, drift, and gain due to programming and the range of possible input signals is not always analyzed. The system must be designed to bound the range of these variables or analyzed with computer models and simulations. In addition, testing of the actual system to establish a safe bounded system must be done. Models and simulations must be confirmed by data from test of the actual system. The systems engineer shall develop and coordinate the testing, the simulation work, the software development, the hardware development, the requirements in the DS, the FMEA, and the safety analysis as part of the control system analyses. A key systems engineering skill required by all systems engineers is control system analysis. Software systems engineers, especially, must have high control system analysis skills.

Control system outputs and feedback

The range of output interface variations and feedbacks are not always analyzed. Cable and wiring can have a dramatic input on the control system characteristics of your system. Cable capacitance can give you response delays or feedback problems as in the critical output interface discussion of operational amplifiers. Cable inductance, especially in step function outputs have caused large L di/dt voltage transients that have exceeded the voltage rating of devices at the interface, causing them to short, overheat, and cause a fire. This is a common problem that has caused fires in test equipment and even in systems that are designed to monitor for fires. The impact of connecting test equipment on your control system can be insidious. The system has different control system characteristics in actual mission use from that in the tester configuration. The analysis of the "high speed clock driver circuit" shows how the test cable changed the clock signal waveform. Consider the possible impact on analog servo drivers that control an airplane wing flap. The system tests good on the tester, but intermittently fails or drifts on the airplane. It is possible to establish a condition where a control system will break into oscillation because of the characteristics of the output load and its response to a step response output (sort of like ventricular tachycardia that can lead to sudden death in the human system).

The packaging of integrated circuits requires a control system analysis of the outputs and their impact on the transfer function of the device. For example, the parasitics of the package and leads provide capacitors and inductance that can affect the transfer function of high-impedance, high-gain, and high-frequency response devices.

The feedback of your system must be known and it must be analyzed. Its relationship to the control transfer function must be known. If it is delayed by the interface or by the control algorithm response, it can change the frequency response and, consequently, the stability of your system.

One of the serious problems with process systems can be that the only feedback they respond to is that feedback that is internal to their system. They have no customer feedback. This condition can cause your system to fail. You need to analyze the feedbacks to your system.

For further reading about control system analysis look into *Introduction to Control System Technology* by Bob Bateson. This book is an excellent introductory to control system analysis.[9]

Introductory System Engineering by John G. Truxal is a control systems reference for modeling, dynamics, optimization, Bode, and Nyquist.[10]

Servo Mechanisms and Regulating System Design by Harold Chestnut and Robert W. Mayer is considered by some to be their bible of control system design.[11]

Control Systems Engineering by C. J. Savant covers basic feedback control, system design, learn by doing, get into the lab, do some empirical modeling, use Bode.[12]

6.2.12 New Technology Analysis

When your system is to use new technology, you must do an analysis that shall include a trade-off study of at least three alternates that prove that the new technology is far superior to prior technology and is worth the risks. The analysis shall include an analysis of production yield and a cost analysis with worst case variance. A backup design shall be baselined in case problems arise during the development process. In addition, at least two suppliers for the technology shall be identified.

6.2.13 System Test Analysis

System testing is an integral part of every system mission; therefore, the analyses of how and when the system is to be tested start as part of the mission analyses. This shall result in testability requirements in the DS and in test plans in the PMP. The test equipment requirements shall be established; a DS for the test equipment shall be developed, and the actual test equipment shall be baselined. The system DS part 2 acceptance test procedure (ATP) shall be developed, and actual testing shall be taking place to confirm the design meets the requirements. The test data must be analyzed and used to confirm production costs and yields. A test analysis specialist may do an independent analysis for the testability of your system. What is the coverage of the testing? Are some failures undetectable but will cause system failure or degradation of the system when it is used to perform its mission?

A staff systems engineer was called upon to resolve a conflict between the lead system engineer and the production engineer. Production had to build, test, and calibrate 700 digital to analog (DA) converters with special output buffer amplifiers. The calibration requirements required a selection tolerance of plus or minus one ohm, which was very costly to do and necessitated expensive test equipment. In the knowledge gaining discussion, the staff systems engineer discovered that the DA converters and the amplifiers were being individually tested and serialized, and had individual test data. This data was being filed as history data to track component supplier quality, which is a very commendable thing to do. The staff system engineer did a requirements accuracy allocation analysis of the main system mission requirement, which was the measurement accuracy of the helicopter pilot's line-of-sight. This analysis showed that the calibration accuracy required could be increased to plus or minus three ohms in the DS test procedure. This reduced the cost of calibration. However, the staff systems engineer's test analysis also included an analysis of the production test process. It was determined that if the incoming test data of the DA and amplifier serialized components was used to pair up amplifiers and DA converters into kits, over 30 percent of the system could meet the specification requirement with no calibration. This is an example of why systems engineers must do test analyses. A book by Roger T. Stevens, *Operational Test & Evaluation—A Systems Engineering Process,* is recommended for further study.[13]

6.2.14 In-depth Specialty Analyses

When you take the system engineering self-assessment skill survey, you discover that you lack in-depth skills in many areas. You must get some help from experts in your skill deficiency areas. You may also require help because development schedules do not give you enough time to personally do all the analyses required to develop the system. Therefore, you need to do an analysis of what specialty skills are needed, when they are needed, and how long they are needed. Then you have to find these important people and schedule their work. You may require, for example, thermal, structural, chemical, materials, reliability, testability, maintainability, EMI, software, power system, market, logistics, producibility, tool experts, and even customers as specialty analysts.

6.2.15 Product Analysis

The product analysis for your system starts with a market analysis that includes the establishment of market size, and what the product or system can be sold for and at what profit. This analysis also includes benchmarking your competition. What percentage of the market can you obtain? Benchmarking in the context here refers to an in-depth study of your number one competitor to establish a standard by which you can be measured. What is your reliability, quality, accuracy, effectivity, customer support, product features, etc. compared to the benchmark standard? What can you do to enhance the salability of the product or system to get more of the market? Can you do better in the service and support of your system? How can you improve on your competition benchmark?

Product analysis includes all of the cost analyses—UPC, top-down, bottom-up, value engineering, and especially life cycle cost (LCC). A producibility cost must be done by the production department to ensure you will get lowest cost and highest yield in the manufacture of your product.

You need to do an analysis of how you intend to support and assist the customer in the use of your product. This includes methods to get timely feedback from the customer on how to improve your product and service. Is your system a quality system? Have you designed it and its production and support processes so it can be constantly improved for greater customer satisfaction?

A recommended reference for product analysis is *Systems Engineering and Analysis* by Benjamin S. Blanchard and Walter J. Fabrycky. This book has a series of chapters which related to designing for operational feasibility.[14]

Recommended study for improved quality are *Quality, Productivity, and Competitive Position* by W. Edwards Deming;[15] *Total Quality Control (TQC)* by A. V. Feigenbaum;[16] and *Juran's Quality Control Handbook* edited by Dr. J. M. Juran.[17]

6.2.16 System Analysis Tools

One of the skills in the system engineering self-assessment skill survey is tools knowledge. The systems engineer needs to know about and recognize the wide range of tools required to do the development and analysis of a system. Tools are hardware, software, and procedural. Hardware tools may be hammers, screwdrivers, saws, wrenches, soldering irons, oscilloscopes, meters, electrocardiograph machines, stethoscopes, laboratories, environmental chambers, test equipment, calculators, computers, and special test jigs. Software tools may be all the computer-aided software programs, computer-aided-design, computer-aided-system engineering, computer-aided-software engineering, operating systems, simulations and models. Procedural tools are all the methodologies and processes you use such as the system engineering process, the design process, the problem-solving methodology, the scientific approach, probability, statistical analysis methods, and diagnostics methods.

You need to be careful that you do not forget that tools are only intended to assist you in your development of the product or system. Some systems designers become tool experts and forget about the system they are designing.

You may be asked to select a tool for the systems engineering process. A test you may use to determine if a specific tool is operational is to request a copy of the system development specification generated by the tool for the tool. If the tool can provide a DS of itself by itself, then it has the potential to do that for your system development. Also ask if the tool can provide all the analyses and all the documentation required to perform the total system engineering process for the life mission of the system. Experience has shown that systems engineers select and use a tool as specifically required at the time of need. We are very pragmatic in our selection and use of tools.

The following are some recommended references you may wish to study to further your knowledge of tools: *System Engineering Tools* by Harold Chestnut. This book is the baseline for the subject of system engineering tools.[18]

Computer-Aided System Engineering by Howard Eisner provides an appendix which lists a sample of microcomputer-based software, which though dated is a good starting point if you are searching for a specific software tool.[19]

Towards Creative Systems Design by Henry C. Lucas is must reading if you are developing or selecting a computer tool.[20] His key premise is that users should design the (tool) system.

6.3 THE ANALYSIS CHECKLIST

A checklist for analyses to perform is provided in chapter SEE-8, "The Analysis Checklist." Note that this checklist is part 3.2.2.9 of the PMP. It is recommended that this checklist be used as part of your system engineering and system engineering management process.

6.4 ANALYSIS PROBLEMS

1. The voltage divider of figure 6.3 is made up of two resistors and gives an output range of 4.195 Vdc to 5.767 Vdc. If you add a 1K potentionmeter with a 3.3 ohms per degree of adjustment capability, you can initially calibrate at room temperature with a stable V_{IN} of 12 ± 0.1% the nominal output voltage to 5.0 ± 0.2%

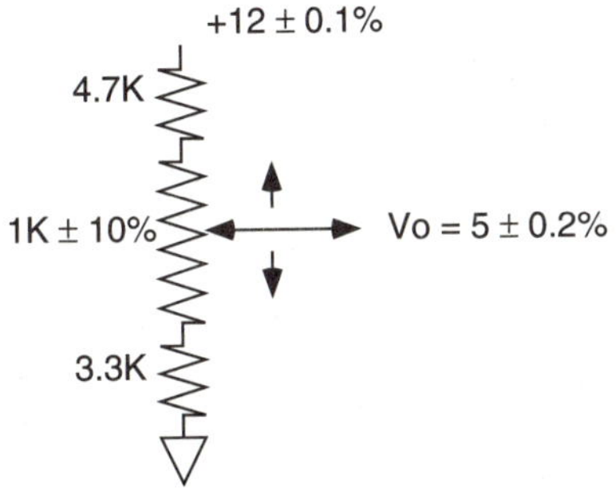

a) What is the range of V_o after calibration given the "In-use" V_{IN} range of 12 ±10% VDC and the resistor ± 5% and pot ±10% tolerances?

b) If the two resistors and potentionmeter were replaced with an eight (8) bit digital to analog converter with an accuracy of ± 1 least significant bit driven by a precision 10 ± 0.1% VDC, what would be the initial calibration of Vo to 5 VDC nominal possible by software programming? What would be the range of software programming? What are the concerns in programming?

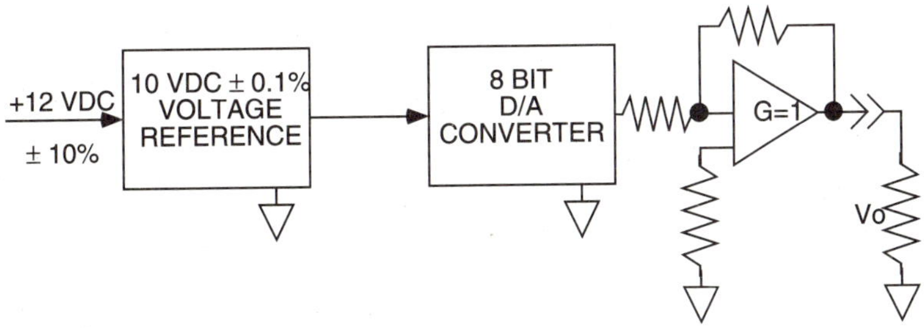

What is the range of Vo after calibration? What do you need to know in order to answer this question?

2. A famous Old English puzzle is called "Dog's Mead." Although it relates to a farmer, his family, and his land, it involves a good deal of engineering mathematics and logic. The problem is to find the age of Mrs. Groosby, Farmer Dunk's mother-in-law. And you must not assume the puzzle was invented this year.

You'll need to know that there are twenty English shillings to the pound sterling, that an acre is 4840 square yards, and that a rod is a quarter of an acre. Also, these hints help: One number in the puzzle is the area of Dog's Mead in rods, but it relates to something in the puzzle quite different from that area. One of the numbers across is the same as one of the numbers down. So here you are:

Across

1. Area of Dog's Mead in square yards.
5. Age of Farmer Dunk's daughter, Martha.
6. The difference between the length and breadth of Dog's Mead in yards.
7. Number of rods in Dog's Mead times number nine down.
8. The year when Little Piggly came into occupation by the Dunk family.
10. Farmer Dunk's age.
11. The year Farmer Dunk's youngest child, Mary, was born.
14. Perimeter of Dog's Mead in yards.
15. The cube of Farmer Dunk's walking speed in miles per hour.
16. Number fifteen across minus number nine down.

Down

1. The value of Dog's Mead in shillings per acre.
2. The square of Mrs. Groosby's age.
3. The age of Mary.
4. The value of Dog's Mead in pounds sterling.
6. The age of farmer Dunk's first-born, Edward, who will be twice as old as Mary next year.
7. The square, in yards, of the breadth of Dog's Mead.
8. The number of minutes Farmer Dunk needs to walk one and one-third times around Dog's Mead.
9. See number ten down.
10. Ten across times nine down.
12. One more than the sum of the digits in the second column down.
13. Length of tenure, in years, of Little Piggly by the Dunk family.

3. Who Owns the Zebra?

1. There are five houses, each of a different color and inhabited by men of different nationalities, with different pets, drinks, and candies.
2. The Englishman lives in the red house.
3. The Spaniard owns the dog.
4. Coffee is drunk in the green house.
5. The Ukrainian drinks tea.
6. The green house is immediately to the right (your right) of the ivory house.
7. The licorice eater owns snails.
8. Carmels are eaten in the yellow house.
9. Milk is drunk in the middle house.
10. The Norwegian lives in the first house on the left.
11. The man who eats marshmallows lives in the house next to the man with the fox.
12. Carmels are eaten in the house next to the house where the horse is kept.
13. The gumdrop eater drinks orange juice.
14. The Japanese eats peppermints.
15. The Norwegian lives next to the blue house.

WHO DRINKS WATER?
WHO OWNS THE ZEBRA?

4. Automobile Yearly Gas Usage

   Brainstorm all the ways you can reduce your yearly automobile gas usage. Write a functional equation. Do an analysis to determine the best three solutions for minimum usage. Do an analysis of the best three solutions based on your personal preferences.

5. Solve the national debt.
6. Do a life cycle cost analysis of your home.
7. Do an attributes analysis of your present automobile: size, weight, LCC, MTBF gas mileage, features, power, expected life, safety.
8. Your system is found to be over cost and under reliability goals. Provide recommendations to improve both cost and reliability.
9. Passengers on airplanes sometimes carry electronic devices such as lap-top computers and audio disk players that cause interference that affects the aircraft electronics. Brainstorm for solutions to the problem. Write a functional equation. Recommend three possible solutions. What do you say to those who say there is no interference problem? They say airline restrictions on use of electronic devices are only intended to get more passengers to buy head phones for the movie and more drinks, so the airline can make more money.
10. Do a failure mode and effects analysis of your home. Recommend three changes that will eliminate possible critical failures or safety hazards.

6.5 References

1. Blanchard, B. S., and W. J. Fabrycky. *Systems Engineering and Analysis.* Englewood Cliffs, NJ: Prentice Hall, 1981, chapters 4 and 17.
2. Eisner, Howard. *Computer-Aided Systems Engineering.* Englewood Cliffs, NJ: Prentice Hall, 1988, chapter 11.
3. Pfeiffer, Fred R. *Control Theory Class Notes.* Minneapolis, MN: Honeywell Microwave Systems, February, 1986.
4. Chestnut, Harold. *Systems Engineering Methods.* New York: John Wiley & Sons, 1967, chapter 7.
5. Breipohl, Arthur. *Probability System Analysis.* New York: John Wiley & Sons, 1970.
6. Dhillion, B. *Reliability Engineering in Systems Design and Operation.* New York: Van Nostrand Reinhold, 1982.
7. Eisner, Howard. *Computer-Aided Systems Engineering.* Englewood Cliffs, NJ: Prentice Hall, 1988, chapter 14.
8. Brooks, Frederick P. Jr. *The Mythical Man Month.* Reading, MA: Addison-Wesley, 1982.
9. Bateson, Robert. *Introduction to Control System Technology.* Columbus, OH: Merrill Publishing, 1989.
10. Truxal, John G. *Introductory System Engineering.* New York: McGraw Hill, 1972.
11. Chestnut, Harold and Robert W. Mayer, *Servo Mechanisms and Regulating System Design.* New York: John Wiley & Sons, 1951.

12. Savant, C. J. *Control Systems Engineering.* New York: McGraw Hill, 1953.
13. Stevens, Roger T. *Operational Test & Evaluation—A Systems Engineering Process.* New York: Wiley–Inter Science, 1979.
14. Blanchard, B. S., and W. J. Fabrycky. *Systems Engineering and Analysis.* Englewood Cliffs, NJ: Prentice Hall, 1981, "Design for Operational Feasibility," chapters 13–17.
15. Deming, W. Edward. *Quality, Productivity, and Competitive Position.* Boston, MA: MIT Press, 1982.
16. Feigenbaum, A. V. *Total Quality Control (TQC).* New York: McGraw Hill, 1961.
17. Juran, Dr. J. M., ed. *Juran's Quality Control Handbook.* 4th ed. New York: McGraw Hill, 1988.
18. Chestnut, Harold. *System Engineering Tools.* New York: John Wiley & Sons, 1965.
19. Eisner, Howard. *Computer-Aided System Engineering.* Englewood Cliffs, NJ: Prentice Hall, 1988, "A Sample of Microcomputer-based Software," Part V Appendices.
20. Lucas, Henry C. *Towards Creative Systems Design.* Baltimore, MD: Columbia, 1974.

Design Reviews

(The Four Rights)

In Chapter SEE-4 I discussed the program start-up review as a review that considers both technical and program baselined requirements. In this chapter I will help you develop the discipline and skills required to perform the technical review process. I will discuss "the four rights" for design reviews: the right reasons, right time, right people, and the right way. Reviews shall be an opportunity to provide for constant improvement, especially when you do the mandatory peer review process. Peer review means that all designers such as electronic hardware, mechanical, software, process and all others who design must review their design with a peer who is a recognized expert on the type of design you are working on. This one-on-one, private, friendly, "I wish to help you," mentoring design review, where the designer and the reviewer spend time together with an appropriate checklist, is the basis for all good design reviews.

To improve your designs you need the design review discipline. Design reviews follow a disciplined methodology that requires the use of checklists and follow-up action. One key design review question asks of *all designs* if three possible solutions have been analyzed. This is called the "three solution rule." The three solution rule is contrary to the thinking of people who believe that discipline will make us noninnovative and those who misuse the phrase "get it right the first time." There are many famous examples of people trying over and over to achieve constant improvement and ultimately to be successful: Thomas Edison in developing the best filament for the light bulb, the Wright brothers in building the flying machine, Abraham Lincoln in losing election after election before winning to serve. Many people involved in medical research, and people in sports follow a disciplined approach resulting in innovation and success. The system engineering process is a disciplined iterative (do it over and over until you get it right) process for constant improvement. You rarely get it completely right the first time; therefore, you need the three solution rule and design reviews to help you achieve innovation and success. The design review is an integral part of the system engineering process (fig. 7-1).

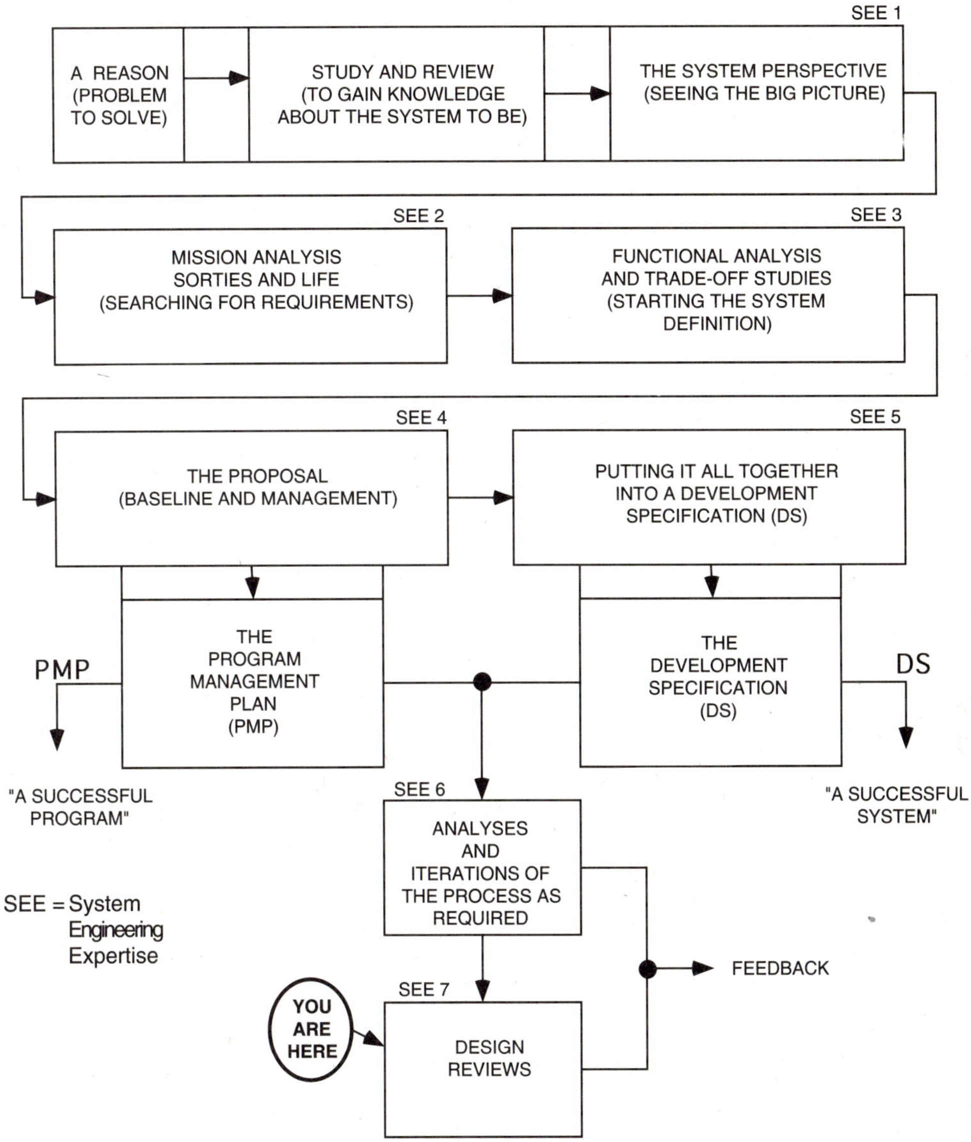

Figure **7-1.** The system engineering process

7.1 Design Review Procedures—The Four Rights

7.1.1 Right Reasons

It is important to decide why a design review is desired. The design review objectives must be clearly communicated to all participants. Some of the right reasons to conduct a design review are baseline concept feasibility, the system development specification, mission capability improvement, cost reduction, or production readiness. The reason or reasons for the design review shall be clearly stated and well understood by design reviewers and those whose designs are being reviewed. It is important for all the participants to believe that the reasons for the design review make sense and that by working together they can address the objectives to improve the design and remove risk.

7.1.2 Right Time

Timing is critical for conducting the design review. The design and documentation must be far enough along to be reviewable. Don't review too early! Required supporting data such as detailed block diagrams, flow diagrams, analysis, cost data, test data, drawings, and specifications must be ready. Data that applies to the reason for the review must be provided to the reviewers sufficiently before the actual review to allow for their study prior to the design review meeting. The data should be applicable data only; don't bury them in paper!

The corollary to reviewing too early is reviewing too late. You could be too late if the design process has progressed so far that design changes will not be possible without dire program cost and schedule impact.

Timely reviews are possible, however, because formalized review procedures and checklists have been established that define the reasons and the optimum time for specific reviews. The following sequence of reviews is generally accepted by the military and industry: Systems Requirements Review (SRR), Systems Design Review (SDR), Preliminary Design Review (PDR), Critical Design Review (CDR), Test Readiness Review (TRR), and Production Readiness Review (PRR). Note: This is not an argument for all these design reviews, but it does encourage the right timed sequence; that is, the right reasons at the right time. If design reviews are timed to occur when adequate data is available and in the correct sequence, "right time" reviews can be held.

7.1.3 Right People

The selection of the right design reviewers for a design review is the most important right. The reviewers not only must be available and have the appropriate expertise, but should have some very important general characteristics:

- **Attitude**: must have a desire to serve, be a believer in the design review process, be dedicated and willing to work, and be willing to follow through and complete the job
- **Attributes**: must love to learn and study, be a digger, and be a good communicator

- **Acceptable**: must be recognized and respected as an expert by those being reviewed

One of the objectives of the design review process is to develop people with these general characteristics.

7.1.4 Right Way

Assuming the right reasons, right time, and right people are established, then the right design review procedures must be utilized. The following procedure is an example of a top-level (level-4) review procedure for very critical designs. It requires a maximum level of effort. It is possible to reduce the required level of effort as you shall see later.

A top-level maximum effort design review procedure (see section 7.2 for reduced effort procedures):

1. The *chief engineer* with concurrence of the *program manager* shall establish the design reviews to be held on the program and the reasons (objectives) thereof.
2. They shall concur on the right time for the reviews.
3. The right people to serve as design reviewers shall be contacted and commitments to serve made. This team of approximately seven priorly successful, appropriate design experts shall be selected from outside your development team. They shall be looking at your system under development for the first time in the first review. You may continue to use this team in all subsequent reviews in the review sequence, but it is very important to get an outsider's reaction and questions as part of your initial reviews. One of the reviewers shall be designated as chairperson for the reviews by the chief engineer.
4. The *engineering manager* and his *lead engineers* shall reserve a conference room, establish the agenda, send out meeting notices, and send out the data package to be reviewed one to two weeks prior to each design review meeting. If the data package is very large, it will be appropriate to have a data package familiarization meeting to help the reviewers expedite their review.
5. Prior to the design review meeting, *the reviewers* shall review the data package; they shall also ask clarification questions of the designers as required. Reviewers and reviewees shall be cognizant of the design review checklists. All designers shall have completed their level 1 "peer" reviews. The lead, staff, and chief engineers shall have established level 2 and 3 action registers.
6. The design review meeting shall be held. The design review chairperson shall make meeting attendees cognizant of the requirements that relate to the design review.

a. The objectives for the review shall be posted.

b. An open, honest, working together, helping atmosphere shall be established.

c. Area of Concern Report Forms (example, fig. 7-2) shall be made available for all attendees. Reviewers, reviewees, and observers should all be allowed to fill out these relatively simple forms to document any design concern that may come to mind during the meeting.

d. Attendees of the meeting shall be given an opportunity to express themselves at the conclusion of the meeting.

e. A formal *action item register (AIR)* shall be established and reviewed during the meeting. This list shall include *all* of the design concerns raised during the meeting, in addition to all other prior uncompleted action register actions. Close out of actions will require sign off by the *director of engineering*.

CONCERN AREA CLASSIFICATION** #

DESIGN REVIEW CONCERN

SYSTEM

CONCERN BY ________

RECOMMENDED ACTION BY ________

ACTION IMPLEMENTED: BY: ________

DATE: ________

**			
MN — MANAGEMENT	SF — SOFTWARE	P — PRODUCIBILITY
SE — SYSTEMS	A — ANALYSIS	H — HARDWARE
ME — MECHANICAL	MAR — MARKETING	FS — FIELD SUPPORT

Fig. 7-2. Example concern report form

7. Five days after the design review meeting, the design review chairperson shall provide a formal report that includes the *Action Item Register*. This report shall be addressed to the director of engineering, chief engineer, lead engineering manager, program manager and all review attendees.
8. Five days after receiving the design review chairperson's report, the engineering manager shall meet with the chairperson to establish plans to address the action items and provide a response report to those on the design review distribution list. This report shall include the *Action Item Register* with action plans recorded.
9. Action item register closeout: The action item register (AIR) will be a periodic review item with the program manager and the appropriate engineering sign offs until all action items are satisfactorily closed out.

7.2 TAILORING YOUR DESIGN REVIEWS TO FOUR LEVELS OF EFFORT

Design reviews are required for all system development programs, but the level of effort can be tailored to the level of risk and complexity of the design to be reviewed. All designs shall be reviewed at the first level of review, which is the peer review. Greater effort may be given to the review by using more people and processes as described by levels two, three, and four. The level of effort and design review plans must be established in the initial program management plan (PMP) as part of the system engineering process. Four possible levels of effort that may be selected are described as follows, but you must always do the peer review. Even if you are designing alone, find someone who will listen to you talk about your design.

Level 1. Each designer will use the appropriate checklist or checklists as listed in section 7.4 and record in his or her design data book the reviews and consultations with peer engineers in order to resolve the checklist questions. The designer shall be prepared to provide this information to higher levels of design review effort. This review is a peer review and should include the name and signature of the reviewing peer or peers.

Level 2. Level 1 and the designer's lead engineering manager will review the peer review process, including checking for appropriate peers and checklists. The manager will establish an action item register and will review and approve the actions taken to resolve design review concerns.

Level 3. Levels 1 and 2 and a single staff or specialist systems engineer shall be assigned as a lead design reviewer to do an independent design review using the appropriate design review checklists plus any other special questions that may apply to the design. The results and action items from this review will be provided to the chief engineering manager to whom the lead engineering manager reports. The chief engineering manager will approve the actions to level 2 and 3 actions taken to resolve design review concerns.

Level 4. Levels 1, 2, and 3 and a review team of approximately seven members—an independent team of experts—who are provided with a data package and follow a process that includes a design review meeting. The level 4 process is provided as an example in section 7.1.

7.3 The Three Solution Rule

The three solution rule forces the system design engineer to become innovative; the rule is, "All engineers shall develop three solutions to every problem." Two checklist questions for all design reviews are: What are the three possible ways this design could be implemented and what are the advantages and disadvantages of each of the three solutions? The designer may work with a mentoring peer as part of the design review to evolve three solutions that may then undergo a trade-off study to select the optimum design.

7.4 Design Review Checklists

The reasons for a specific design review and the time of the review (such as system requirements review or preliminary design review) will govern the appropriate checklist of questions for each review. Some generic questions are asked at every design review to force discipline. The three solution rule question, for instance, and also, "have all prior design review action items been completed? How much does the design cost? How safe and how reliable is the design?" There are also many checklists that have been generated by specific companies, industries, and the government. Use those checklists required by your customers, but also develop your own questions and checklists that are specific to the design you are reviewing. In this book I provide a typical design review checklist to get you started.

7.5 A Typical Design Review Checklist

The following design review checklist asks generic design review questions that should be asked at all design reviews. It also asks specific questions that deal with the following reasons for the design review: Is the initial system concept feasible? Is the system development specification (DS) adequate to initially lead the development process? Is the program management plan (PMP) adequate to start the management of the development? Have all risks been identified and have risk elimination plans been developed? This checklist is basically a "program start-up review" checklist, which is really a review of the adequacy of the initial system engineering process (SEP) and system engineering management (SEM).

A System Design Review Checklist Example

Generic questions

- Have all prior design review action items been completed?
- Do you have an adequate system development specification (DS)?
- Do you have an adequate program management plan (PMP)?
- Do you have an active process of review and improvement of your DS?
- Do you have an up-to-date PMP, an active action item register process, and a periodic program-brief status reporting process?
- Is the data provided for the design review complete and adequate for the stated reasons and objectives for the design review?

Specific questions

- Has mission analysis been done?
- Has a system perspective analysis been done?
- Have team building sessions been held?
- Where are the results of the system functional analysis recorded?
- Have three possible solutions to the system mission been analyzed?
- What trade-off studies were done and where recorded?
- Is the baseline system completely and consistently defined in the DS?
- Does the design review data include a complete system DS that provides a baseline system capable of performing the mission and PMP plans to ensure the development of the system within schedule and cost?
- Do all the system designers know their technical, schedule, and cost requirements including program cost allocations and product design-to costs?
- Are the system technical attributes established?
- Are requirements traceable to the mission?
- Are the specific requirements clearly and briefly stated?
- Are program schedules, work breakdown structure—including assignment or responsibility—and program cost allocations established?

- Do the DS and the PMP follow a complete table of contents outline structure that includes all the possible generic elements plus all the special system specific elements?
- Are the technical requirements established so the three solution rule can be applied by the designers? That is, do the designers have room to be innovative?
- Have all risks, both technical and programatic, been identified and plans made for elimination been established?
- Does the development team have any skill deficiencies? What is being done to overcome these areas where specialty skills are required?
- Have market, producibility, and profitability analyses been done?
- What is the status of the technical performance monitors (TPMs)? Which ones are under technical risk monitor?
- Are all interfaces defined and requirements established?
- Has the power distribution system been defined and analyzed?
- What is the power used profile?
- Have all critical special circuits been identified and given special analysis?
- Have reliability, test, product, control system, new technology, specialty, and software analyses been done?
- Has a UPC model been developed?
- Has a balanced system attributes study been done?
- Is computer-aided design (CAD) usage required or planned?
- Has the impact of all the environmental requirements been analyzed?
- Has an accuracy apportionment analysis been performed?
- Do you have a baseline production build and test model established?
- Has failure mode and effect analysis (FMEAs) been done?
- Has a safety and hazard analysis been done?
- Has a development specification tree been established? Are people assigned responsibility for the subsystem specifications? Does the system development specification establish top-level requirements for each of the subsystem specifications?

The preceding questions are not intended to be all that will be asked but are intended to get the review started. It is expected that many more questions will be asked as the reviewers study the specific design being reviewed.

7.6 THE DESIGN REVIEW CHAIRPERSON

If you are given the responsibility to be the design review chairperson, you need some guidance. The design review chairperson must do some homework in accessing each situation: What are the four rights? Who are the designers? Who is the customer? What are the customer requirements? How much time and money has been allocated for the review? What is the level of review required? And what are the management issues?.

Two overriding goals should be to establish *peer review process* as an integral ongoing part of the process and to establish a *working together* relationship between designer and reviewers. If these two things are established, you can achieve good design review process.

One method you may try is to establish working sessions of the peer reviewer and designer, where they review the concern sheets and then jointly agree on actions to be placed on the action item register. The systems designer and reviewer, the hardware designer and reviewer, the software designer and reviewer, etc. could, in a workshop format, work together and then together report to the total assembled group their agreed upon recommendations.

Note that *all* concern sheets shall be functionally collected into categories, where they may then be combined, consolidated, and prioritized. *Never* throw away a concern. Find a place for it. This is very important to establish that all people involved in the design review feel that their concerns are addressed. In some design reviews, concerns and brainstormed ideas have been "voted upon," and a lot of concerns and good ideas were in effect thrown away. This is an area where your SEE-3 functional analysis skills can be used to ensure you utilize all the concerns and ideas.

Training design review chairpersons requires that they first become familiar with the SEE-7 chapter, which assumes they have previously gone through SEE-1 through SEE-6. They should also ideally be established as the design review chairpersons for the life of the project/program/product. This will establish continuity and complete understanding of the project/program/product/people and their designs.

The design review chairperson must also be universally accepted as one with prior, successful, applicable experience relating to the specific project/program/product. The chairperson shall have been involved in prior good design reviews based on the methodologies herein described. Then the design review chairperson must fit the criterion of "right people" with the right attitude, attributes, and acceptability.

The design review chairperson is self-trained by doing. When you lead that first design review as chairperson, remember the words from a song sung by Gracie Allen in 1936, "Stiff upper lip—stout fellow—we can muddle through." Then when you do your second and third design reviews, as you learn from your experiences, each will be better and better.

7.7 System Engineering Design Review Themes

When a system is being design reviewed there are some general themes that all design review participants must think about.

- Development specifications are alive and used.
- Thou shall consolidate not proliferate; that is, avoid complexity.
- Plans are made and followed.
- Your system is really only a subsystem.
- An action register exists and is used.
- Systems must be designed for constant improvement.
- Unbalanced systems will drift to cutoff or saturation.
- One person's component is another's system.
- We must all combine SEP and SEM; each of us must do both.
- What does it cost?
- What is the power profile?
- What are the critical circuits: highest frequency, highest gain, most power dissipation, largest dynamic change, highest impedance, new technology, and highest cost?
- Timing analysis: turning the system on, mission analysis of all sub systems and critical components, power supply timing impact of all random events, turning the system off.
- Skills and experience of the system design team members.
- Interfaces—all of them.
- Safety—is the system safe?

7.8 Design Review Problems

1. Develop a design review checklist for one of the following process systems:
   a) A jury trial process
   b) A system engineering design review process (note we are talking about the process not a specific design)
   c) A hospital emergency treatment process
   d) A fast food restaurant process
   e) A slow food restaurant process
   f) A student grading process

2. Develop a design review checklist for one of the following systems:
    a) An electronic home security system
    b) An automotive antilock brake system
    c) An airplane all-weather instrumental landing system
    d) The Mississippi river water control system
    e) The Hubble Observatory Space System
    f) A robotic storage and retrieval library system
3. Develop a design review checklist for a preproduction design review for one of the following products:
    a) A new plow
    b) A laptop portable personal computer
    c) A microwave oven
    d) A patio door
    e) A line of battery operated carpenter tools
    f) A jet engine
4. Develop some specific design review questions in the following areas:
    a) Management
    b) Software
    c) Producibility
    d) Systems
    e) Analysis
    f) Hardware
    g) Mechanical
    h) Marketing
    i) Field support

Flying on Your Own

(Now it is time for you to fly over your mountain)

8.1 SEEING AS AN EAGLE

You have been working on developing your "seeing the big picture skills" by doing system perspective exercises. You recognize that every system is really only a subsystem, and you need to be very much aware of the bigger system above and its mission. Now as you read the newspaper, watch television, and work on your job, you will start to ask a lot of questions and also have a desire to apply the system engineering process to solve problems.

The maglev system

A newspaper article tells how much fun it will be to ride the new magnetic levitation (maglev) train from Albany to Manhattan so you can attend the opera. It will take one hour one way at 300 mph and only cost $160 round trip. If you want to go from Boston to Manhattan it will cost $320 round trip and take two hours one way because you have to go from Boston to Albany, then from Albany to Manhattan. If you wish to commute from Albany to Manhattan daily, it will cost $800 per week, and you will spend over two hours a day commuting. The so-called good news from federal, state and local governments indicates that maglev will be financially self-sustaining, will reduce pollution, will provide jobs, will take advantage of the technological groundwork that has been laid by France, Japan, and Germany, will provide tremendous growth for Albany, will benefit everyone, will be fun to ride, will be too costly for small time criminals, will take ten years to develop, will only cost $6.7 billion, and the only downside may be that you will not be able to manage your growth.

Give me a break! The system engineer in me makes me ask questions. If it is self-sustaining, how many people can afford to spend close to $40,000 a year commuting to work? How many people will take evening pleasure trips that will cost $300 per person (train, meal, opera tickets plus miscellaneous)? How many times a year will the average person ride this train? What will be the impact of high-power magnetic fields on people and the environment? How will it affect the ecology for ten years while under construction? Who pays for the jobs generated during the ten years of construction? What if you only get one-half the number of predicted passengers so it is not self-sustaining? Who pays for those operating jobs? Is it worth it?

In Europe high-speed rail systems have been torn up because of a lack of business. In America we love our cars. How will this system impact other transportation systems? How many commuter airplane companies will be put out of business and how many jobs will be lost? How about safety? Let us do some failure mode and effect analyses: What if the train crashes at 300 mph or breaks down in a blizzard in the Berkshire mountains with a full train on December 23? Crime possibilities are the same as airplanes. Maglev could be a target for a terrorist's bomb or a crazy person with a gun. How about the social aspects of such a system? Will we generate a new class distinction, those that can afford to ride this train and the rest of us? We all know that if the government predicts that it will only cost $6.7 billion that it will definitely exceed $14 billion, and guess who gets to pay the bill for the next ten years before the train even runs? Where is the development specification and the program management plan for maglev? Is it totally baselined? Requirements established? Do we have plans for its development? What trade-off studies have been done? What are the alternate systems that we could postulate?

Instead of developing more high-speed transportation systems, what can we do to reduce travel? What if we did most of our work in our home and commuted only once or twice a week to our office on an allocated schedule? Can we expand the concept of subcontracting cottage industries? Can we better utilize our present systems? Can we enhance our present ride-share computer networks to increase car pooling?

By use of the system engineering process could you propose an alternate integrated system to maglev? The first question prompted by that question is, what is the problem? Why do we need maglev? Is it to generate jobs or is there a lower-risk way to generate jobs? Do we need to show the rest of the world that we are better at technology than they are? If so, why? Do we really need high-speed transportation? Are we really trying to reduce pollution, or is maglev just needed to improve the economy of New York and Massachusetts at the expense of the rest of the country?

Let us postulate a problem statement for a bigger system above: Present nationwide transportation needs are expanding such that they are causing high cost, loss of time, and pollution; therefore, we shall propose a new integrated transportation system that will slowdown expansion as well as accommodate growth in transportation capability with a reduction in cost, lost time, and pollution. This new system shall be developed by use of mission analyses which will study all the reasons people travel. It shall consider alternates to travel such as two-way visual telephone, etc. This new improved integrated system shall be proposed by the generation of a development specification and a program management plan.

This problem statement provides a mission for you to use the system engineering process. You shall now do a functional analysis of our mission to develop a new integrated system. You shall brainstorm by asking, "What are all the actions of persons and things required to solve this problem?" Your brainstorming will ask what knowledge, what skills, what experience, and what specialists are required to develop this new system? Do you have the skill, knowledge, and prior, successful applicable experience to develop a new system to solve this problem? You need a skilled team, so you may decide to develop a self-assessment skill survey to use to assemble the ideal small group of systems engineers to work on this program.

You can assemble a team of experts; then have the team do the skill survey exercise to ensure that the team has all the skills to do the job. This exercise provides team building, ensures the team knows each other's areas of expertise and also knows what skill deficiencies the team has. Your team must have some inexperienced learners, which the skilled team members are to mentor. The mentoring process establishes dialogue that forces the mentors to think as well as the mentored to learn.

Your team now has a stated problem, which it has discussed. It has sufficient knowledge and skills to evolve the required system. It can now have a team attributes workshop as discussed in chapter SEE-1, followed by the system engineering process as outlined in this book. It will provide a development specification and a program management plan that will propose a solution to the problem.

System developments by Tracy Kidder

In his books *House* and his Pulitzer Prize winning *The Soul of a Machine,* Tracy Kidder describes the typical methods used to design and build a house and a computer "machine." In *House,* the skilled team consists of the home buyers, their architect, the contractor/carpenters, and a number of subcontractors. This team did, in an ad hoc way, follow the system engineering process. Architects can use the system engineering process.

In the *Soul of a Machine* an unskilled team takes too long, works long stressful days, has too many problems, and spends too much money in developing a new computer. They needed the system engineering process. In contrast, in 1971 a highly skilled computer development team was assembled at a small company in Bloomington, Minnesota, and in a few months, it successfully developed and put into production a new innovative computer. This was accomplished within budget, on time, and with very few problems. A highly skilled team and the system engineering process made the difference.

Products

The system engineering process can be applied to all products such as personal computers, tools, automobiles, windows, flowers, marketing cups of candy, or a tooth brush.

Processes

The system engineering process can be applied to all processes such as a chemical process, a financial process, a marketing process, a gymnastics training process, candy in a cup retailing process, or a manufacturing process.

The media

Be very careful what you say to reporters because they will misquote you. We need to improve the clarity, honesty, and accountability of our reporters on television, in newspapers, magazines and books. The concept "freedom of speech and expression" does not provide license to libel, propagandize, misquote, insinuate, cause emotional pain, or in any way bring about harm. The media system needs stronger feedback to force accountability. Communication process experts can use the system engineering process to improve media systems.

The ecology

Problem 3 in chapter SEE-2 discusses the ASWAN dam on the Nile. Many problems could have been avoided if the mission analyses had included studies of the dam's impact on disease, society, insect life, aquatic life, water purification, and the fishing industry. The water all over the world is being seriously affected by dams, fertilizers, and industrial wastes. All systems engineers must determine the impact of their system on the environment. Civil, agricultural, industrial, and ecology systems engineers must use the system engineering process, with special sensitivity to environmental impact.

The children

Improvement in family and education systems requires the application of the system engineering process by educators and social systems engineers. The life and sortie missions of children must be analyzed so we can provide improvements in family life and mission-driven education.

Medical

In the introduction to this book, I talked about Doctor Plummer of the Mayo Clinic and why I felt that he was not only an excellent doctor but also a systems engineer. So, too, doctors can apply the system engineering process. In addition, control system engineers know that the dynamics of many human control systems are quite fast, such as blood pressure, blood chemistry, and heart tachcardia; therefore, we have been evolving patient monitor systems with the ability to constantly monitor (reference workshop problem 3 of this chapter, "The Generic Monitoring and Analysis System"). The application of the system engineering process to the medical profession is wide ranging. It can be applied to solving very difficult problems by using the functional analysis methodology, "layered analysis methods" discussed in chapter SEE-6. It can also be used to develop or improve specific process systems such as emergency room, specific surgery procedures, billing and insurance procedures, patient records, lab testing, and doctor competency improvement processes.

The flea bites

I have been told that the systems engineering process can be used to avoid conflicts, eliminate wars, and bring about a peaceful coexisting world. Study has evolved an interesting corollary between conflicts around the world in places such as Palestine, Northern Ireland, and Yugoslavia and the life cycle of the flea. This corollary also seems to exist in social relationships between people and also in certain diseases.

The adult flea lays eggs. They hatch. The larvae grow and molt. The larva spins a cocoon and pupates, and they live in nests that could be the living room rug. When the flea is ready to emerge, it remains in the pupae case until a host animal approaches. The vibration of the room or ground associated with this approach releases the emergence behavior. In rooms that are not used for long periods, quite a large brood of fleas can accumulate. Starved by the long waiting period, they viciously bite to get blood. These new adult fleas then start the cycle over by laying eggs. This cycle has brought about the spread of plague epidemics among humans, but it is the thought-flea in the mind of humans that is the source of the evil attack conflict mission that we need to eliminate.

The system engineer studies the evil mission cycle and then develops ways to break the cycle. The cycle of the flea can be broken by an obsession with cleanliness. We can also spray to kill the adults/larvae/eggs/pupae when they emerge as adults. We treat the host with a chemical in the blood stream that causes the flea to die if he bites the host (this chemical obviously must not injure the host). Then we consistently break the process longer than the longest possible cycle. The flea in the mind can be the desire for revenge for past grievances. How do we clean the mind and kill this thought-flea? Perhaps it is as easy as to learn how to forgive and forget for the sake of our children and grandchildren. Many people in history have been instrumental in breaking the cycles of evil. Many people daily do things that help break the evil cycles of conflict. We need to use the system engineering process to evolve conflict resolution systems.

The judicial process

The jury process is an example of the many judicial processes that could be improved if the system engineering process was applied. Present methods of selecting juries, especially for long trials, result in some serious distortions of the concept of "a jury of your peers." The selection process is long and tedious. The population of people available for jury duty is limited to those who are not working or can afford to be away from their jobs for a long period of time. Many jurors are selected specifically because they have no knowledge, skills, or experiences that can relate to the defendant. How can they understand the defendant? Jurors are restricted in how much information they are given, may request, or can obtain on their own. Information on prior arrests are not divulged, for example. Long jury deliberations develop an incentive to vote in any way just because the jurors want to go home. Even worse the jurors may be afraid of possible revenge from the defendant.

These problems, real or perceived, are worthy of the application of the systems engineering process to bring about jury process improvement. Lawyers and Judges can apply the system engineering process to improve judicial processes.

Your turn

Now it is your turn to apply your skills in the system engineering process to solve problems in your specific area of expertise.

8.2 SKILL SURVEYS

The self-assessment skill survey is a very powerful tool for building teams. Each system type requires a specific body of knowledge, skills, and prior applicable successful experience within the system engineering development team. I have devised and used a skill survey for electronic control systems, which was described in chapter SEE-1. You may develop your own specific skill survey for your type of system by brainstorming with a diverse team of experts to select skills required. You may then pattern the survey, the skill definitions, and the scoring process after the one described in this book.

The skills that may be generic to all system types are marked with an ☛ in front of the following skill definitions. You may expedite your skill survey for your type system by starting with these generic system engineering skills.

You may wish to develop your own specific skill survey if you are working on a system type other than an electronic control system, such as an educational system, a medical system, a judicial system, a civil engineering system, or any specific process type or specific product.

Skill Definitions for Electronic Control Systems

Accuracy. System techniques are available to enhance the system accuracy. The systems engineer should be familiar with these techniques and use them when applicable. Analog-to-digital (A/D) convertor ground measurement provides a zero offset correction; a reference voltage measurement provides a span correction; ratiometric measurements can eliminate common mode signals; the equation for balanced differential amplifier network analysis can be used to optimize accuracy for designing balanced measurement channels. Heuristic methods, software accuracy enhancement algorithms, and models can also be used.

Acoustics. The systems engineer must know what the system will sound like as it is running. Will it be objectionable? The systems engineer should be able to calculate sound level and know how to design for reduction.

Analog circuit design. The design of electronic circuitry that operates in the active region. The design and application of amplifiers such as audio and operational, active filters, and analog control systems are all examples of analog circuit design. A "3" skill level would include one who has experience in applying analog amplifiers in specific analog circuits; a "6" skill level would include one who has designed analog-to-digital (A/D) and digital-to-analog (D/A) circuitry or is one who has designed analog control systems; the "8/9" skill level includes someone who had done all of the preceding.

Architecture. A style and method of design and construction. The development of the optimum architecture for a system is a key skill for a systems engineer. For example, the systems engineer may select a dual computer with a shared, commonly mapped memory area as an architectural feature that will ensure adequate throughput, workable algorithm partitioning, and high-speed coprocessor data transfer. The systems engineer is familiar with architectural features and is capable of performing trade-off studies thereof.

Breadboard building. Mounting and connecting components to form a functional version of an electronic circuit. The objective is to prove functionality of the design and to obtain test data for circuit analysis; skill in layout, wiring, soldering, heat sinking, sketching, drawing, completeness, electrical noise decoupling, and spurious element control are required.

☛ ***Communicator.*** One of the top nine skills for a systems engineer. The systems engineer must work with internal and external interfaces in order to establish complete understanding. Always ask questions to establish clarity. The system development specification (DS) is the key communication tool used by the systems engineer to communicate with the system designers and ultimately the system producers, service people, and customers.

The top nine system engineering skills:

1) Communicator, 2) Working with others, 3) Finishing a job, 4) Mission analysis, 5) Functional analysis, 6) System DS process, 7) Reliability analyses, 8) Control system analysis—system analyst, 9) Cost estimator.

Company procedures. The Engineering Operation Procedures System (EOPS), Program Management Operating Procedures System (PMOPS), design process procedures, configuration management procedures, are all examples of procedures with which the systems engineer should be familiar. The company way of doing your job (procedures) are, it is hoped based on years of successful performance. The systems engineer knows how to get things done within the procedures system, including the use of applicable level of effort, tailoring of requirements, expediting, appropriate bypassing, proper response to mandatory procedures, and also knowing how to improve procedures.

Computer-aided design (CAD). These are software tools used for design analysis such as "SPICE" and "HICAP," for circuit design; "STARDYNE," for structural analysis; and various other tools for all engineering disciplines (mechanical, electrical, reliability, systems).

Computer-aided manufacturing and integrated documentation (CAMAID). These are computerized drafting tools used to generate engineering drawings. The Computer-Vision system and an IBM PC with Futurenet software are examples of the hardware/software that provide CAMAID.

Computer hardware. The circuitry that constitutes the physical equipment for a computer. The associated skill for computer hardware is the ability to design and/or use computer hardware. The use skill level is about a "3," while the design skill level is about a "7." Computer hardware design includes: central processor unit (CPU), computer architecture, computer input/output design, microprogramming, hardware debugging, and all the computer peripheral designs. Design them all successfully and you are a "9" level.

Computer software. The knowledge of computer software programs that are available and applicable to your projects. You also need to know where to go to get information on what is available in software for design/ desktop publishing, word processing, business, etc.

Computer technologist. The *combination* of both hardware and software skills in the application of computers.

Hardware: CPU design, computer architecture design, computer I/O design, computer microprogramming, documentation, debug, etc.

Software: Machine language programming, a number of high-order languages (HOLs), hierarchial structuring, modularization, executive, assemblies, compilers, documentation, debug, available computer software programs, etc.

☛ ***Control systems analysis***. The understanding of basic control system relations, feedback, stability and transfer functions. The control system description and analysis using LaPlace transformations and transfer functions, Root-locus, Bode Plots, Nyquist diagrams, sampled data systems, Z-transforms, nonlinear systems, frequency response, transient response, and Kalman filters. This also includes system analysts such as those that analyze and develop algorithms for navigation, telemetry, queing, star searching, gyro coning, etc. This is one of the most important skill areas for the systems engineer.

☛ ***Cost estimator***. Developing good cost estimating skills is very important. Developing a parametric cost model for estimating the cost of doing a job, projecting how costs will change in the future, optimizing the cost of a product, recognizing the impact of maintenance costs, etc., are all skill aspects of cost estimating. Unit product cost (UPC), life cycle cost (LCC), value engineering, and work breakdown structure (WBS) all require cost estimating skills. This is one of the most important skill areas for the systems engineer.

☛ ***Customer liaison***. Communicating and cooperating with the customer with the objective of establishing a "win-win" relationship in providing a successful system, subsystem and/or component. The systems engineer has the responsibility to establish the *technical liaison* with the customer. The development specification is a key tool used to establish precise and concise understanding.

☛ ***Data analysis***. The systems engineer must know how to interpret data. Question the source and method of obtaining data. Question the objective of the data. Know how to develop a mathematical definition of the data, and do an analysis of what the data indicates. In addition, the interrelationships between various data sets are developed.

For example: This system nominally responds to temperature variation by accuracy and response time changes of a certain plotted characteristic. It also responds to component variations which change the accuracy and response time variations by different characteristics. The summation of these effects result in a family of curves, that can be defined by the following mathematical expressions . . .

Data busses. Skills in applying, designing and analyzing data busses are very important in digital systems. Both parallel and serial busses must be understood. A general knowledge of all the standards such as Mil–Std–1553, ARINC–429, IEEE–488, RS–232, MULTI–BUS, PI BUS, SDLC, TM STDS, UNIBUS, and on and on (as required), must be developed to apply these standards to your systems. Design and analysis skills are: software programming protocols, transmission line theories, lumped-constant equivalent circuits of distributed parasitic elements, the impact of specific parasitic elements at critical points in circuitry, ground return design, frequency response, pulse shrinkage, common-mode-rejection, differential balanced networks, optical coupling, strip line, EMI considerations, nuclear impact, timing diagrams, data formation, ground bounce, and on and on (as required).

☛ ***Design reviews***. A design is reviewed by an expert or a team of experts to ensure that the design will be successful. A design can be of a system, a circuit, a procedure, a management plan, etc. There are four rights for a design review: right reasons, right time, right people, and right way. Military Standard 1521 (Mil–Std–1521) and many engineering procedural systems provide instructions on how to perform design reviews.

☛ ***Development specification (DS) writing***. Development specification writing is a powerful tool to guide the development of a system. The DS is a document that clearly and accurately describes the essential technical requirements for systems, subsystems, components, products, items, processing, materials, and services, including the procedures by which it will be determined that the requirements have been met. This is one of the most important skill areas for a systems engineer.

Digital circuit design. The design of logic circuitry. Circuits that have two states: on or off, 1 or 0, saturated or cutoff. Circuits have rise times, fall times, propagation delays, parasitics, and temperature coefficients. Timing analysis skills are a must. Digital circuits are used in digital computers to provide logic, control, communications, memories, and arithmetic functions. Digital circuit design also includes the application of complex digital functional elements such as microprocessors, analog-to-digital converters, digital-to-analog converters.

Drafting (or computer graphics). The ability to provide a complete set of drawings for a system is very important to its development. Drawing the detailed system block diagrams, the functional diagrams, the schematic diagrams, the interconnect diagrams, the mechanical drawings, the partitioning diagrams, and, especially, ensuring the completeness of the drawing package is a system engineer's job. The system engineer normally does not draw all these drawings but should be capable of doing so. He or she will establish, probably by sketches or computer graphics, the entire drawing package. The systems engineer will be called upon to check all the system drawings to ensure completeness and correctness.

Electromagnetic interference (EMI) analysis. Electromagnetic interference analysis requires the SE to recognize EMI generating sources and also the EMI susceptibility of the circuitry. Radio Frequency (RF) design, packaging techniques, parasitic components, modeling, current distributions, field configurations, filtering, filter design transmission lines, coaxial cables, shielded cables, twisted pair cables, connectors and wave guides are all areas that should be understood. The EMI analysis, and especially EMI testing, require the SE to be supported by EMI specialists.

Engineering specification (ES) writing. The engineering specification (ES) is an internal company specification written for production build and testing of a subassembly, such as a printed circuit assembly. The design systems engineer tells production how to test the design in production to confirm that it is functional.

Error analysis. This analysis is a detailed worst case analysis to ensure the design fulfills the error apportionment criteria. The design can be produced in production with high confidence that most of the units produced will meet the error apportionment criteria.

Error apportionment. In the development of the DS requirements, the systems engineer must establish error allowances apportioned among the various error-contributing modules. If a measurement accuracy of 1.0% is required, for example, 0.02% may be given to the input filter, 0.01% to the input MUX-SW, 0.04% to the A/D converter, and 0.03% held in reserve for test equipment measurement inaccuracy.

☛ ***Failure mode and effects analysis (FMEA)***. The failure mode and effects analysis is an analysis usually performed by a reliability engineer with support from the systems engineer. The effect of all possible hardware failures and software misprogrammings are analyzed. When a failure mode is found that may cause a system malfunction or serious safety hazard, the system is redesigned to reduce the effect of the potential failure. This skill should be an inherent part of the design process.

The systems engineer can, through judicial partitioning, utilization of built-in test, redundancy techniques, and parts selection criteria, reduce the possibility of potentially critical failures.

☛ ***Finishing a job***. The systems engineer, by nature of his ability to see the big picture perspective, sees what is required to finish a job to ensure the outputs from one job flow in as inputs to a future job. The systems engineer is not only a self-starter but also a complete finisher. This is one of the most important skills for the systems engineer.

☛ ***Functional analysis***. The systems engineer does an analysis of the system to determine functions required in order to accomplish a mission. For example, a commercial airplane functional analysis would include functions such as propulsion, flight controls, communications, navigation, and electrical power. Functional analysis is hierarchical in nature: a functional analysis on the top-level function "electrical power" leads to such functions as power takeoff, generation, storage, distribution, conversion, etc. Functional analysis is a skill that is applicable to all engineers and is therefore one of the most important skills for a systems engineer.

☛ ***Human factors***. The human element of a system and the interface of the human to the system. Consideration must be given to human physiological factors (e.g., reaction to environmental forces), psychological factors (e.g., need, expectation, attitude, motivation), anthropometric factors (e.g., human physical dimensions), and their interrelationships.

Hydrodynamics. The systems engineer many times needs to provide interface to mechanical actuators, which are many times hydraulically driven. Therefore, it is helpful to have some knowledge of the dynamics (response time, nonlaminar flow, hardover input bounce, etc.) of the hydraulic system.

Illumination engineering. It is valuable for the systems engineer to know light spectral characteristics and effect on human operators. The emergence of optic coupling is another subject of which the systems engineer should have some knowledge.

☛ ***Innovator***. The skill of introducing something new and developing the "something new" to prove feasibility. This skill starts with mission analysis and brainstorming for ideas. It then carries the idea to demonstrated practice. The innovation skills of a systems engineer are very important. The systems engineer constantly asks and seeks a better way, a better solution!

Integrated circuit technology. Very high-speed integrated circuit (VHSIC) and very large-scale integration (VLSI) are both terms that relate to integrated circuit components. Very high-frequency pipeline processors are an example of VHSIC, while microprocessors, which provide a total complicated computer on one integrated circuit, are an example of VLSI. Programmable array logic (PALs) and application specific integrated circuits (ASICs) are also sometimes referred to as VLSI devices. The systems engineer should be knowledgeable of past and present integrated circuit devices. In addition, the systems engineer should be knowledgeable of integrated circuit design and process methods. Solid state physics, materials, masks, geometries, interconnects, packaging, and, especially, parasitics in applications all are areas the SE should know.

☛ ***Lead engineer***. Very high technical skills are baseline for the development of lead engineer skills. A lead engineer assumes technical responsibility for a system or subsystem and then performs as a work director, design reviewer, and coordinator of design/development activities. Excellent technical communication capability is a key attribute of the lead engineer.

☛ ***Life cycle cost.*** Systems can be designed to be ultrareliable by using the most expensive parts and ultraconservative design: field maintenance and repair can then be negligible. Systems can also be designed with the cheapest parts, resulting in a lot of field maintenance and repair costs. The equations, study, and analysis of these cost factors is referred to as life cycle costs (LCC). The basic elements of a life cycle cost equation are: LCC = [(Research and Development Cost) + (Production and Construction Cost) + (Operation and Maintenance Support Cost) + (Retirement and Disposal Cost)] / number of systems produced.

☛ ***Maintainability design***. In order to put in a new gasket, you have to remove the entire engine on your car. Poor maintainability! Designing equipment to provide ease of maintenance is a skill. Both mechanical and electrical ease of access, with failure isolation and repair scenarios optimized, is required for good maintainability design. Testability is a subset of maintainability design.

Mechanical packaging. The mechanical design of the system, including the placement, mounting, interconnection, and environmental aspects of all the system components. Thermal and structural design and analysis of the system are performed by the mechanical packaging engineer.

Microprocessors. Knowledge and application experience of current microprocessors. This experience should include actual hardware design into a system and associated software development.

☛ ***Mission analysis***. A time line study of system function is the mission sortie. The systems engineer should also perform life mission analyses including a time line analysis of the life of a specific system and also the analysis of the system through all its life cycle phases. This is an application of the system perspective. *Mission Analysis is a skill that is applicable to all engineers*; it is, therefore, one of the most important skills for the systems engineer. See "Time Analysis" skill definition, which is a lower-level application of mission analysis, and, as such, is how systems engineers initially develop their mission analysis skills.

☛ ***People systems***. The systems engineer will do an analysis of the program to assess the skills required for the people assigned. The systems engineer will also be familiar with the skills capability of the specific people under consideration. Matching the skills required to the skills available is part of the people systems.

Another part of the people systems is the development of skills through mentoring, on-the-job training, and formal training courses. Interpersonal relationships, team building, disciplined structure with the correct, specific "room-to-do" sensitivity, and effective listening-conflict resolution are all required for this very important skill.

☛ ***Planner***. Plan your work, then work your plan! Plan your work means to make up "to do" lists, schedules, cost estimates, contingency actions, risk analysis, matching skill needs of work to be done to skills of people available, etc. Then also work your plan by monitoring, measuring, and modifying to make your plans ensure the work will be done as desired. Every plan defines: inputs from, activities to be performed, outputs to, who, when, how, where, and how much.

Power supply design. The skills required to select the appropriate type of power supply design for a specific system. The size, weight, efficiency, EMI characteristics, cost, reliability, load profile, handing characteristics, overcurrent characteristics, overvoltage characteristics, transient response, and power-up and power-down characteristics are all features that should be known. In addition, the ability and skills to actually design power supplies are included in this skills category. Modeling, computer-aided design (CAD), power supply architectures, and power supply interfaces (both to variable input power and loads) are skills that relate to power supply design.

Power system design. The skills required to select a power supply for a system as in power supply design, plus the skills needed to develop the requirements for the power supply. This includes load analysis, accuracy, load profile analysis, power factor correction, distribution of energy storage, and system response to input power and/or load transients. The power supply system design includes the ability to construct the system power model and utilize analysis to optimize the power system, including power source interface, power supply, and power distribution. The power distribution design includes knowledge of grounding methods and ground current distribution, which requires an intimate understanding of the hardware operation and its load profile, including the possible impact of the software. EMI filtering is another aspect of power system design.

☛ ***Probability***. Probability concepts and theory are important to the systems engineer for various system analyses. The universe (population) and the sample, random events, mutually exclusive events, independent events, the addition theorem, the multiplication theorem, probability distribution models (binominal, Poison, uniform, exponential, and normal) are all familiar terms to a systems engineer with probability analysis skills. The fundamental definition of reliability is heavily dependent on the concepts derived from probability theory.

☛ ***Production design***. Design the system so it can be produced at a profit. This includes unit product cost analysis, production process analysis, test analysis, yield analysis, value engineering, and worst case analysis. The systems engineer should also do production walk-throughs as part of the life mission analysis. The production readiness review (PRR) process should be understood.

☛ ***Program management***. These are skills in managing the business aspects of a program. Customer interface, cost estimating, scheduling, team building, risk analysis, variance analysis, negotiations, program management plans (PMPs), system engineering management plan (SEMP), status reviews, technical objectives, and interface communications are all skills to develop. The establishment of appropriate plans, the monitoring thereof, and both external and internal excellent communications, on a program demonstrate good PM skills.

☛ ***Proposal writing***. Proposal writing is an especially important skill for a systems engineer. Applying customer liaisons, the system perspective, mission analysis (both sortie and life), functional analysis, trade-off studies, and partitioning and developing specification process very quickly/adequately to lead a proposal technical team should be one of the system engineers strong skills. The systems engineer works with the program manager and the customer to provide the proposal baseline system definition.

Propulsion. Systems engineers are becoming more involved in total integrated systems; therefore, knowledge of propulsion systems is helpful. The interaction of the thrust vector and other surfaces and response times thereof are needed to do a complete mission analysis of vehicles.

☛ ***Reliability analysis***. Calculations of the mean time between failures (MTBF), component stress exposure, redundancy impact, and reliability model development are examples of reliability analysis. This is one of the most important skills for a systems engineer.

☛ ***Report writer***. Make an outline, organize your material and thoughts in a systematic way. Write a theme, establish your objective, then write and rewrite your report. Now imagine yourself as the report audience, read it from its perspective. Is it concise or too wordy? These are report writing skills, but they also apply to any formal presentation, oral as well as written.

☛ ***Risk analysis***. The systems engineer reviews specific projects for risks regarding cost, schedule, and technical aspects. The systems engineer also applies the system perspective and mission analysis to the risk analysis for future activity on related follow-on projects, competition, ultimate market, and producibility. The systems engineer develops plans for overcoming the risks and works them into the overall long-range plan for the system. The systems engineer establishes cost performance measures (CPM), schedule performance measures (SPM), and technical performance measures (TPM), and then ensures that they are monitored and met.

***Safety analysis*.** This analysis is usually performed by the reliability specialist with support from the systems engineer. The analysis is to ensure that the product or system under analysis is free from hazards that may endanger the users and handlers during the total life mission of the product.

***Sketching (or fast computer graphic)*.** A picture is worth a thousand words, and a good sketch (or a fast computer graphic) in the system specification is an excellent communication tool! System specifications with a number of good sketches are smaller and easier to understand. Block diagrams, outline drawings, perspective drawings, an installation sketch, cabling drawing, specification tree, partitioning diagrams, timing diagrams, mechanization drawings, control system diagrams, etc. are all needed. The ability to do fast and understandable sketches or computer graphics is a good systems engineering skill to have.

***Software coding*.** The line-for-line development of a set of steps to be followed by a computer system or other device. Software coding is also commonly referred to as "programming." The software system has been designed into working functional modules that are presented to the software coder for software coding (programming). This skill also includes designing and developing test methods and debug processes for software module validation.

***Software design*.** Given a modularized software definition and requirements from the lead systems engineer, design the software by developing the hierarchial structure, the startups/shutdown process, executive, functional module definition and requirements, flow diagramming, timing analysis, memory sizing analysis, and software development monitor and test process. Tool selection, operating systems, and compilers for software development are also considered a subset of software design.

***Software systems*.** The performance of a digital computer-based system is software dependent. A hardware digital control system can be made to perform differently merely by programming the digital computers differently. Therefore, the software engineer has the wherewithal to develop a system. Synergism between the lead systems engineer and the software systems engineer is imperative, and must be established by the systems engineering top-level system development specification iterative process. It is better yet to realize that skills in software systems are a special skill area that should be practiced by all systems engineers. They can make the interface to other skill areas, such as hardware systems engineer, analyst systems engineer, mechanical systems engineer, and program manager systems engineer, easier. The software systems systems engineer has strong software skills but also has adequate skills in other areas to interface with systems engineering peers.

***Solve difficult problems*.** The skill to solve difficult problems can be developed. *First*, the basic problem-solving methodology must be learned: study, ask questions, analyze, define the problem in its simplest terms, brainstorm for possible solutions, write a "function of" equation, prioritize to most likely to succeed, document possible solutions, establish trade-off criteria, do trade-off and decision chart analysis, apply optimum solution, take data, analyze results, iterate with as many try-agains as required. *Second*, apply this methodology to problems until it becomes second nature.

☛ ***Standards (industry, military, etc.)*** The systems engineer must have knowledge of standards that apply to the system being developed. This includes reviewing and questioning their applicability with the customer. Exceptions to requirements can be negotiated. The skill and perseverance to review, question, and negotiate standard requirements is one of the most important skills a systems engineer can have. Some specific Military Standards of interest are:

MIL-STD-499 — engineering management. The system engineer management plan (SEMP)

MIL-STD-490 — specification practice

MIL-STD-1521A — design reviews

MIL-STD-881 — work breakdown structure

MIL–STD–704C — electrical power

Statistics. The knowledge and understanding of *descriptive statistics* (analysis techniques directed to the description of sample data) and *inferential statistics* (deals with sampling and the drawing of an inference about a population from the sample) are systems engineering analysis skills.

Stress analysis. This analysis is performed on all the components in the system. Are voltage, wattage, or current ratings being exceeded? Is the local environment developing "hot" spots where the power dissipation may be exceeding the derating criteria? This analysis must be done by each design engineer, with support from the reliability engineer, the mechanical engineer, and the thermal analysis specialist.

System analyst. (See Control systems analysis)

☛ ***System design***. System design requires: customer liaison, study, problem definition, brainstorming, system perspective application, sortie mission analyses, life mission analyses, functional analyses, trade-off studies, baselining, partitioning, system development specification process, design analyses, test analyses, reliability analyses, product analyses, producibility, control system analyses, new technology analyses and specialties analyses. System design entails the following SE functions: management (SEM), system analysis tools (SAT), system definition function (SDF), system implementation function (SIF), system verification function (SVF), and system support function (SSF).

☛ ***System partitioning***. The apportionment of the system into hardware and software modules. These modules have to be sized within constraints. They must be producible and interface logically within a hierarchial structure. The inputs, functional activity, outputs, feedback and detailed requirements must be specified for each module.

☛ ***System spec writing***. This iterative *process* is used to timely develop the documentation of the system definition and detailed requirements specifications. This skill is "Job No. 1" for the lead systems engineer. All lead system engineers—top-level lead, hardware lead, software lead, test lead, and mechanical lead—must have this skill.

Testability design. Testability is a subset of maintainability design. It deals with designing systems to provide adequate accessibility to signals and data to provide for the isolation of faults to replaceable modules. Built-in test (BIT) (software) and Built-in test equipment (BITE) (hardware) are designed-in features relating to testability design. The proper location and buffering of test points and the generation of adequate and confirmed test procedures for test and fault isolation reflect good testability design skills by the designer.

Thermal analysis. The systems engineer must have some basic knowledge of the heat dissipation profiles within the product and must be able to relate them to the external mounting and external environment *throughout the life mission* of the system. In addition, the systems engineer should have sufficient knowledge to understand heat transfer mechanizations, thermal resistances, impact of the buildup of unclean surfaces, and be capable of doing simple heat transfer calculations. The detailed thermal analysis *requires* support by thermal analysis *specialists* with modeling computer program analysis tools.

☛ ***Time analysis***. Systems engineers must have a complete understanding of the value and critical nature of time. Their skills should include the ability to develop time sequences. In addition, they must be able to do an analysis of the impact of events out of desired sequence. They should develop the ability of doing time variance and worst case timing analysis. Increments of time to systems engineers are relative, and they can deal with the analysis with least significant elements of picoseconds or years. This skill is the basis for mission analysis; systems engineers all develop skill in doing timing analysis in an ad hoc way, but when they see the relationship of timing to system sortie and life, then all the sub-missions timings are put into perspective as part of mission analysis.

☛ ***Tools knowledge***. Knowing about tools and/or developing the *appropriate* tools for the job. Tools can be hardware, software, procedural, and combinations of hardware/software/procedural. Examples of hardware tools are logic analyzers, oscilloscopes, meters, calculators, special test jigs, and circuits. Software tool examples are operating systems, design tools such as CAD and those used in QFD, as well as analysis support programs. Procedural tools are those methodologies such as the problem-solving methodology, decision charting, PERT diagramming, matrix reduction methods, functional analysis process, and flow diagramming. Combination tools are represented by computers used in a test or simulation configuration where hardware tools are provided to run software tools to perform a test sequence procedure.

☛ ***Training***. Have you taught others formally or informally? Systems engineers are expected to lead the technical development of the system; they, therefore, are mentors and trainers. Likewise, specialists engineers are expected to mentor others in their area of specialty of design or analysis.

Travel. Do you travel? Do you plan your trips with objectives established ahead of time? Do you avoid unnecessary trips? Do you try to combine trips? Do you keep trip costs down? When on a trip do you develop rapport with customers? Do you know all the tricks of traveling (best time of the week, etc.)?

☛ ***Unit product cost***. UPC is the cost of a device as produced (including material, production labor, and production test). It also entails the development of a specific equation for a specific product, which defines its cost elements. The ability to develop both top-down (mission driven) and bottom-up (design driven) UPC models and tracking methodologies, (also see life cycle cost).

Use test equipment. The systems engineer should have skills in the laboratory and at field test sites to use test equipment such as logic analyzers, oscilloscopes, meters, telemetering equipment, software support equipment, etc. The accuracy of the equipment and also the loading of the system by the equipment should be understood. Related skills to use for test equipment data are data analysis, error apportionment, accuracy enhancement, testability design, and tools knowledge.

☛ ***Value engineering***. The process of analysis of cost reduction redesign activity. This very ordered and disciplined process determines *all* costs, and has shown that sometimes hidden costs make some so-called cost reductions more costly than present design. Value engineering has, however, been used to cut costs from devices by concentrating on reducing high cost drivers, which could be material, components, or process cost elements. The systems engineer must know what the remaining planned production quantity is, the present cost, and the cost to do the redesign. If the quantity to be produced is high enough, the reduction in cost big enough, and the cost to redesign low enough, you may be able to do the redesign and thereby cut cost.

☛ ***Variance analysis***. Variance is a measure of dispersion: it is the difference from a mean. The analysis of this range of dispersion or deviation from the means is a very important task for a system engineer. It becomes especially important when a number of items are combined so that the interrelations of the variances must be established. Worst case analysis and central-limit theorems root of sum of squares (Rss) analyses are subsets of variance analysis.

Wire lists. To the systems engineer, a wire list is much more than just a point-to-point listing. It is an opportunity to do a mental analysis of the interfaces. As the systems engineer develops or analyzes the wire list, he or she looks at source impedance, distributed capacitance and inductance, relationships between lines (crosstalk possible?), fanout, total distributed load, possible noise, etc. This analysis starts with the initial baseline for the system as part of the interface definition.

☛ ***Working with others***. Getting along and "getting to know you" so we can work successfully together is a key desire of a systems engineer. Teamwork wins, and if the team wins, we all win. We all experience "people problems," but the skill to overcome them separates success from failure.

Worst case design. Worst case design is sometimes called "worst case analysis" or "design margin analysis." This analysis is a mathematical testing of the design. Examples of Worst Case design are:

1. Summation of the minimum propagation delays through two independent paths to a common device to ensure that the logic circuitry will not cause a momentary logic function error.
2. Analysis of the initial tolerance range, plus temperature effect, plus aging effect of the components of a filter to ensure these factors do not change its frequency response out of the desired min/max range of the specification.
3. The specification range of the electronic device parameters does not cause the circuit to be out of the required specification.
4. Software nesting will not cause a "lockup" or incorrect jump condition.
5. Software provides for correct input/output response times worst case or even better "enhances" system response as a function of hardware temperature. (Also see Error analysis, Variance analysis and Accuracy enhancement.)

Scoring your skill survey

Using the templates, add up your scores in each of the classifications: HASP, H, A, S, P, M, and TOTAL. Record these additions on the survey analysis report, calculate the percentages for each of the classifications, and record the percentages. Select the highest percentage of H, A, S, P, and M, and place 1.0 in the "normalized" column for the highest percentage. Now ratio the percentages for each of the other classifications against the highest percentage and record in the normalized column. You may get greater then 1.0 for HASP if your skill percentage for common system engineering is greater then your highest of H, A, S, P, or M skills. Plot these on the profile chart. Rank your H, A, S, P, and M percentages, from highest percentage = 1 to lowest percentage = 5. I have programmed the prior process into my computer. You may wish to do the same.

The survey reviewer comments should list the number of years experience including applicable years prior to your college degree and what you consider yourself as indicated on the first page of the survey. Also record in the comments: the total skills percentage, the skill profile, the highest skill classification, and the lowest skill classification. Now scan the grades of the specific skills, looking for expert (9) areas and for low skill (0.1, 2) areas. Record expert and low skill areas. The last item to place on the survey reviewer's comments is from the last page of the survey: do you wish to be a systems engineer or just wish to improve your system engineering skills?

Recommendations can now be made based on the data collected in the comments. Provide a consolidated statement such as, "You are a well-rounded program manager systems engineer with expert hardware skills. You are an expert in working with the customer. To become a better systems engineer you could improve your knowledge and skills in development specification process, data analysis, software systems, and also analysis skills—reliability , life cycle cost, and errors. This should flatten your skill profile and also raise your total skills."

If you wish only to improve your system engineering skills, the recommendations would be different because you wish to improve all the "HASP" skills and do not worry about developing a flat profile. The ideal systems engineer has one high-expert skill classification, a high-flat profile for all the other classifications, a total skill level of over 50%, and at least ten years experience.

SKILL SURVEY SCORING TEMPLATES

Skill	M	H	A	S	P
Mission Analysis	-	H	A	S	P
CAD	M	H			
CAMAID	M	H			
Drafting	M	H			
Proposal Writing					P
Control System Analysis			A		
Data Busses		H			
Architectures		H			
Sketching	M	H			
Company Procedures	M				P
Tools Knowledge	M	H	A	S	
Working with Others	-	H	A	S	P
Training					P
Risk Analysis	-	H	A	S	P
Travel					P
Software Systems				S	
Production Design	M	H			
System Design	-	H	A	S	P
Wire Lists	M	H			
Program Management					P
Lead Engineer					P
Solve Difficult Problems	-	H	A	S	P
Planner					P
Cost Estimator					P
Innovator	-	H	A	S	P
Report Writer					P
Communicator	M	H	A	S	P

Skill	M	H	A	S	P
Analog Circuit Design		H			
Digital Circuit Design		H			
Software Coding				S	
Software Design				S	
DS Writing		H		S	
ES Writing		H			
Breadboard Building	M	H			
Use Test Equipment		H			
Data Analysis	-	H	A	S	P
Functional Analysis	-	H	A	S	P
Design Reviews	-	H	A	S	P
Customer Liaison	-	H	A	S	P
Reliability Analysis			A		
Unit Product Cost	M	H			
Finishing a Job	-	H	A	S	P
Human Factors	-	H	A	S	P

Skill	M	H	A	S	P
Acoustics	M	H	A		
Worst Case Design	M	H	A		
Value Engineering	M	H	A		P
FMEA			A		
Maintainability Design	M	H	A		P
Safety Analysis			A		P
Life Cycle Cost			A		P
System Spec Writing	-	H	A	S	P
System Partitioning	M		A		P
Standards	M	H			P
Error Analysis		H	A		
Error Apportionment			A		
Testability Design	M	H			
Accuracy Enhancement			A		
Power Supply Design	M	H			
Power System Design	M	H	A		
People Systems					P
Mechanical Packaging	M	H	A		
Stress Analysis	M		A		
EMI Analysis	M		A		
Thermal Analysis	M		A		
Integrated Circuit Technology		H			
Time Analysis		H	A		P
Microprocessors		H		S	
Computer Technologist		H		S	
Computer Hardware		H			
Computer Software				S	
Probability			A		
Statistics			A		
Hydro-Dynamics	M	H	A		
Variance Analysis			A		
Propulsion			A		

CODE: HASP = COMMON SE SKILL

H = HARDWARE SKILL

A = ANALYSIS SKILL

P = PROGRAM MANAGEMENT SKILL

M = MECHANICAL SKILL

Maximum Scores:

14	—	HASPS	X9	=	126
32	—	H	X9	=	288
25	—	A	X9	=	225
8	—	S	X9	=	72
17	—	P	X9	=	153
25	—	M	X9	=	225
75	—	TOTAL	X9	=	675

SYSTEM ENGINEERING SKILLS SURVEY

SURVEY—ANALYSIS REPORT:

System engineering skill survey scores for ______________________________

SKILL	SCORE	%	NORMALIZED	PROFILE	SE TYPE RANK
Common SE Skills	____ /126	____	____		
Mechanical Skills	____ /225	____	____		____
Hardware Skills	____ /288	____	____		____
Software Skills	____ /72	____	____		____
Analysis Skills	____ /225	____	____		____
Program Manager Skills	____ /153	____	____		____
Total Skills	____ /675	____		0 0.5 1	

These are your personally developed scores. They only establish your perception of your skills. They are intended to assist you in your systems engineering career development.

Survey reviewers comments: ______________________________

____

SYSTEM ENGINEERING SKILLS SURVEY

Recommendations:

8.3 SAMPLE WORKSHOP PROBLEMS AND WORKSHEETS

To help you fly on your own I have provided some workshop problems and worksheets. Problem 1 relates to an automobile and problem 2 relates to a house. Most of us live in houses and drive automobiles, so you should have enough knowledge to get started. These problems relate to my background and experience, but you can use the methodology and the worksheets for your system type. Most of the examples in this book also relate to electronic hardware systems, but the system engineering process can be applied to other types of systems. In chapter SEE-2, I talked about how writing this book used the system engineering process. In chapter SEE-3, I showed the development of the engineering department process system EOPS. I indicated that you could remove the word engineering from the resulting specification tree and substitute practically any process system such as post office, county government, university, or medical clinic. In chapter SEE-5, I showed you some examples for a software tool system CA-CES.

If you are interested in electronic control systems, then do problems 8.1, 8.2, or 8.3 by using the worksheets. If you have a different specific system type problem, then write your own problem statement, develop an applicable skill survey, and then use the worksheets.

System Engineering Workshop Problems

Many different problems have been used for SEE workshops, some examples of which follow. Part of the process of teaching "SEE" requires that the instructor must *tailor* the problem to each specific six- to seven-person team. This selection of appropriate problems must be done based on analyses of the student's skill survey data. The workshop methodology has been used to baseline many systems for very specific teams working on real development programs. It is especially helpful for proposal efforts. The workshops can be defined by the following action function element.

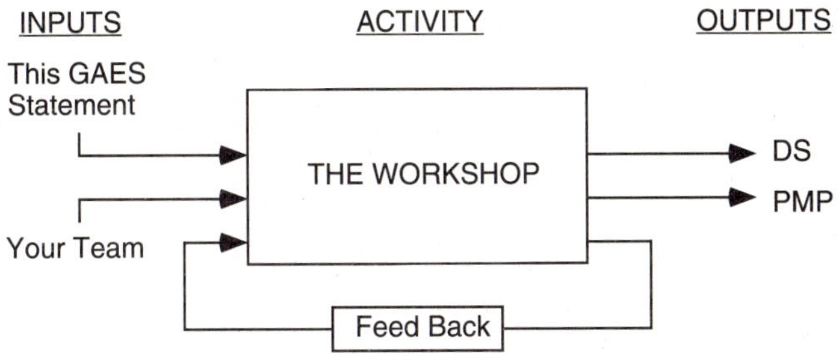

I have also used the worksheets to lead individual students in extended twelve-week courses, where students work in applying system engineering process and management to their specific present real job or problem. It is highly recommended, however, that all students become involved in a team workshop to develop skills in working with others.

System Engineering Expertise Workshop—Example Problems

1 Generic Automotive Electronics System (GAES)

Statement of the problem

Automotive electronics has developed in piecemeal fashion—engine control, cruise control, brake control, warnings, navigation, communication, displays, radio, etc. Usually each of these are separately developed, produced, and then kludged together. We wish to integrate all the electronics into a single system. Conceptually, this type of integrated system has been demonstrated by the Total Avionics Control System (TACS) of chapter SEE 4. GAES shall be modularized, thus allowing for add-on options. It shall be cost effective but shall incorporate innovative features that will enhance its salability.

Improved automobile performance and reduced automotive cost shall result from use of this system. Maintainability, reliability, and safety shall all be improved over that of present automobiles. The automobile propulsion will be conventional gasoline or diesel engine. The GAES shall utilize a computer with flexible software modules for wide range of specific system installations. Enhancement options shall include security, driver fatigue monitor, collision avoidance, intercom/phone, safety features, and other options which will enhance the marketing of GAES.

We shall develop this GAES sufficiently to provide a demonstration and validation model. We shall baseline the system sufficiently to perform producible and profitability studies. You have one year and a maximum of two million dollars to provide a prototype system that can be the base for a full-scale development program starting twelve months from now.

Your task

Provide a preliminary development specification (DS) that *defines* a baseline system and *requirements* thereof. Provide a preliminary program management plan (PMP) that will *plan*, *monitor*, and *control* this one-year development program. The plan and specification for this system development will be presented at a kick-off meeting for the people who will be working on the program. Also present at this meeting will be people who represent all the interfaces to this system and to this development program.

Your methodology

1. Team building
2. System perspective
3. Mission analyses, sortie and life
4. Functional analyses
5. Baseline system definition
6. DS process

7. PMP process, including the establishment of technical performance monitors (TPM), cost performance monitors (CPM), schedule performance monitors (SPM), and selection of specific analyses to be done.

8. A design review

2 Home Environmental Control System (HECS)

Statement of problem

In the past we have functioned for the most part as a component supplier to the private home environmental control business. To improve our competitive position we are contemplating the development of a Home Environmental Control System (HECS) whereby we would provide a total system made up of compatible components: furnaces, air conditioners, air cleaners, solar panels, water heaters, etc., and a common system controller. It is desired that the system be modularized so various combinations of components may be purchased for specific installations. Conceptually, this type of integrated system has been demonstrated by the Total Avionics Control System (TACS) of chapter SEE-4. The controller shall be computer based with flexible software modules for a wide range of specific system installations. The software modules shall be developed to provide a library of firmware modules that may be purchased for a variety of installation types and a multiple of system enhancements. Enhancements shall include capability of auxiliary functions such as home security, smoke and fire alarms, intercom/radio, interface with personal computers and computer networks, phone jacks, television antenna outlets, and home entertainment systems.

We shall develop this HECS sufficiently to provide a demonstration and validation model. We shall baseline the system sufficiently to perform producible and profitability studies. You have one year and a maximum of two million dollars to provide a prototype system that can be the base for a full-scale development program starting twelve months from now.

Your task

Provide a preliminary development specification (DS) that defines a baseline system and the requirements thereof. Provide a preliminary program management plan (PMP), which will plan, monitor, and control this one-year development program. The plan and specification for this system development will be presented at a kick-off meeting with the people who will be working on the program. Also present at this meeting will be people who represent all the interfaces to this system and to this development program.

Your methodology

1. Team building

2. System perspective

3. Mission analyses, sortie and life

4. Functional analyses

5. Baseline system definition
6. DS process
7. PMP process, including the establishment of technical performance monitors (TPM), cost performance monitors (CPM), schedule performance monitors (SPM) and selection of specific analyses to be done.
8. A design review

3 Generic Monitoring and Analysis System (GMAS)

Statement of problem

Develop a system to assist in the monitoring and analysis of intermittent failures. Intermittent failures can occur in all systems such as: electronic, human, mechanical, civil, and process. Intermittent failures of systems are difficult to troubleshoot and very costly to repair. Some intermittents occur only once every three months. The combination of variables that cause them to occur can be difficult to determine and difficult to duplicate. Consequently, intermittents are solved by component replacement trials—catch it when it happens—exercising your system by varying inputs and/or output loads and/or exposure to environmental excitation, and a lot of analysis. These methods are costly and time consuming.

Intermittents can be caused by momentary failure of a component under a special combination of conditions, or they can be caused by the combination of marginal components at a unique environmental condition. The cause may even be a basic design error that occurs under a unique combination of inputs/outputs/environment. Many times the intermittent failure is the result of a momentary random event; therefore, it is very difficult to "catch" sufficient data to pinpoint the cause of the failure.

Many times we take data on the working system in an attempt to gather data of marginal conditions or trends. This data usually is inconclusive; it tends to confuse and, sometimes, tempts the "expert" to generate an unproveable theory. Sometimes we just let the system fail intermittently until it ultimately dies or just fails completely.

Some examples of intermittent failures:

1. Navigation system velocity take-offs (see chapter SEE-6, "Functional Analysis Layered Analysis Methods"). This system says you're moving when you are stationary!
2. Sudden human physical health problems. Your system reaches the failure threshold.
3. Fuel valve on a furnace locked up. Fumes fill the house.
4. Camper furnace cutout. It gets cold.
5. Integrated Propulsion Control System (IPCS) Disengaged to Backup.

6. Advanced Digital Optically Coupled Control System—memories will not load.
7. Radio and television sets. They cut out and then come back on.
8. Semitruck electronic brakes lock. People are killed!
9. Electronic spark control in an automobile. Car hops down the road.
10. Egyption land navigator—display lockup. You're moving, but your display says you're stationary.
11. Arlington Heights smoke detector. It detects a moth and gives false alarms.
12. Fires in test systems. System dynamics cause shorts that cause fire in your tester.
13. Cold start injector on a Volkswagon automobile would not function below a certain wind-chill. Your car doesn't always start in very cold weather.

We shall develop GMAS sufficiently to provide a demonstration validation model for a selected specific system intermittent and to baseline the system sufficiently to perform producible and profitability studies. You have one year and a maximum of two million dollars to provide a prototype system that can be the base for a full-scale development program starting twelve months from now.

Conceptually, this type of system has been demonstrated by prior flight control systems with integral Built-in test equipment (BITE) hardware, Built-in test (BIT) software, and stimulus. The use of a histogram memory has been demonstrated by embedded computer support equipment, and expert system modules have been demonstrated by the Computer-Aided Cost Estimating System (CACES) as defined in chapter SEE-5. A histogram memory is a memory that is used to continuously store the last few minutes (or some length of time) a number of selected system internal measured parameters. Data older than the last few minutes is discarded. If a failure condition is detected by GMAS, it then saves the data in the histogram as an aid to analyzing the cause of the intermittent failure. Expert system software is used automatically to perform this analysis and provide a report to the GMAS user. Those who have worked in attempting to solve intermittents know that if we had more data, especially leading up to the intermittent, we could easily determine the cause of failure.

Your task

Provide a preliminary development specification (DS) that defines a baseline system and the requirements thereof. Provide a preliminary program management plan (PMP), which will plan, monitor, and control this one-year development program. The plan and specification for this system development will be presented at a kick-off meeting with the people who will be working on the program. Also present at this meeting will be people who represent all the interfaces to this system and to this development program.

Your methodology

1. Team building
2. System perspective
3. Mission analyses, sortie and life
4. Functional analyses
5. Baseline system definition
6. DS process
7. PMP process including the establishment of technical performance monitors (TPM), schedule performance monitors (SPM), cost performance monitors (CPM) and selection of specific analyses to be done.
8. A design review

Day 1—Session 1

System engineering team attributes

- Vocally interview each team member while the team is gathered around a table:
  - — Name and address
  - — Schools and degrees (if applicable)
  - — Years experience (include applicable prior to degree)
  - — Experience (single most prominent)
  - — Specialties (pick top 3):

    H = Hardware   SE = System Engineer
    SW = Software   M = Mechanical
    A = Analyst   O = Other
    PM = Program Management

  - — WT (Record your weight)   HT (Record your height)
  - — New and good (What is new and good for you within the last year? Just married, new baby, etc.)
- Total (team weight in pounds, team height in inches, # years experience, most predominate experience type, top three team specialties, team's most predominate new and good)
- Does your team have any skill deficiencies?

- Is your team "right brain" (intuitive), "left brain" (rational), or "composite"?

Day 1—Workshop Session 2

SEE-1—System perspective

- The team shall, while gathered around a table, fill out the System Engineering Perspective Worksheet:
  - You are a subsystem of what bigger system and what are the major components of your system?
  - What time phase are you in?
  - List and briefly describe the five physical parts of your system as you now perceive it
  - What (if any) are constraining conditions?
  - What actions are to be performed to develop your system?
  - Establish specific cost and date goals

Day 1—Workshop Session 3

SEE-2—Mission analyses

- The team shall, while gathered around a table, develop mission analysis time lines:
  - A sortie mission for the major system above your subsystem
  - A sortie mission for your subsystem
  - A life mission for your subsystem
- Write specific requirements on your sortie time lines as they occur to you such as environmental, operational range timing constraints, accuracies, etc.
- Show design reviews, development specifications, cost analysis, etc. on your life mission time line

SEE TEAM # ________ **ATTRIBUTES** DATE: ________

NAME ADDRESS	SCHOOL DEGREE	YEARS EXPR	EXPERIENCE	SPECIALTIES (PICK TOP 3)	WT / HT	NEW AND GOOD
TOTAL						
AVERAGE						

NOTE: PLACE THIS WORKSHEET IN YOUR PMP AS SECTION 6.1 TEAM ATTRIBUTES

THE SYSTEM PERSPECTIVE

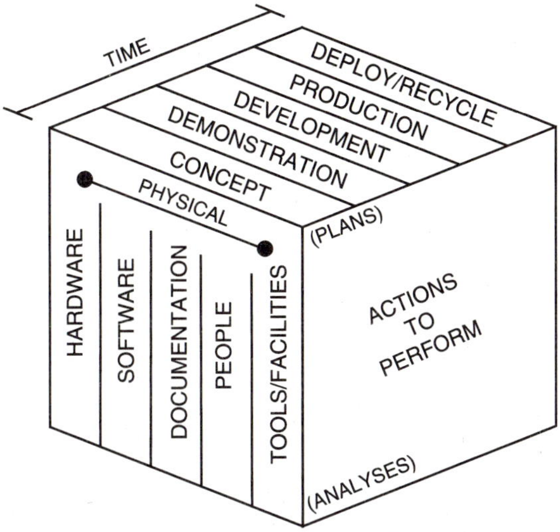

FRONT VIEW

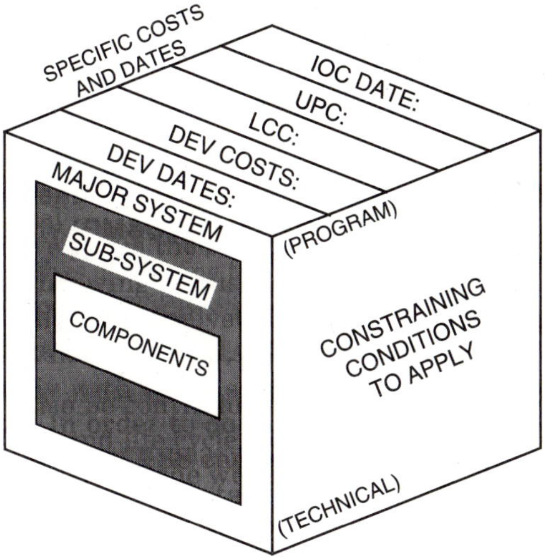

REAR VIEW

IOC = INITIAL OPERATIONAL CAPABILITY

UPC = UNIT PRODUCT COST

LCC = LIFE CYCLE COST

DEV = DEVELOPMENT

NOTE: PLACE THIS WORKSHEET IN YOUR PMP AS SECTION 6.2 SYSTEM PERSPECTIVE

SORTIE MISSION ANALYSES

________ ________ (MAJOR SYSTEM)	
________ ________ (OUR SYSTEM)	

SUB-SYSTEM: ____________ FOR ____________ MAJOR SYSTEM

NOTE: PLACE THIS WORKSHEET IN YOUR DS AS PART OF 3.1 SYSTEM DEFINITION

LIFE MISSION ANALYSIS

				← SORTIES →		
CONCEPT	DEMONSTRATION	DEVELOPMENT	PRODUCTION	DEPLOYMENT	PRE-PLANNED PRODUCT IMPROVEMENT	DISPOSAL AND RECYCLE

SUB-SYSTEM: ________________ **FOR** ________________ **MAJOR SYSTEM**

NOTE: PLACE THIS WORKSHEET IN YOUR PMP AS SECTION 6.3 LIFE MISSION ANALYSES

Day 2—Workshop Session 4

SEE-3—Functional analyses

- The team shall perform functional analysis of your subsystem sortie mission time line down to two levels.

  A) Brainstorm: What are the actions of persons and things required to perform the mission?

  B) Collect into a hierarchial diagram (add additional items as they occur to you, also establish "attributes," avoid team conflicts by assigning trade-off studies and analyses).

  C) Repeat A) and B) to a lower level which shall result in the functional control block diagram for your subsystem.

Day 2—Workshop Session 5

SEE-4—Baselining

- The team shall by assignment of specific tasks to each team member develop a baseline system definition as follows. (All of the items on the list below must be addressed, in some system-types some of the items may not be applicable.)

  — Scope (statement for the development specification)

  — Time line of system mission

  — Functional block diagrams—hardware and software

  — Interface drawing

  — Specification tree

  — Mechanical sketch

  — Installation sketch

  — Detailed block diagram

  — Attributes table

  — Bill of material

  — Other

NOTE: THIS BASELINE WILL BECOME PART OF DS SECTION 3.1 SYSTEM DEFINITION

FUNCTIONAL ANALYSIS — TOP LEVEL HIERARCHIAL

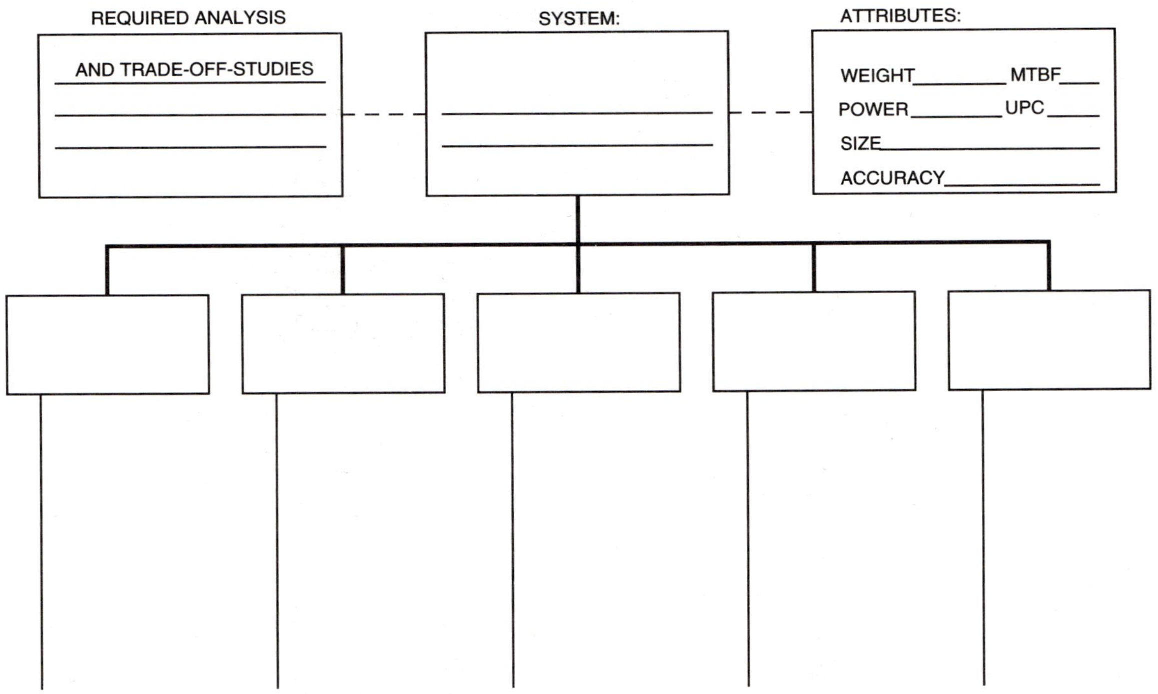

SUB-SYSTEM: ____________ **FOR** ____________ **MAJOR SYSTEM**

NOTE: PLACE THIS WORKSHEET IN YOUR PMP AS SECTION 6.4 FUNCTIONAL ANALYSIS - HIERARCHIAL

FUNCTIONAL ANALYSIS — CONTROL BLOCK DIAGRAM

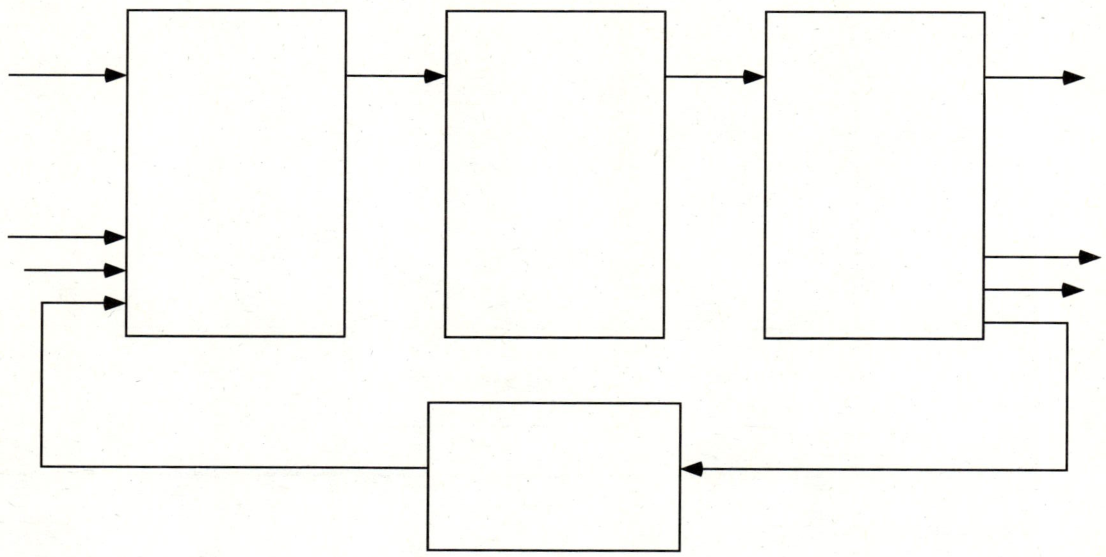

SUB-SYSTEM: ____________________ FOR ____________________ MAJOR SYSTEM

NOTE: PLACE THIS WORKSHEET IN YOUR DS AS PART OF SECTION 3.1 SYSTEM DEFINITION

Day 3—Workshop Session 6

SEE-5—System development specification (DS)—process

(Assign specific parts of the following to each team member)

- Provide System DS—Refer to chapter SEE-5 for examples)

  1.0 Scope—From problem statement and system perspective cube worksheet

  2.0 Applicable Documents—Problem statement plus other DS applicable documents

  3.0 Requirements

    3.1 Definition—Your baseline plus add write ups to describe your figures.

    3.2 Specific Requirements—Use outline, each team member shall write at least one specific requirement!

  4.0 Quality

  5.0 Delivery

  6.0 Notes

Day 3—Workshop Session 7

SEE-4—PMP

(Assign specific parts of the following to each team member)

Provide PMP

1.0 Scope—Use problem statement and SE perspective cube worksheet. Write PM scope statement, also fill in program brief.

2.0 Applicable Documents—List problem statement plus other PM applicable documents.

3.0 Go through PMP outline as a checklist, insert any data you may have generated. If your system type is not applicable so indicate by initials NA.

Specifically Provide:

3.1.1 Work Breakdown Structure (WBS), staffing and program cost allocations.

3.1.2 Program top-level schedules.

3.2.2 System Engineering—Comment on SEMP, reliability, specifications, and support.

3.2.2.2 Design Reviews—Discuss (also include a program start-up review).

3.2.4 Test Plans—Give top-level plan.

3.2.5 Production Plans—Reference UPC model. Also provide top-level flow diagram.

Respond as time allows to other PMP areas of specific interest to your team or problem.

Day 3—Workshop Session 8

SEE-6—Analyses

- Establish analyses to be done and assign responsibilities.
- Review the checklist: "Analyses Required and Technical Performance Monitors."
- Each team member shall sign up for responsibility for some of the analyses required.
- If your system type does not seem to relate to some of these analyses, still do a review of the list and use "other" and indicate why not applicable on items you can't apply to your system.
- Attach this Checklist as PMP section 3.2.2.9 analyses required and technical performance monitors (TPM).

SYSTEM DEVELOPMENT SPECIFICATION NO. DS

PART ______ OF ______

TITLE:

SIGNATURES	TITLE	DATE
PREPARED BY		
CHECKED BY		
APPROVED BY		

☐ CUSTOMER FUNDED | CONTRACT NUMBER

☐ COMPANY FUNDED | PROJECT NUMBER

REVISIONS

LTR	DESCRIPTION	DATE	APPROVAL	LTR	DESCRIPTION	DATE	APPROVAL

PAGE																									
REVISION																									

PAGE																				PAGE OF
REVISION																				

Development Specification Table of Contents

System ______________________________

AUTHOR:

__________	1.0 Scope
__________	2.0 Applicable Documents
__________	3.0 Requirements
__________	3.1 System (Your Baseline)
__________	3.2 Specific Requirements*
__________	3.2.1 Performance Requirements
__________	3.2.2 Interface Requirements
__________	3.2.3 Mechanical Requirements
__________	3.2.4 Environmental Requirements
__________	3.2.5 Hardware Design Requirements
__________	3.2.6 Software Design Requirements
__________	3.2.7 Reliability, Maintainability and Safety Requirements
__________	3.2.8 Testability Requirements
__________	3.2.9 Documentation Requirements
__________	3.2.10 Produceability Requirements

__________ 3.2.11 Installation and Logistics Requirements

__________ 3.2.12 Other Requirements

__________ 4.0 Quality Assurance

__________ 5.0 Preparation for Delivery

__________ 6.0 Notes

*__________ Team ______ Initials Name

Initials	Name
__________	______________
__________	______________
__________	______________
__________	______________
__________	______________
__________	______________
__________	______________

*The items under "Specific Requirements" should reflect your specific system type; so if needed, change headings as they apply to your system. 3.2.1–3.2.12

PROGRAM MANAGEMENT PLAN (PMP)

FOR ______________________________

	SIGNATURES	TITLE	DATE
PREPARED BY			
CHECKED BY			
APPROVED BY			

☐ CUSTOMER FUNDED — CONTRACT NUMBER

☐ COMPANY FUNDED — PROJECT NUMBER

REVISIONS							
LTR	DESCRIPTION	DATE	APPROVAL	LTR	DESCRIPTION	DATE	APPROVAL

PAGE																								
REVISION																								

PAGE																				PAGE OF
REVISION																				

Program Management Plan Table of Contents

Program ____

Responsible Persons

____________ 1.0 Scope of PMP and Program Brief

____________ 2.0 Applicable Documents

____________ 3.0 Program Requirements

____________ 3.1 Program Definition

3.1.1 Work Breakdown Structure (WBS), staffing assignments, and program funds allocation.

3.1.2 Program schedules

3.2 Specific Program Requirements

____________ 3.2.1 Program management process—monitor and control

3.2.1.1 Team building, internal and external

3.2.1.2 Communications, internal and external

3.2.1.3 Risk analysis and plans

3.2.1.4 Monitoring: Schedule performance monitor (SPM), cost performance monitor (CPM), technical performance monitor (TPM), and variances

3.2.1.5 Special contract requirements

_______________ 3.2.2 System Engineering Management

3.2.2.1 System engineering management plan per contract

3.2.2.2 Specifications

3.2.2.3 Standards

3.2.2.4 Reliability, maintainability, safety

3.2.2.5 Support specialties integration

3.2.2.6 Configuration management

3.2.2.7 Design reviews

3.2.2.8 Engineering parts management

3.2.2.9 Analyses required and Technical Performance Monitors

3.2.2.10 Logistics

_______________ 3.2.3 Software Development Plan

_______________ 3.2.4 Test Plans

_______________ 3.2.5 Producibility Plans

_______________ 3.2.6 Subcontract Management Plan

_______________ 3.2.7 Deliverables and Program Closeout Review

____________ 4.0 Quality Control Plan

____________ 5.0 Preparation for Delivery

____________ 6.0 Notes

____________ 10.0 Appendices (as required)

|

|

Responsible Persons:

Initials	Name	Mail Stop	Phone
______	____________________	____________	________
______	____________________	____________	________
______	____________________	____________	________
______	____________________	____________	________
______	____________________	____________	________

PROGRAM BRIEF

SCOPE

PROGRAM NAME ________________ TYPE ________________

CUSTOMER ________________ CONTRACT # ________________

LIFE CYCLE PHASE (FILL IN BAR: [hatched] — PRIOR ACTIVITY, [solid] — THIS PROGRAM, [cross-hatched] — FUTURE GOAL)

CONCEPTUAL	DEMO/VAL	FSD	PRODUCTION	DEPLOYMENT	PRODUCT IMPROVEMENT	DISPOSAL

SCHEDULE: START DATE ________________ (INDICATE YEARS AND FILL IN BOTTOM)

PLAN			
ACTUAL			

COST ________________ SHARING ________________

PLAN				
ACTUAL				

→ | ← SCALE — (CHECK ONE) 1K ___ 10K ___ 100K ___ 1M ___ 10M ___

PROGRAM INPUTS	ACTIVITIES	OUTPUTS

PEOPLE

PROGRAM MANAGER ________________ ADMINISTRATOR ________________

CONTRACTS ________________ MARKETING ________________

ENGINEERING MANAGER ________________ SYSTEMS ENGINEER ________________

PRODUCTION ________________ QUALITY ________________

CUSTOMER ________________

PEER CONTRACTORS ________________

SUB CONTRACTORS ________________

DESIGN ________________

DRAFTING ________________ DEVELOPMENT/EVALUATION LAB ________________

RELIABILITY ________________ LOGISTICS ________________

OTHER ________________

PROGRAM NAME ____________________

COMMUNICATIONS

INTERNAL ____________________

EXTERNAL ____________________

FACILITIES AND TOOLS

____

RISKS: (SHADE IN RISK IN APPROPRIATE SECTION)

HIGH

RISK DIAL

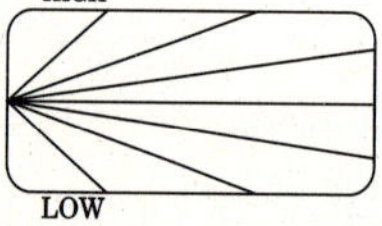

AVE

LOW

GAINS: ____________________

STATUS: (SHADE IN APPROPRIATELY)

TECHNICAL		SCHEDULE		COST		
	○		○		○	RED
	○		○		○	YELLOW
	○		○		○	GREEN

ADDITIONAL COMMENTS ____________________

ACTION ITEMS REGISTER (AIR) ITEMS ____________ (# ACTIVE AIR ITEMS) ____________

PROGRAM START-UP REVIEW (PSR) ____________________

TECHNICAL REQUIREMENTS ESTABLISHED ____________________

TECHNICAL SPECIFICATION ____________________

UNIT PRODUCT COST (UPC) ____________________

PRODUCTION READINESS REVIEW (PRR) ____________________

DESIGN REVIEW (DR) LEVEL ____________________

DR ACTIVITY ____________________

PRINTED WIRING ASSEMBLY PROCESS # ________ ON SCHEDULE ________ BEHIND ________

SOFTWARE PROCESS: # ________ ON SCHEDULE ________ BEHIND ________

POWER SUPPLY ____________________

SYSTEM ENGINEERING ____________________

TECHNICAL PERFORMANCE MEASURE PROCESS: # ________ REQUIREMENT MET ________ BEHIND ________

REPORTER ____________ MONTH ____________ YEAR ____________

PROGRAM REPORT # ____________

Analyses Required and Technical Performance

Monitors for ________________________________ System

(3.2.2.9 of the PMP)

Analyst *

__________ 1.0 Technical Performance Monitors

Weight __________ Power __________

Cost __________ Reliability __________

Volume __________ Maintainability __________

Accuracies __________

____

____

Environmental Impact __________

____

____

Other __________

__________ 2.0 System Design Analysis

Interface __________

Power Distribution __________

Electromagnetic Interference (EMI) __________

* Responsible analyst

Special Circuit to analyze ____________________

____

____

____

Technical Risk____________________

____

____

Other ____________________

__________ 3.0 Specific Reliability Analyses

Mean Time Between Failures (MTBF) __________

Failure Mode and Effect Analysis (FMEA) __________

SAFETY ____________________

STRESS ____________________

Failure Recording and Corrective Action (FRACA) ______

Other ____________________

__________ 4.0 Specific Test Analyses

Testability ____________________

Built-In Test Equipment hardware and Built-In Test Software (BITE and BIT)____________________

Test Generation ______________________________

Data Analysis ______________________________

Other ______________________________

______________ 5.0 Specific Product Analyses

Producibility ______________________________

Market ______________________________

Value Engineering ______________________________

Logistics ______________________________

Maintainability ______________________________

Other ______________________________

______________ 6.0 Control System Analyses

Models ______________________________

Simulations ______________________________

Stability ______________________________

Redundancy ______________________________

Interfaces ______________________________

Timing ______________________________

Other ______________________________

__________ 7.0 New Technology Analyses

Risk/Gains __________

Viability __________

Usability __________

Need Ability __________

Costs __________

Other __________

__________ 8.0 In-Depth Specialty Analyses

Thermal __________

Structural __________

Electromagnetic Interference (EMI) __________

Materials __________

Other __________

__________ 9.0 Specific Software Analyses

Functional __________

Control System __________

Hardware Characteristics __________

System Timing __________

Memory and Throughput __________

Language and Tools ______________________

Development Process and Documentation ______________

____

Process Tailoring ______________________

Support and Test Environment ______________

Configuration Management and Quality

____

Other ______________________

______ 10.0 Other specific system type analyses required ______

____

Instructors Review Worksheet

6 CRITERIA FOR DS's AND PMP's

	DS	PMP
1 — CONSISTENCY		
2 — SPEAKS TO THE DEVELOPERS		
3 — ATTRIBUTES BASELINES		(SPM/CPM)
4 — STRUCTURE		
5 — SUFFICIENTLY ESTABLISHES THE REQUIREMENTS	(TPM)	
6 — GIVES ROOM TO DESIGN/MANAGE		

8.4 As with Wings as Eagles

You have studied this book and learned the system engineering process; now fly on your own. I'm effectively kicking you out of the nest, my friend.

Index

C

D

E

F

G

H

I

J

L

M

N

O

P

Q

R

S

T

U

V

W

Z